# GALILEO'S FINGER

# GALILEO'S FINGER

## *The Ten Great Ideas of Science*

### PETER ATKINS

OXFORD

UNIVERSITY PRESS

# OXFORD
UNIVERSITY PRESS

Great Clarendon Street, Oxford OX2 6DP

Oxford University Press is a department of the University of Oxford.
It furthers the University's objective of excellence in research, scholarship,
and education by publishing worldwide in

Oxford New York

Auckland Bangkok Buenos Aires Cape Town Chennai
Dar es Salaam Delhi Hong Kong Istanbul Karachi Kolkata
Kuala Lumpur Madrid Melbourne Mexico City Mumbai Nairobi
Sao Paulo Shanghai Singapore Taipei Tokyo Toronto

Oxford is a registered trade mark of Oxford University Press
in the UK and in certain other countries

Published in the United States
by Oxford University Press Inc., New York

British Library Cataloguing in Publication Data
Data available

Library of Congress Cataloging in Publication Data
Data available

ISBN 0 19 860664 8 (hb)

10 9 8 7 6 5 4 3 2 1

Designed & typeset by Pete Russell, Faringdon, Oxon
Printed in Great Britain
on acid-free paper by T. J. International Limited, Padstow. Cornwall

The middle finger of Galileo's right hand.

The finger was detached from Galileo's body on 12 March 1737
when his remains were transferred to the main body of the church of
Santa Croce, Florence. It is currently in the Museo di Storia della Scienza.
The vessel that contains the finger has a cylindrical alabaster base,
with the following description:

*Leipsana ne spernas digiti, quo dextera coeli*
*Mensa vias, nunquam visos mortalibus orbes*
*Monstravit, parvo fragilis molimine vitri*
*Ausa prior facinus, cui non Titania quondam*
*Sufficit pubes congestis montibus altis,*
*Nequidquam superas conata ascendere in arces.*

Do not look down upon the relic of a finger, by means of which
a right hand measured paths in the heavens and revealed to mortals
celestial bodies never seen. By preparing a small piece of fragile glass it
first dared a feat which long ago was beyond the powers of young
Titans, who piled mountains high in a vain attempt to
ascend to lofty citadels.

# CONTENTS

# THE EMERGENCE OF UNDERSTANDING

**W**HY HIS FINGER? Galileo marks the turning point, when the scientific endeavour took a new direction, when scientists—an anachronistic term at the time, of course—rose from their armchairs, questioned the efficacy of the preceding attempts to come to grips with the nature of the world by thought in alliance with authority, and took the first faltering steps down the path of modern science. In the process they rejected untested authority, and while not letting go entirely of armchair speculation and inner contemplation, forged a new, more powerful alliance with the technique of publicly verifiable experimental observation. We see this aspect of Galileo's finger dabbling in all our current scientific pies. We see it in physics, where it first began to stir; in chemistry, where it found its way at the beginning of the nineteenth century; and in biology, especially since biology ceased being merely a source of wonder during the nineteenth and twentieth centuries.

In short, this book celebrates the effectiveness of Galileo's symbolic finger for the winkling out of truth. That only Galileo's physical finger is preserved but the descendants of his techniques thrive is also symbolic of the transitoriness of personal existence in contrast to the immortality of knowledge. Galileo's finger, then, represents that misty concept the 'scientific method'. He was not alone, of course, or first, in introducing this approach to the discovery of knowledge, but he is sufficiently prominent in the history of ideas for it to be reasonable to adopt him as a symbol of its introduction. One aspect of this amazingly potent method for disinterring truth about the world that distinguishes science from its principal rival—namely, impressively expressed but ultimately idle speculation—is the centrality of experiment. Going out into the world and making observations under carefully controlled conditions minimizes the subjective component of our perception and, in principle, opens up the observations to public scrutiny.

Galileo also developed the art of simplification, the isolation of the essentials of a problem, the peering in his thoughts through the clouds that in real systems conceal the underlying simplicity, just as he looked through his actual telescope and saw the complexity of the heavens. He set aside the creaking cart pulled through mud; instead, he considered the simplicity of a ball rolling on an inclined plane, a pendulum swinging from a high support. That isolation of the core phenomenon from the creaks and confusions of reality is a key part of the scientific method. Scientists see the pearl in the oyster, the jewel in the crown.

Some, of course, will claim that here lies weakness. True understanding, they claim, comes from appreciating the hurly-burly of reality: the cart bogged down, the lover lamenting, the lark ascending. That scientists examine a butterfly to study its mechanism is an abnegation of understanding, they claim. We have to see this objection in perspective, not reject it out of hand. Most scientists, being human, accept that sentiment is a wonderful component of our interaction with the world, but few would accept that it is a reliable route to truth. They prefer to disentangle the awesome complexity of the world, examine it piece by isolated piece, and build it up again, with deeper understanding, as best they can. They study the behaviour of a ball on an inclined plane in order to understand the cart on the hill; they study the pendulum in order to understand the swing of the athlete's leg. Their opponents will cry that understanding the physics of vibration does not illuminate the joy of music and that picking apart a symphony into a bag of notes destroys our understanding of its composition. The scientist replies that we must first understand what it is to be a note, then move on to understand why it is that some chords are agreeable and others dissonant, and then—maybe not for decades—try to understand the psychological and artistic impact of a sequence of chords. Science aims for thoroughness of understanding, never losing sight of the ultimate goal and not rushing to it impatiently half-baked. Whether scientists will ever comprehend our joy of comprehending the world, of acting out our lives within it, and all the other great questions that philosophers, artists, prophets, and theologians consider to be their territory of discourse is a matter for idle speculation. And we all know how useful that has been.

By a great idea, I mean a simple concept of great reach, an acorn of an idea that ramifies into a great oak tree of application, a spider of an idea that can spin a great web and draw in a feast of explanation and elucidation. I have had to be selective, and I have no doubt that others could come up with other super-spiders that would capture other juicy flies of science. Here, though, is my choice.

I have concentrated on ideas rather than applications. I have written little about black holes and space travel, and hardly anything—except in my reflective Epilogue—about that wonderful shift of paradigms of explanation that we are currently experiencing in the form of information technology and computation. My aim has been to identify the *ideas* that illuminate and, in most cases, provide the foundation for technological advance. That imaginative intellectual descendant of Galileo, Freeman Dyson, distinguishes between *concept-driven science* and *tool-driven science*. Almost the whole of my account is concept-driven. In this distinction Dyson echoes another distinguished thinker, Francis Bacon, who classified ideas as *fructifera*, bringers of fruit, and *lucifera*, bringers of light. I concentrate on the latter. It is arguable whether molecular biology and the consequences of knowing the structure of DNA are *lucifera* or *fructifera*, and whether they are concept-driven or tool-driven, and therefore whether they should be included here. I have opted for the former in each case because no other discovery has contributed so much to our comprehension and application of biology and it would be absurd to exclude it. Perhaps in molecular biology we are seeing the merging of *lucifera* and *fructifera* into a science of unprecedented dynamism.

Scientific exposition is not like reading a novel, where events unfold in a simple linear fashion. To understand a scientific idea, you might need to read quickly first time, jumping the bits that seem too demanding or (heaven forbid) too boring. Indeed, although I consider that there are natural sequences of presentation, such as climbing up from dark fundamentals into the daylight of the familiar or drilling down from the familiar to the more fundamental (I adopt the latter), the chapters are more or less independent and can be read in any order.

The second aspect to keep in mind is the drift into abstraction that characterizes modern science. Abstraction is another important face of Galileo's finger, and we need to be alert to its role and its importance. First, abstraction doesn't mean useless. Abstraction can have enormous practical significance because it points to unexpected connections between phenomena and allows thoughts developed in one field to be used in another. Most importantly, though, abstraction is a way of standing back from a set of observations and seeing them in a broader context. One of the most satisfying *eureka!* moments in science, and in reading about science, is the Cortez-like experience of seeing oceans merging into a single whole, and realizing the connection between phenomena that had seemed disparate. My intention is for us to travel to the high ridges of science where we can experience this awesomely satisfying merging, and as we travel gradually to unfold the pleasures of ever greater abstraction. Thus, I start with monkeys and peas, lead past atoms, into beauty, then through spacetime, and culminate with that awesome apotheosis of abstraction, mathematics. If you read the chapters in sequence, then you should find that each successive chapter deepens your comprehension of what has gone before.

We are about to set out together on a challenging but ultimately deeply satisfying journey. Science is the apotheosis of the spirit of the renaissance, an extraordinary monument to the human spirit and power of comprehension of the puny human brain. My principal hope is that as the journey progresses and I lead you carefully to the summit of understanding, you will experience the deep joy of illumination that science alone provides.

# ONE

# EVOLUTION

## THE EMERGENCE OF COMPLEXITY

**THE GREAT IDEA**
*Evolution proceeds by natural selection*

*Nothing in biology makes sense except in the light of evolution*

THEODOSIUS DOBZHANSKY

**L** IFE is such a precious thing that for long it was thought to require its own special creation, for how could something so extraordinary and so special emerge spontaneously from lifeless slime? Indeed, what is the crucial component of things that endows them with life? The answers to hugely important questions like these emerged in two waves. There was first the wave of empirical explanation, when observers, most of them naturalists and geologists in the nineteenth century, inspected the outward forms of nature and drew far-reaching conclusions. Then there was the second wave, in the twentieth century, when moles with scientific eyes burrowed below the surface of appearance and discovered the molecular basis of the web of life. The first of these approaches is the subject of this chapter; the second, which greatly enriches our understanding of what it means to be alive, is the subject of the next.

The ancient Greek philosophers, as ever, had their own views about the nature of living things and, as in most of their well-meaning pronouncements, they were utterly but engagingly wrong. The self-styled god Empedocles (*c*.490–430 BCE), for instance, shortly before he unwisely chose to demonstrate his own divinity by hurling himself into the crater of Mount Etna, had supposed that animals are built from a universal kit of parts which, conjoined in various combinations, gave elephant, gnat, horned toad, and man. That the world is populated by these familiar combinations rather than flying pigs and asses with the tails of fish is due to only certain combinations being viable. Nature had presumably experimented with other combinations in its anticipation of the *Island of Doctor Moreau*, but after a brief hobble, flutter, or wallow these experimental creatures had flipped over and died.

Nearly two thousand years later the same view was echoed, but on a molecular scale, by the Comte de Buffon, George-Louis Leclerc (1707–88), who considered that organisms emerged spontaneously from aggregations of what we would now call organic molecules, and that the number of possible species is the number of viable combinations of these molecules. Buffon thought he should know: his great work the *Histoire naturelle, générale, et particulière* (which started to appear in 1749) was planned to extend to fifty volumes, and of the thirty-six he completed, nine were devoted to birds, five to minerals, and eight (published posthumously) to cetaceans, reptiles, and fishes.

But where did all these creatures, all living things in fact, come from? There is an awful lot of them, with a couple of million species recorded and possibly 10 million or more to be found. Aristotle, ever magnificently intellectually fertile and magnificently wrong as usual, had supposed that animals had fallen from the stars or had originated spontaneously, fully formed. The Yahuna Indians of the Amazonian Basin took a neo-Aristotelian view and considered that manioc arose from the ashes of the murdered and cremated Milomaki. The Cahuilla Indians of California had a similar belief, with vines springing from his cremated stomach, watermelons from the pupils of his eyes, and corn from his teeth. Less agreeably, wheat sprang from the ashes of his lice eggs and beans from his semen.

Other religions have offered seemingly simple accounts in which all creatures great and small were made by a god, and that's that. However, even some of the church fathers found it difficult to come to terms with all the statements in the Bible. The erudite Gregory of Nazianzus (*c*.330–89; Nazianzus was somewhere in Cappadocia, Asia Minor), for instance, considered that God must have done some of his creating after Noah's flood, for the latter's little ark was too small to have housed representatives of all species.[1] The Archdeacon of Carlisle William Paley (1743–1805) considered it indisputable that he had identified the origin of creatures in the book he published in 1802, sportingly

---

[1]  Built from gopher wood, the ark was 300 cubits long, 50 cubits wide, and 30 cubits high; a cubit is the length of the arm from the elbow to the finger tips, and is about 45 centimetres (about 18 inches).

entitled *Natural theology, or evidence for the existence and attributes of the deity collected from the appearances of nature*, in which he famously argued by analogy with a traveller coming upon a watch, remarking on its intricate design, and having no doubt that a watchmaker was behind it. Thus, anyone coming across the intricacies of nature must inevitably conclude that God had had a hand in its design and construction. Anaximander of Miletus (*c*.610–545 BCE), though, contributing to Western philosophy when it was the newest of green shoots, had in fact already glimpsed something of the truth when he supposed, purely speculatively and as part of his, Thales', and Anaximenes' philosophical programme to account for beings, and being in general, that species of animals could be transformed into one another.

As so often in science, the first step towards true understanding rather than fanciful speculation is the collection of data, which in this case means identifying and classifying all the types of organisms that constitute the biosphere, or at least as many as patience, persistence, and providence allow. The most useful names recognize relationships, as in the convention of family members retaining their surnames. By the middle of the eighteenth century, with international maritime trade established, even stay-at-homes had become familiar with the plethora of organisms and oddities that populated the world, and had realized that simple names such as cow and dog were inadequate, just as inhabitants of Lapland would find their language inadequate in Uganda. The first universally accepted system of nomenclature was devised by the Swedish botanist Carl von Linné (1707–78), better known to us by his Latinized name, Linnaeus. Linnaeus set out his system of nomenclature in his *Systema naturae*, published in 1735, and the systematic nomenclature of plants is commonly taken to stem from his *Species plantarum* of 1753. In these works, Linnaeus introduced the hierarchy of membership (Fig. 1.1), with kingdom near the top, and the pyramid broadening to ever more particular types as we descend through phylum, class, order, family, genus, and species. This scheme has since been elaborated by the inclusion of various intercalated layers, such as subfamily and superfamily. Thus, we humans are classified (ironically, some would argue) as the species *Homo sapiens*, of the genus *Homo*, in the family Hominidae, in the superfamily Hominoidea, of the infraorder Catarrhini, of the suborder Anthropoidea, of the order Primates, of the subclass Eutheria, in the class Mammalia, in the superclass Tetrapoda, which is a member of the subphylum Vertebrata, of the phylum Chordata, in the kingdom Animalia, of the domain Eukarya, in the empire of organisms.

A deficiency of the Linnaean system is that it is based on recognizing similarities rather than, as is scientifically more appropriate, on identifying under-

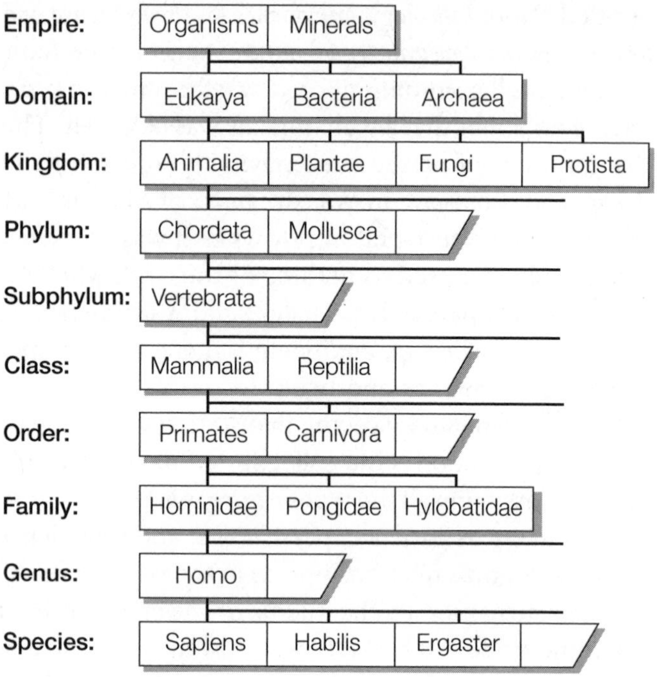

**Fig. 1.1** The Linnaean classification originally consisted of eight ranks (domain, kingdom, phylum, class, order, family, genus, species) organized a bit like the Roman army. Since then, the tree of classification has acquired many intercalated ranks, a few of which are shown here. This tree shows how humans fit into the extended Linnaean system. Where a particular taxonomic level shows only some of the taxons stemming from the superior taxon, the row ends in a sliced rectangle. The classification scheme remains controversial at just about every level: some, for instance, prefer to think in terms of five Kingdoms (with the bacteria included in that rank).

| Empire: | Organisms | Minerals | | |
| --- | --- | --- | --- | --- |
| **Domain:** | Eukarya | Bacteria | Archaea | |
| **Kingdom:** | Animalia | Plantae | Fungi | Protista |
| **Phylum:** | Chordata | Mollusca | | |
| **Subphylum:** | Vertebrata | | | |
| **Class:** | Mammalia | Reptilia | | |
| **Order:** | Primates | Carnivora | | |
| **Family:** | Hominidae | Pongidae | Hylobatidae | |
| **Genus:** | Homo | | | |
| **Species:** | Sapiens | Habilis | Ergaster | |

lying relationships. Moreover, the precise definitions of classes, phyla, and so on, are difficult to give and actually not of any particularly deep fundamental significance. The current fashion in taxonomy is *cladistics* (where *klados* is Greek for a young shoot) which scrutinizes the descent of organisms from a common ancestor and identifies the different branches, or *clades*, of the bush of life (Fig. 1.2). Cladistics was introduced by the German taxonomist Willi Hennig (1913–76) and elaborated in his *Phylogenetic systematics* (1966). According to Hennig, classification should reflect genealogical relationships and organisms should be grouped strictly on the basis of their descent from a common ancestor. Unlike the light-heartedness of theoretical physicists, who adopted everyday words like 'spin' and 'flavour' for their schemes, Hennig burdened taxonomy with Greek, and cladistics deals with symplesiomorphies (characteristics shared by more than one creature), synapomorphies (shared derived features), and so on; fortunately, we shall not need this burdensome language as we shall mostly use the Linnaean system. However, cladistics is very powerful, logical, and useful as it is based on the genealogy of organisms, which is arguably the only rational basis of classification.

Immediately, though, we run up against a profound problem which will pervade the rest of the discussion and troubles even the newer systems of classification: what do we really mean by a *species*? Even today, there is a lot of argument about the precise definition of the term. The argument has little practical significance, but because the concept is so central to historical arguments about the origin of species, we need to touch on it at least briefly. It might, in fact, be better to accept the impossibility of devising a universally valid definition, regarding the term 'species' as intrinsically fuzzy, and not to impose inappropriately rigid walls to box it in.

The common-sense definition of a species, the one adopted by what are sometimes called *typological taxonomists*, is a group of organisms that looks —that is, possesses identifiable morphological characters—unlike another group of organisms. Plato had much the same idea with his concept of *eidos*, or

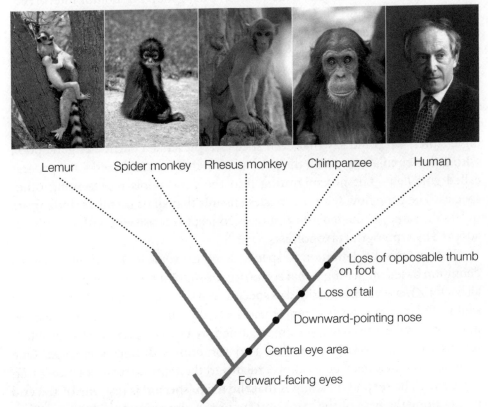

Lemur    Spider monkey    Rhesus monkey    Chimpanzee    Human

Loss of opposable thumb on foot

Loss of tail

Downward-pointing nose

Central eye area

Forward-facing eyes

**Fig. 1.2** In a cladistic classification, the tree, a *cladogram*, branches at the emergence of each significantly unique characteristic. Formally, we say that the classification is based on *synapomorphies*, which are shared derived homologies; a *homology* is a character inherited from a common ancestor. This cladogram indicates how humans fit into the scheme.

'perfect form', an ideal, a true essence, represented only imperfectly by actual beings. We have little difficulty in distinguishing a sparrow from a blackbird by their 'identifiable morphological characters' and regard them as different species of bird. We have no difficulty, we think, in recognizing the essential 'birdness' of the two creatures and seeing that it is distinct from the 'plantness' of a turnip, and we can distinguish the 'sparrowness' of one from the 'blackbird-ness' of the other.

A somewhat more sophisticated definition is the *biological species concept*, in which a species is defined as a group of organisms that breeds internally but is reproductively isolated from other such groups. According to this view, a species is an isolated island of vigorous reproductive activity, not unlike Mykonos in midsummer. This definition captures sparrow and blackbird as different species because they breed within each group but do not interbreed. Reproductive isolation can arise in many different ways. For instance, the groups of organisms might be geographically isolated—that is one reason why islands are so important in the history of evolutionary ideas—or breed at differ-ent times of the year. The groups might find each other repulsive (or, at least, unattractive), or find it frustratingly physically impossible to copulate however lustful they might feel towards each other.

If we anticipate the mechanism of heredity to be elaborated in the follow-ing chapter, we could say that each species represents a particular gene pool, with genes circulating within the pool as its members interbreed—the process called *gene flow*—but not migrating into the gene pools representing other species. The gene flow within a species ensures that all its members look more or less the same, so the biological species concept is consistent with the criteria adopted by typological taxonomists.

Then why is the definition of species so controversial? One problem with a definition based on mating is that some organisms don't mate. For example, not all bacteria mate yet are classified as species, and there are plenty of examples of multicellular organisms that reproduce asexually (such as the common dande-lion, *Taraxacum officinale*), yet are regarded as pukka species. This problem reveals that the word 'species' has two sometimes distinct meanings. One meaning, the one referred to above, relates to the reproductive isolation of an organism. The second meaning is that the term 'species' is just one of the end points along the base of the taxonomic pyramid, the ultimate unit of classifica-tion of a group of organisms regardless of whether or not they are capable of mating with other organisms. That is, a species is just a *taxon*, a unit of classifi-cation. The use of 'species' to denote simply a taxon is common in palaeontology, where a single lineage may be ascribed different names at different stages of its development and even though its successive members never had the option to consider mating. Thus, *Homo erectus* evolved into *H. sapiens*, and never walked out together: they are examples of what are sometimes called *chronospecies*.

The recognition of these difficulties has motivated alternative ways of defining species, definitions that sometimes span the boundaries of and conflict with the biological species concept. For instance, one way of classifying organisms is *phenetically*, in which organisms are put into the same groups according to purely objective measurements, including discrete measurements, such as giving 1 for 'has wings' and 0 for 'has no wings'. The 'know your partner' games in newspapers, magazines, and dating agencies are loosely phenetic. The advantage of the phenetic approach is that it is purely objective, and does not rely on making subjective judgements about the appearance of an organism or trying to guess whether, given the opportunity, one organism—perhaps now extinct—could mate with another. One problem with this scheme is that although phenetically identified groups of organisms look almost identical, they nevertheless might be unable to reproduce with each other. So, although they are the same phenetic species, they are distinct biological species. An example is the fruitfly *Drosophila* with its two (non-interbreeding) categories *D. pseudoobscura* and *D. persimilis*. These two organisms are virtually indistinguishable phenetically, so they form one phenetic species, but because they do not interbreed they constitute two biological species.

There are other definitions of what it means to be a species, and the application of the criteria they adduce muddies the water even further. The *ecological species concept* recognizes the importance of the role of the environment and the resources and dangers it supplies. It defines a species as a group of organisms that exploit a single ecological niche. The *recognition species concept* takes note of the ability of an organism to recognize a potential mate. An advantage of this definition, which is closely related to the biological species concept, is that whereas the ability to interbreed must often be inferred, recognition can often be observed directly. It may be the case that a new species emerges when one group of organisms fails to recognize as potential partners its former mates. That recognition need not be by appearance: plants and animals communicate in a variety of ways, including by sound and, more discreetly, or even unconsciously, to our senses, by the emission and detection of the chemicals we call *pheromones* and which humans sometimes incorporate, for ultimately similar reasons, into their perfumes and lotions. Finally (but only in this brief survey, for there are other definitions), there is the *phylogenetic species concept*, in which a species is defined as a group of organisms that have a common ancestor but differ in at least one characteristic. According to this definition, the members of two different phylogenetic species could differ in as little as a single characteristic and be able to interbreed.

There is no doubt that species have evolved and are evolving still. The evidence for past evolution is the fossil record, which provides an extraordinary sequence of images of the population of the Earth through time. The record is incomplete, just as at present no museum—a museum usually takes much better care of its possessions than the raw Earth—holds an example of every extant species, but it is sufficiently complete for us to be able to trace back through time the ancestry of living things, including our own origins in the—the cliché 'dim and distant' is inappropriate— surprisingly bright and recent past.

The science of the fossil record and its interpretation in terms of the history of life on Earth is called *palaeontology*. The word 'fossil' derives from the Latin word *fodere*, to dig up, via *fossile*, that which is dug up. The early fossil hunters subscribed to the Platonist view that a fossil was the image of an ideal form that had been created by the action of some kind of *vis plastica*. However, we now know that a fossil consists of the mineral parts of skeletons (bones are principally calcium phosphate together with the protein cartilage) and teeth (also calcium phosphate with various hard coverings). Fossils are found in sedimentary rocks, rocks that have been formed as minerals have been deposited and compressed, such as limestone. Igneous rocks, rocks that have oozed out on to the surface from deep below, are never populated by fossils. Some fossils are found in metamorphic rocks, which are sedimentary or igneous rocks that have been modified by high temperatures and pressures. Some fossils are organic material, such as wood, that has become mineralized as water has percolated through and filled internal cavities with stony deposits. The organic parent has gone entirely by the time we stumble upon it, and the fossil we unearth is a three-dimensional mineral facsimile of the original. Shells are often preserved, but the aragonite form of calcium carbonate from which they are formed is converted to the harder, denser form known as calcite. Organic materials are not preserved in this way, but imprints of feathers (a rigid kind of protein) and fleshy parts (composed of soft kinds of proteins lubricated by fats) are often found preserved in the rock in which the fossil is embedded. Some tiny creatures are preserved intact in the solidified resin we call *amber*. Bigger creatures, such as mammoths, have been found preserved in glacial ice.

The ground beneath our feet is alive in the sense that there is a ceaseless upwelling from the molten regions below to produce new regions of the lithosphere, the outermost solid encasing of the inner molten Earth. A rising plume of magma causes the lithosphere to spread from the upwelling zone and then to dive below again far away at a subduction trench. Embedded in this conveyor belt are the cow-pats of crusts we call the continents, which therefore migrate around the surface of the globe. These *plate tectonic* processes, originally proposed to a scornful world by the German geologist Alfred Wegener (1880–1930) and argued in his *The origin of continents and oceans* (1915), but since about 1960 accepted through work that showed how the hitherto presumed immobile, rigid sea floor could spread, have transformed the appearance of the Earth (Fig. 1.3). They have also caused local buckling of the continental crust with effects ranging from orogenesis (mountain building) to the formation of chasms, foothills, and valleys.

Amid this great stirring and crinkling of the countryside it is not surprising that in some cases mixing of geological strata has taken place, and that in places here and there a fossil of one generation has been brought below the fossil of another, and older fossils have been washed in from elsewhere to mingle with their descendants. We can usually detect these apparent inconsistencies by tracing the shape of the strata and seeing that it has buckled. In fact, when we consider the power of tectonic events in alliance with the tumultuous effects of climate, when oceans froze into ice ages, glaciers scraped to and fro, and then great hundred-metre-high tsunamis of meltwater replenished the oceans when the ice retreated, it is astonishing that there is any record at all of our distant past. Global war—Earth against organism, organism against organism—has raged on the relics of life, and we are lucky to find even a single tooth.

But we have found much more than a tooth. Luck in death, if there is any such thing, let a dinosaur die where it would not become a dinner, let it sink into the mud, be overlain with sediment, and in due course poke through into the daylight when erosion ablates the surface. The richest fossil record consists of marine invertebrates with hard skeletons that lived in shallow waters. The least well represented are organisms without skeletons, and easily damaged creatures such as birds. Some fossils occur in huge numbers: chalk hills are mounds of the fossilized remains of the single-celled algae known as coccolithophores

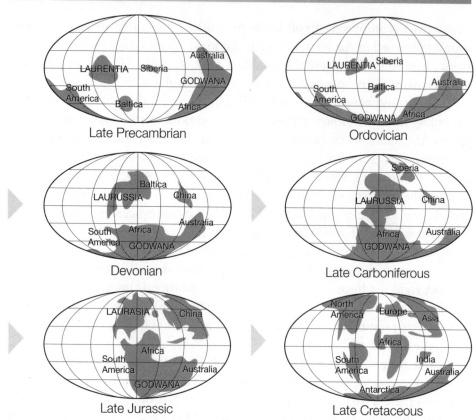

Late Precambrian

Ordovician

Devonian

Late Carboniferous

Late Jurassic

Late Cretaceous

**Fig. 1.3** Our vision of Earth, with its familiar distribution of continents, takes on an entirely different aspect if we regard it in the long term. On a time scale of millions of years, the surface is fluid, with continents drifting over the sphere as matter upwells from the interior and returns at distant subduction zones. On this sequence of diagrams we see the gradual emergence of the modern Earth over the past billion years (for the age corresponding to each named period, see Fig. 1.9). The diagrams mark the regions that were destined to become our present continents and countries.

(Fig. 1.4). These fossils are being laid down today, for about 1.4 billion kilograms of coccolithophores are deposited each year. Their presence in seawater is partly responsible for its opacity. Indeed, in the summers of 1997 and 1998, the entire Baring Sea changed colour from deep blue to aquamarine as it bloomed with billions of coccolithophores in their brief but muted moment of enjoyment of life en route to becoming uplands of the future.

The fossil record, glaringly incomplete as it is, is highly suggestive of evolution, with species coming and going, one species evolving into others, others going extinct, the whole resembling a bush with branches that branch, twigs that die, and the current biosphere as its leaves. The record seems to show the bush-like history of the biosphere, with sometimes ambiguous but plausible

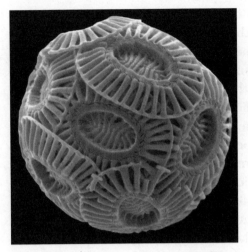

**Fig. 1.4** A scanning electron microscope (SEM) image of a common species of coccolithophore, *Emiliana huxleyi*. Each hub-cap shaped object is a separate coccolith. Our chalk and limestone uplands are built from dead and compressed coccolithophores.

lines of descent. However, there are alternative explanations of the fossil record, and because the matter is so important for an understanding of our place in nature, we need to examine them.

One alternative to evolutionism is transformism; we deal with this proposal later. Here we consider another alternative, *creationism*, in which each species is forever unchanging, except for minor variation. According to creationists, a species springs into existence ready-made, cleverly designed, with life breathed into it by an uncreated, omnipotent, worship-worthy Creator. Perhaps the species will be eternal but perhaps it will become extinct, stepping aside to make room for a new manifestation of the Creator's inscrutable whim. The Creator has an unbounded capacity for designing and building animals with seemingly inexhaustible variations on themes built round the ability to torture, maim, and kill one another. Among these dripping fangs lies, of course, the Creator's free-willed image, Man.

Creationism, including its transparently camouflaged variant 'Intelligent Design', is not science: it is an untestable assertion pursuing and impelled by an anti-science, religiously motivated agenda. To be maximally charitable, creationism serves much the same purpose as Galileo's Simplicio, a literary device for showing that a scientific explanation, in this case evolutionism, provides superior explanations. It is important that science is constantly required to provide explanations; the problem with creationism is that its proponents do not realize that they are but Simplicios and that their incessant petulant pestering and, even worse, distortion of the evidence, is time-wasting, tiresome, and in danger of sealing the eyes of the young to the actual glories of creation.

So, what are the arguments against creationism? There are so many that it would overburden this chapter to state them all. I will give a hint of their flavour by giving only three. First, numerous new species have been formed in recent times, which strongly suggests it has gone on in the past and that fossils are not just bin-ends, a record of extinction rather than evolution. Second, it is sometimes argued that evolution has no predictive power, therefore cannot be tested, and therefore is no more a variety of science than creationism is. That assertion is not true. The fact that evolution has occurred emerged from observations on the remains of and extant varieties of macroscopic creatures. In the twentieth century it became apparent that evolution could be traced at a molecular level. The effective prediction is that the details of molecular evolution must be consistent with those of macroscopic evolution. That is found to be the case: there is not a single instance of the molecular traces of change being inconsistent with our observations on whole organisms. Third, one of the legal tests of copyright infringement is to note whether a product replicates the mistakes introduced, sometimes intentionally, into the product that is copied. Mapmakers sometimes introduce tiny errors—an additional house, for instance, in a landscape—to trap plagiarists. There are two kinds of plagiarized mistakes in biology. In one, evolution starts off in a stupid direction (it has no foresight), and then has to live with the consequences. The mammalian eye is an oft-quoted example, for as the eye evolved it trapped itself into a potty design, as might be adopted by a Potty Designer, with the blood vessels in front of the retina and therefore having to leave the eye by poking through the retina and leaving a blind spot. This design has been followed ever since. The other kind of mistake occurs at the molecular level, in the form of *pseudogenes,* for instance, which are duplicated non-functional strips of mutant DNA, the equivalent of false houses in maps.[2]

Let's return to science and the established fact of evolution. *Microevolution* is the process of developing tiny modifications. *Macroevolution* is the generation of new species and higher taxa (orders, families, and so on) as a result of the accumulation of the changes brought about by microevolution, the process called *phyletic gradualism.* As we have remarked, the experimental evidence for this gradual evolution is clouded by the presumed incompleteness of the fossil record, which often lacks the transitional forms we might expect. There are two possible explanations. One is that the transitional forms did exist, but have vanished without trace. The alternative is that phyletic gradualism is incorrect, that the fossil record is more complete than people had thought, and that speciation

---

[2] For a detailed account of this evidence, see http://www.talkorigins.org/faqs/molgen/.

(the formation of new species) occurred in spurts of a few thousand year following a long period of quiescence, or 'stasis'. This highly controversial theory of *punctuated equilibrium* was proposed by Niles Eldridge and Stephen Gould (1941–2002) in 1972. In this theory, it is supposed that a small isolated community underwent a burst of modification in the process of *allopatric speciation* ('allopatric' simply meaning that variation occurs in a different geographical region from its ancestor). The ancestral site is therefore unlikely to contain a record of the intermediate forms, and fossils of the new species will be found at the ancestral site only if the fully evolved new species spreads back into it again: the understandable absence of intermediate forms enhances the impression of the abruptness of the transition between the two forms.

Phyletic gradualism and punctuated equilibrium, in the form it was first proposed, are probably best regarded as the opposite ends of a spectrum of possibilities. It is not appropriate to regard them as competing models of evolution, but more as marks on a meter that indicates the speed at which speciation occurs. Some events, the emergence of some species, correspond to a pointer reading close to gradualism, and other events, the emergence of other species, correspond to a pointer reading close to punctuation. It is extraordinarily difficult to distinguish the rate of evolution of a species and to be sure that the fossil record is complete. That is not to say that the more recent versions of punctuated equilibrium are not controversial, for it has been elaborated beyond the simple 'fast–slow' counterpoint of its earliest manifestation, in part by the proposition of mechanisms for the maintenance of stasis and the switch to episodes of rapid variation. The philosophical stance of the theory is also controversial, for whereas Darwinism proposes that speciation is the accumulation of changes representing adaptation, punctuated equilibrium views speciation as the driving force for adaptation. That such controversies exist should not be interpreted as a failure of the theory of natural selection (and certainly not the fact of evolution): they are signs of a vigorous debate about the details of one of the most important processes in the world.

There is one further point to emphasize. Evolution does not necessarily lead to greater sophistication: the direction of evolution is not always up. An organism may find that it can accelerate its reproductive activity, and thus more successfully populate the Earth, if it discards a lot of social or anatomical baggage. Why bother with a lot of socially contrived activities if one can get down to the central job of reproduction without them? Moreover, the habitat may change, and the remnant members of an unsuccessful species might suddenly find that their hour has come, and under the changed conditions can out-reproduce their hitherto more successful rivals. The tunicate, the sea squirt (*Ciona intestinalis*), has another solution, and is the ultimate couch potato. This little chap is a motile hunter in its larval form and therefore needs a brain. However, once it has found a suitable niche to which it can anchor itself to become sessile, it no longer needs to think, so it eats its own energetically burdensome brain. Brains are great consumers of energy, and it is a good idea to get rid of your brain when you discover you have no further need of it.

How does all this rich variety of life originate? William Paley, as we have seen, knew that he knew, for he was confident that each species was God's creation, and that was that. Jean Baptiste Pierre Antoine de Monet, Chevalier de Lamarck (1744–1829) also thought he knew, and is intellectually more admirable than Paley in so far as he struggled with the problem of finding a mechanism. Lamarck, first soldier, then bank clerk, then later assistant botanist, and finally professor of insects and worms, spent a lifetime in poverty, his last few years completely blind. Poverty dogged him even in death, for he was buried in a rented grave, to be turned out when his tenure expired after five years to make way for a new incumbent, and his remains dispersed. His name is now more associated with scorn than respect, yet he deserves respect as the founder of invertebrate biology (a term he coined) and for at least trying to find an explanation of the existence of species. He began to publish his speculations—they were no more than that, they were certainly not scientific theories—on the mechanism of evolution in 1801, but his most complete account was presented in his *Philosophie zoologique* (1809).

Lamarck supposed that all organisms are engaged in a metaphysical quest towards perfection and become transformed from an original protist seed that contains some kind of Platonic essence of the species. This quest is driven by 'nervous fluids' of various ill-defined kinds which nourish organs that are exercised and starve organs that are not. He also speculated—and it is this feature for which he is best remembered, although he probably regarded it as a minor component of his overall thesis—that characteristics once acquired are inherited. His best-known example is the elongation of the neck of the giraffe as it strives to reach higher leaves and become an ever more perfected giraffe, with the elongation achieved in one generation being inherited by the grateful next.

We might scoff at the simple naivety of the idea, but before molecular biology ruled out any possible mechanism for such inheritance, it was difficult to disprove the concept. Lamarckian views, which are referred to as *transformism* rather than evolution, persisted well into the twentieth century. Jocular disproofs were common but irrelevant: that the circumcision of many successive generations of Jews had not led to the atrophy of the foreskin is not an argument, because the boy child was not striving to lose his foreskin. In a celebrated series of disagreeable experiments the influential German biologist August Weismann (1833–1914) cut the tails off many successive generations of mice, and found no diminution of the length of tails of subsequent generations. All such mutilation experiments—and there have been many, both accidental and contrived—although relevant to disproving that acquired characteristics are inherited, are irrelevant to the central aspect of transformism, Lamarck's view that *striving* is central, for only then do the transforming juices run.

In *The vestiges of Creation* (1844), the publisher Robert Chambers (1802–71) had caught a glimpse of a possible explanation. He recognized the importance of mutations, but argued that new species spring capriciously from the accident of monstrous birth. Thus, if a fish were unaccountably born with wings, feathers, and a beak, then the biosphere would acquire something resembling its birds. At about the same time, the *Bridgewater treatises*, a collection of works sponsored by the will of the Reverend Henry Egerton, eighth and last Earl of Bridgewater, 'to demonstrate the power, wisdom, and goodness of God as manifested in the Creation illustrating such work by all reasonable arguments as, for instance, the variety and formation of God's creatures', was a vehicle for the expression of a number of contemporary ideas. The contributions included 'The adaptation of external nature to the moral and intellectual constitution of man' by Thomas Chalmers (1833) and 'The adaptation of external nature to the physical condition of man' by John Kidd (1837). From a modern point of view, both contributions represent the exact opposite of what we now believe to be the case.

At this late stage of the chapter, the hero of evolution, Charles Robert Darwin (1809–82), finally walks diffidently on to the page. Darwin's success at identifying the origin of different types of organism can be traced to his immersion in the natural world from 1831 to 1836, when he served nominally as gentleman's companion but actually as naturalist on board HMS *Beagle*, under Captain Robert FitzRoy, an illegitimate descendent of King Charles II. FitzRoy wanted the company of a gentleman on the lengthy, lonely journey, not least to avoid the fate of his predecessor on the ship, who had shot himself, and in fear of what may have been a disposition, for a few years earlier his uncle, the Home Secretary Viscount Castlereagh, in a fit of depression had slit his own throat.

Immersion in a plethora of seemingly overwhelming data is often the prelude to seminal discovery, with the subconscious beavering away in its own back-room, seeking patterns, and then finally erupting into conscious thought to generate that most precious of personal scientific events, a *eureka*.

During the five years of his voyage, Darwin spent many months on land, usually welcoming it as a relief from the seasickness that rarely left him when on board the tiny vessel.[3] His most famous sojourn, for five weeks from 15 September 1835, was on the Galapagos Islands (the 'islands of tortoises'), off the coast of Ecuador in the Pacific Ocean, where the *Beagle*, like so many before her, had called to collect the characteristic huge tortoises for fresh meat to see her through the next leg of the voyage. All the tortoises on the larger islands have since been hunted to extinction; some species on the smaller islands survive. The Galapagos are a series of volcanic islands that Herman Melville, another visitor at a different time, referred to, with far less insight than Darwin was to exercise, as 'five-and-twenty-heaps of cinders dumped here and there in an out-side city lot'. But even Darwin did not appreciate the significance of his visit until the islands were well astern, for he recorded that it was hard to imagine 'tropical islands so useless to man'. The swirling fogs and shifting currents that surround the islands gave rise to the nickname *Los Encantadas* ('the enchanted ones'); as indeed they were, for the metaphorical fog that hitherto had shrouded the origin of species began to clear as Darwin, tucking with relish into the flesh of the tortoises he had helped to slaughter, reflected on the differences between the corpses of the birds collected from different islands (he visited only San Cristóbal, Floreana, Isabela, and Santiago). As he was to report,

Several of the islands possess their own species of the tortoise, mocking-thrush, finches, and numerous plants, these species having the same general habits, occupying analogous situations, and obviously filling the same place in the natural economy . . . [it] strikes me with wonder.

As we have remarked, islands were crucial to the identification of the theory that Darwin was in due course to call *natural selection*. Not only do they simplify the ecosystem and so render differences more easily observable, but they effect-ively isolate populations and so allow variation and adaptation to develop.

Filled with tortoise and struck with wonder as he was, Darwin still lacked the spark to bring his thoughts into the light. The spark sprang, he subsequently claimed, on 28 September 1838, while he was still reflecting on the abundance of information he had jackdawed away on his lengthy voyage. Reading, for amuse-

---

[3]  The *Beagle* was 27 metres long and 7 metres wide amidships, equivalent to about 15 × 4 'darwins'.

ment, Malthus's *Essay on the principle of population* (1798), in which the elegant and refined Reverend Thomas Malthus (1766–1834), a professor of political economy employed to teach economics to employees of the East India Company, argued that humanity was doomed because populations increase faster than food supplies, consequently humanity would inevitably outgrow its resources. Darwin later recollected:

being well prepared to appreciate the struggle for existence which everywhere goes on from long-continued observation of the habits of animals and plants, it at once struck me that under these circumstances favourable variations would tend to be preserved, and unfavourable ones to be destroyed.

As Thomas Huxley (1825–95), Darwin's bulldog, was later to say, 'How extremely stupid not to have thought of that'.

For well nigh twenty years Darwin reflected upon this observation, gradually building his theory of natural selection, accumulating illustrations, never quite losing his belief in the Lamarckian inheritance of acquired characteristics, and fearing the consequences of publication. He began to write an account of his ideas in 1856, intending it, like George Eliot's Dr Casaubon, to be huge and authoritative. But his plans were interrupted, for there were other island-visiting readers of Malthus. Darwin was appalled to receive a manuscript from Alfred Russel Wallace (1823–1913) entitled *On the tendency of varieties to depart indefinitely from the original type*. Wallace was a distant descendant of the Scottish hero William Wallace, and had travelled extensively in Amazonia as a professional specimen collector from 1848 to 1852. With little in prospect in Europe, he decided to resume his increasingly lucrative but arduous collecting career, and selected the Malay Archipelago (the Indonesian Archipelago) as his destination, arriving in Singapore in 1854. In February of 1858, after years of travel and collecting, and while suffering from an attack of malaria in the Moluccas (the precise island is uncertain, but it was probably either Gilolo or Ternate), he realized—like Darwin—that Malthus's ideas held the key to the explanation of evolution.

Darwin was in a quandary, for his was an idea that he had already nurtured for two decades and priority was about to be dashed from his grasp. He consulted his friends Sir Charles Lyell and the botanist Joseph Hooker. Without being able to consult Wallace, they decided to present the latter's essay and Darwin's accumulating notes at the next meeting of the Linnaean Society in London on 1 July 1858. From that moment, natural selection was out of the bag. Darwin forsook his magnum opus, and frantically and mercifully abbreviated his planned account, which was published in November 1859 as *On the origin of species*, or more precisely, with its full Victorian gothick decoration restored, as *On the origin of species by means of natural selection or the preservation of favoured races in the struggle for life*. Even Darwin thought this title a little cumbersome, and in subsequent editions (there were five more) he went as far as dropping the *On*.[4] As Darwin remarked:

I had two distinct objects in view; firstly to show that species had not been separately created, and secondly, that natural selection had been the chief agent of change.

Our attention must now focus on Darwin, the generally acknowledged discoverer of natural selection. But it would be wrong to ignore Wallace completely in this connection, not least because of the nobility with which he effectively ceded priority to Darwin. However, there are certain features of Wallace's later, very long life that reduce his stature in the field. He could never accept that humans could evolve without a little divine puff and guidance, and he sought to confine natural selection to the evolution of form, leaving consciousness to be moulded by something higher. By extension, much to the horror of his friends, later he also lost himself in the endless murky byways of spiritualism itself.

Natural selection is such a simple idea, yet it is so complex to apply because the considerations it is necessary to make require great circumspection. In short, no tortoise is an island, and to consider the role of natural selection for a species of tortoise, we must consider its response to all the biota in its vicinity as well as the physical state and climate of its location. The evolution of a tortoise will also have consequences for its competitors and predators, which in turn will impact back on the tortoise. Unlike simple linear systems in which influences travel down a simple chain of command, the biosphere is an extraordinarily rich non-linear system, with changes in an organism reflecting back on the organism as its evolution modifies its environment. The development in time of non-linear systems is very difficult to predict, and it is no wonder that evolutionists are unable to predict the future of the biosphere, that apotheosis of non-linear complexity. Here, I will outline some of the ideas characteristic of the *modern synthesis*, or *neo-Darwinism*, which emerged as ideas relating to genetics were

---

[4]   The word 'evolution' does not appear in the *Origin*, except as the final word 'evolved'; nor does the book deal explicitly with the origin of species, which is still a highly vexed question.

found to underpin ideas relating to observational natural history during the early part of the twentieth century. Indeed, natural selection was not fully accepted until the 1930s and the establishment of the modern synthesis. As I have already indicated, in this chapter I am confining myself largely to phenomenology, leaving the molecular basis of evolution to the next.

Natural selection depends on three principles:

### 1.  *There is heritable genetic variation.*

That is, members of a given species are not identical clones; there is genetic noise within a species. Darwin had no conception of the mechanism of inheritance, and favoured a 'blending' theory in which the characteristics of the copulating parents went into a kind of mixing pot. This ignorance about the true mechanism and inclination towards a mechanism which, as his critics were fast to point out, could not result in evolution, was one of the main stumbling blocks for the acceptance of his ideas. The story might have been different if Darwin had bothered to open and read a letter from an obscure monk, Gregor Mendel, who had effectively handed him the key.

### 2.  *Parents over-proliferate.*

That is, in an echo of Malthus, parents produce more offspring than can survive. Some species, such as the elephant, produce only one offspring, and it might die; others, such as frogs, produce thousands, of which perhaps only one survives. Offspring over-proliferation occurs less in big, complex organisms that have to commit themselves to years of parental care, like elephants and, perhaps, middle-class parents in Western countries.

### 3.  *Successful offspring are the ones best adapted to the environment.*

'Success' is more than just survival: it is the ability to go on to reproduce. This principle is the spring of the nineteenth-century right-wing libertarian Herbert Spencer's unfortunate and much misunderstood phrase 'the survival of the fittest', which he coined (in about 1862) in connection with the development of his *Social Darwinism*, in which he crudely extended the subtle ideas of natural selection to the dynamics of societies and opened the door to eugenics, the elimination of virtually all forms of state intervention, and racism. Like all good slogans, 'survival of the fittest' is memorable, and Darwin was seduced into using it in later editions, but it debases the subtlety of the underlying idea.

When considering natural selection, we must keep in mind that it is wholly local in space and time. Natural selection is totally committed to the present

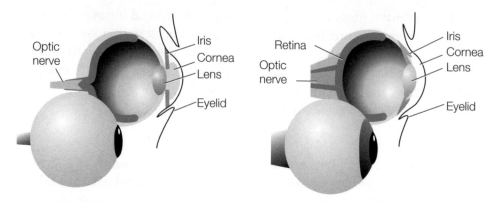

**Fig. 1.5** The diagram on the left shows the general layout of a mammalian eye. Note how the blood vessels are on the front of the light-sensitive retina and have to find a way out through the retina, so leaving a blind spot. The diagram on the right shows the seemingly more sensible arrangement in a squid, with the blood supply at the rear of the retina. Although evolution stumbled blindly into each arrangement, neither could be reversed as the survival value of light sensitivity—which evolved into vision—was so great. Incidentally, there does seem to be at least one advantage of the mammalian arrangement: the flow of blood in this arrangement may help to lower the incidence of disease.

and is wholly without foresight. If an adaptation now turns out to be regrettable in the future, then hard luck on the future: natural selection cannot anticipate that it is driving a species into an evolutionary *cul de sac*; indeed, it cannot anticipate anything at all, not even the following day. Natural selection lives for the moment, it is the ultimate in *carpe diem*. The mammalian eye is an example that we have already mentioned: by a quirk of evolution, the original photosensitive patch that in due course was to evolve into our principal perceptive organ began with blood vessels on the side of the patch that would in due course result in them overlaying the retina (Fig. 1.5). Photosensitivity is such a powerful weapon of predation and avoidance that it was more valuable for an organism to persist with this unfortunate arrangement than discard the advantage by retrenching, inverting the order with a view to better eyesight a million years later. The eye of a squid is more perfect in this respect (but not in others), for it developed along an evolutionary path in which the blood vessels happen to lie behind the photosensitive retina. Another example is the inconvenience in the arrangement of the pipes inside our mouth, where our respiratory and food passages intersect and thereby open up the possibility of choking. The passages intersect because in an early lungfish ancestor, the air opening the fish used for breathing at the surface was very appropriately located at the top of the snout and led into a common space shared by the food passageway (Fig. 1.6). There was no going back from this arrangement, even though it has its dangers. The

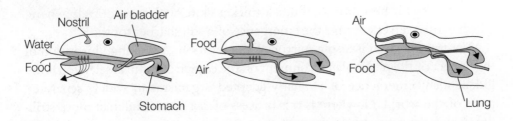

**Fig. 1.6** Another instance of unfortunate lack of foresight is the blind evolution of the mammalian respiratory and digestive systems. The diagram on the left shows the layout of a typical fish. The nostril (formally: *nares*) leads to a closed cavity and is used mainly for olfaction. Oxygen is extracted from the water which enters through the mouth and is expelled through the gills. The air bladder is used to control depth, like ballast tanks on submarines. The middle diagram shows the arrangement in a lungfish, the ancestor of modern mammals. The nostril opens into a passage into the mouth cavity, but is still used only for olfaction. Air is gulped through the mouth and enters the air bladder. It is only a short evolutionary step to the typical arrangement in a mammal, shown on the right, with the nostrils now used to import air. Unfortunately, air and food share a chamber before the former travels to the lungs through the trachea and the latter to the stomach through the oesophagus. This cobbled-together, cheapskate, penny-pinching, but evolutionarily comprehensible arrangement holds out the danger of choking.

seemingly unhygienic economy of using the penis for both copulation (including, particularly in humans, its accompanying rituals) and urination has a similar evolutionary basis and, moreover, the tube leading from testis to penis loops on the wrong side of the tube leading from kidney to bladder.

Natural selection is essentially unpredictable because it is the outcome of sometimes competing tendencies, and adaptations that at first sight might be advantageous remain unachievable. A minor example is the human appendix. For us, an appendix is a danger, for it can become diseased and result in death. Appendicitis results when infection causes swelling, which compresses the artery supplying blood to the appendix. A steady blood flow into the appendix protects against bacterial growth, so any reduction of the flow aids infection, which leads to more swelling. If the blood supply is cut off completely, the bacteria thrive and the appendix bursts. A small appendix is more susceptible to this chain of events than a big appendix, so appendicitis applies the selective pressure that maintains a large appendix in the sense that it is more dangerous to start to shrink than to carry on with what we have got. Consequently, despite its dangers, it is extremely difficult for evolution to eliminate an appendix.

Natural selection is an arms race. The *Red Queen hypothesis* is the idea that predators and prey are engaged in a constant battle, with predators evolving better predation strategies and techniques and prey doing likewise. (The Red Queen instructed Alice to go on running faster in order to stay still.) A sharpened tooth here brings about a thicker skin or fleeter foot elsewhere, which in turn can encourage the emergence of a still sharper tooth.

Natural selection is also a mirror of the location. A striking illustration of the impact of the physical environment on the course of natural selection is the independent emergence of similarly adapted organisms in widely separated parts of the world. Nowhere is this process of *convergent evolution* more striking than in the emergence of marsupial versions of placental mammals: in the former the embryo develops principally in an external pouch and in the latter it develops principally *in utero*. Thus marsupial versions of mammals evolved when Australia broke off from Antarctica during the Cenozoic era about 65 million years ago and cruised north, like a huge Noah's ark, with its isolated ecosystem (Fig. 1.7). The North American wolf (*Canis lupus*), a placental mammal, is similar in appearance to the marsupial Tasmanian wolf (*Thylanicus cynocephalus*). Natural selection's exploration of the available niches has led to many analogous solutions (Fig. 1.8): the mammalian ocelot (*Felis pardalis*) resembles the marsupial tiger cat (*Dasyurus maculatus*), the flying squirrel (*Glaucomys volans*) apes the honey glider (*Petaurus breviseps*), the woodchuck (*Marmota monax*) shadows the wombat (*Vombatus ursinus platyrrhinus*), and the common mole (*Scalopus aquaticus*) looks like the marsupial mole (*Notoryctes tryphlops*). Even the house mouse (*Mus musculus*) has its marsupial *doppelgänger*, the yellow-footed mouse (*Antechinus flavipes*).

We can begin to grasp the interconnectedness of it all by noting that when the volcanic Panama landbridge between the North and South American landmasses, which were separate fragments of Laurasia and Godwana, respectively, emerged about 3.5 million years ago, not only did it lead to interspecies battles as the mammal populations of the North flooded south and battled for survival with the predominantly marsupial populations of the South, but the landbridge

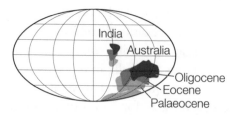

**Fig. 1.7** Australia broke away from the rest of Godwana about 60 million years ago (for the times of the epochs, see Fig. 1.9), and floated north-east with its cargo of isolated animals, which underwent isolated evolution in this one huge island. At much the same time, India floated north, smashed into the continental landmass, and the ensuing crumpling produced the Himalayas.

Flying squirrel *Glaucomys volans*

Honey glider *Petaurus breviceps*

House mouse *Mus Musculus*

Yellow-footed marsupial mouse *Antechinus flavipes*

**Fig. 1.8** Although Australia (and South America) was isolated, evolution had to confront similar problems and came up with similar solutions. Here are two examples of mammals and their marsupial equivalents.

upset the circulation of the oceans and induced the onset of an Ice Age which modified the biota across the planet.

However, evolutionary war is only one component of the driving force of change, for changes in the physical environment also play a central role in driving evolution. These changes include the clearing out of niches by mass extinction and thereby allowing new bubbles of population to develop. *Catastrophism*, the idea that the world is subject to sudden upheaval, as dramatized in the myth of Noah's flood, was the force for change favoured by the highly influential French anatomist and founder of vertebrate palaeontology, Baron Georges Léopold Chrétien Frédéric Dagobert Cuvier (1769–1832), who had as many forenames as there are geological eras, but it fell out of favour when geology became established. The rationalization of geology started in the hands of James Hutton (1726–97) in his *Theory of the Earth* (1795) and was propagated forcefully by Sir Charles Lyell (1797–1875) in the three volumes of his *Principles of geology* (1830–33; Darwin carried a copy with him on the *Beagle*). Hutton and Lyell favoured *uniformitarianism*, in which the physical nature of the Earth is regarded, on the basis of a great deal of evidence from the analysis of strata, as having undergone a slow, steady transformation. However, we now know that there have indeed been catastrophes, most famously the asteroid impact which all but eliminated the successful but insufficiently genetically flexible dinosaurs. These great creatures were doomed by the lack of vegetation that the artificial dust-shrouded night brought or may have been incinerated in a world in which the atmospheric oxygen concentration was possibly significantly higher than it is now. Their demise opened up the world to a surge of mammals.[5]

We shall need to refer to some of the geological eras and periods into which the history of our plastic planet has been divided (Fig. 1.9). Their names have been assigned capriciously, but Wales and the West Country (of England) have done rather well: Cambria (for *Cambrian*) is an ancient name for Wales, Ordivices and Silures (for *Ordovician* and *Silurian*) the names of pre-Roman Welsh tribes, and Devon (for *Devonian*). The names of the epochs into which some geological periods are divided have a slightly unhinged air: they include Palaeocene ('old recent'), Eocene ('dawn of recent'), and Oligocene ('few recent'). I will add, in parentheses, the etymologies of the other names as they arise, except for the remnants of an early attempt to name the periods systematically, as for the Triassic, Tertiary, and Quaternary periods.

The dinosaur-doom extinction at the end of the Cretaceous ('chalky') is only the best known of at least five major events. In the catastrophic event— what-ever it was—that brought the Permian period (Perm, in Eastern Russia) to a close, more than 95 per cent of marine animal species vanished. The Ordovician period was terminated abruptly 440 million years ago, the Devonian 350 million years ago, the Permian period 250 million years ago, the Triassic 205 million years ago, and the Cretaceous 65 million years ago. The causes of most of these extinctions are still largely unknown, but there is no shortage of ideas, including asteroid impacts and drastic falls in sea levels accompanying global cooling. These extinctions are traumatic, but life is very resilient and species diversity rebounds very quickly: within 5–10 million years the diversity equals and often exceeds pre-extinction levels. Extinction events sweep away competitors, open up niches that are ripe for colonization, and are best regarded (except for the extinctees) as glorious opportunities. However, although extinctions are important, we should not exaggerate them. A typical animal species persists for about 2 million years, but extinction bursts occur typically every 20–30 million years, so most species never experience extinction by catastrophe. The

---

[5] If, as is increasingly thought probable, birds are descendants of dinosaurs, then the dinosaur format has proved amazingly resilient and long-lived.

| | Era | Period | | Epoch | Headline news |
|---|---|---|---|---|---|
| Now | | | | | |
| 0.01 | | Quaternary | | Holocene | |
| 2 | | | | Pleistocene | Ice ages, extinction of large animals |
| 5 | Cenozoic | | Neogene | Pliocene | Early hominines (or hominids) |
| 25 | | | | Miocene | |
| 35 | | | Palaeogene | Oligocene | |
| 55 | | | | Eocene | |
| 65 | | Tertiary | | Palaeocene | Early mammals |
| 145 | | Cretaceous | | | |
| 205 | | Jurassic | | | First birds and mammals |
| 250 | Mesozoic | Triassic | | | First dinosaurs |
| 290 | | Permian | | | Extinction of invertebrates |
| 350 | | Carboniferous | | | First reptiles |
| 400 | | Devonian | | | First amphibians, first forests |
| 440 | Palaeozoic | Silurian | | | First air-breathing animals, land plants |
| 500 | | Ordovician | | | First vertebrates |
| 540 | | Cambrian | | | |
| 700 | | | | | First animals |
| 3400 | Precambrian | | | | First organisms |
| 4600 | | | | | Formation of the Earth |

*Millions of years ago* (vertical axis label)

**Fig. 1.9** The geological ages of the Earth, with the names given to the eras, periods, and epochs into which each is divided. Some of the major events are given in the column on the right. The numerical ages are only a guide, and vary from source to source.

dinosaurs' misfortune was that they were so successful: they survived long enough to be extinguished.

Currently, we seem to be in the midst of a novel kind of mass extinction, with human activity rendering the biosphere uncongenial to much of the biota it has to share it with, and possibly to itself. Self-induced extinction of this kind may be an ineluctable concomitant of 'progress', for in an ultra-pessimistic neo-Malthusian viewpoint, it may be that the ability to annihilate oneself inevitably outstrips the development of intelligence. The most gloomy view is that although societies can survive when individuals can kill only a few thousand at a blow (as throughout human history until now), no society can survive when technology has developed to the point at which a single individual has the power to kill tens of millions. Human society may just have arrived at such a point. If it is a general rule for societies on all planets, then there is little hope that we will ever fulfil the cosmic aspirations of humanity that optimistic science fiction so imaginatively inspires. But, at least our own extinction will give opportunities to cockroaches.

For all the rich interplay between geography and genes, the romping ground of natural selection, there remain several central questions. One is the nature of the entity on which natural selection is acting. Is it acting on gene, individual, or species?

We can rule out species as the unit of selection. Organisms do nothing on behalf of their species. Just as natural selection is blind to the future, so it is blind to the clan. An individual competes with other individuals and is driven to seek its own success regardless of the good of the aggregation of organisms that constitute the species. The reproductive drive of an individual invests in selfish behaviour and has no conception of *altruism*, unconscious behaviour that results in self-sacrifice on behalf of others.[6] That is not to say that many kinds of behaviour do not look altruistic—it is only when we examine them carefully that we discover that they are wolves in sheep's clothing and that the altruism is truly selfishness red in tooth and claw. In *reciprocal altruism*, in a variation of the social contract that governs ideal human society, an organism indulges in selfishness by cooperating with other organisms, in large measure because in times of stress the helper might be helped.

---

[6] Note that altruism and selfishness in human behaviour are normally conscious activities; in genetics they are classifications of unconscious, instinctive, preprogrammed behaviour.

At a deeper level, we need to understand that members of a species share genes, and that by helping an apparent competitor to reproduce, an organism is covertly facilitating the propagation of its own genes. This type of altruism is called *kin selection*. Thus, the theoretical biologist J. B. S. Haldane (1892–1964) expressed the view that he would gladly drown if by doing so he saved two siblings or ten cousins. Each of his siblings would share one half of his genes; his cousins would share one eighth (so saving eight cousins would be quits, and saving ten would be to his genes' advantage). The grip of our genes on our behaviour suggests that we should look below the level of species, below the level of individual, and right down into the depth of the genes.

A problem with this view is that there is rarely a one-to-one correspondence with behaviour. Not only is the biosphere a conspiracy of complexity, but so too is the manifestation of *genotype* (the genetic makeup of an organism) in the *phenotype* (the physical characteristics of the organism). Some organisms will deny themselves the joy of reproduction but still contribute to the future by helping their close relatives to reproduce instead. Their queen's genes are so close to their own that they achieve the propagation of their own genes by facilitating her reproduction instead: she can churn out replications of their genes without them having to go to the trouble of doing it themselves.

Another problem is to keep track of the consequences of competition at one level (individual, for instance) for the higher level (the species). It could be that an advantage to an individual is deleterious to the group. Because an individual has no evolutionary foresight, it will disregard the consequences of its own behaviour for the group. When food is scarce, some individuals will continue to breed and pass on their genes to later generations; they will not abstain on behalf of the species. As a result, the species will evolve in the direction indicated by the gene flow of the selfish replicators. In modern evolutionary biology, *group selection*, selection at the level of a species or comparable group of individuals, is frowned on: natural selection takes place at a lower level, and all evolutionary trends that appear to indicate selection between species can normally be traced to a consequence of selection at a lower level. In fact, provided we exclude the special case of kin selection, there are no definite examples of adaptations that unequivocally benefit the group: there is no content to the slogan 'for the benefit of the species'.

The problem of the unit of selection can be expressed in a different way, for selection is maximal at a certain level. At the lowest level of being, at the level of atoms, it doesn't matter who butchers whom, because all atoms survive murder, mayhem, and massacre. At a much higher level, let's take the kingdom

Animalia, it also doesn't matter who butchers whom, because the kingdom survives regardless of its changing composition. The effect on survival is much more significant when we reach the level of individuals and their genes, for the distinction between killer and killed is now vitally important. Edging up the scale a little brings us to species: the death of an individual certainly affects the future of the species, because it is usually better to have as many reproducers around as possible and your survival is a contribution, provided you are reproductively capable. The class Mammalia is also a bit more likely to survive if the diner is a mammal and the dinner is not, but dog-eat-dog—in general, mammal-eat-mammal—is almost neutral. Edging in the opposite direction, down the scale from individual, we encounter its genes, which are the blueprint for the individual and the species. Is a dinner of someone else's genes more or less important than a dinner simply of someone else?

One approach to determining the unit of selection is to identify what entity is potentially immortal. Atoms are immortal, but they are the representatives of the empire of minerals, not the empire of organisms. The components from which the double-helix of DNA is formed (the 'nucleotide bases', which we discuss in Chapter 2) are intrinsically lifeless, just as the letters of the alphabet are not literature. Even if these components were immortal, they would not properly be regarded as alive. The human genome, the entire complement of DNA in each cell, is also not immortal, because it is chopped and changed in a process called *meiotic recombination* when sexual reproduction takes place, with one strip—a gene—replaced by another (see Chapter 2 for a discussion of this process too). But we have jumped a level: the gene, a reproductively active strip of DNA. A gene is potentially immortal—until it undergoes a mutation—for it is transferred from genome to genome, from mouse to mouse, virtually intact.[7] Is it then the unit of selection? In his book *Adaptation and natural selection* (1966), George Williams argued that a gene should be regarded as any portion of chromosomal material that potentially lasts for enough generations to serve as a unit of natural selection. In his justly famous book *The selfish gene* (1976), the Oxford zoologist Richard Dawkins (b. 1941) developed this idea ruthlessly and explored how by acting selfishly, a gene spreads into the biosystem and propagates its own survival.

I mentioned in the Prologue that science typically deepens its insights and extends its reach by adopting greater levels of abstraction. That trend is discernible in biology. Natural selection is a natural compost heap for the cultiva-

---

[7] I say 'virtually intact' because even if the random breaks in DNA that occur during meiosis occur in the middle of the gene, the recombination step reconstitutes the gene in the new genome.

tion of abstraction, and the identification of the gene as the unit of selection is a major step in this direction. Thus, Dawkins looks for natural selection taking place at the lowest level of all, the gene, and regards an organism as a throw-away vessel that the ruthlessly selfish (I emphasize, in a technical sense) gene employs to ensure its own propagation. The unconscious gene unconsciously moulds its vessel, its phenotype, to adapt as well as possible to its environment, for the best-adapted vessels will ensure that the gene proliferates.

There is a still lower level of selection, one that is even more abstract than the gene, one that is potentially even more immortal. A gene encodes pheno-typic information, such as information about the layout of the body, its col-oration, or the physiological modifications needed for amplifying the loudness of a roar. The gene is a physical entity which has to be renewed as metabolic processes copy the strands of DNA and ensure that replicas are passed to every cell and to the next generation. As such, as a physical entity, even the gene is not immortal, for the physical gene must be constantly rebuilt. The fact that the information is encoded in DNA is a detail, an implementation not a foundation. When we regard a gene as the unit of selection, we are actually focussing on the *information* it conveys, and just as the organism's body is a disposable vessel for the gene, so the sequence of bases in DNA is a disposable physical realization of the information the gene contains. The really immortal component of life is not the physical gene, it is the abstract information it contains. Information is immortal, and information is ruthlessly selfish. Genetic information is probably the ultimate unit of selection, with DNA its realization and a body its discard-able, subservient vessel.

The living world emerged when inorganic matter stumbled on a way of passing on intricate, unpredictable information, and found that it could achieve immortality for that information by its ceaseless replication. Here lies another furiously running Red Queen, for permanence is achieved only by perpetual replication. In the same spirit, our own nominally civilized, cultivated, intelli-gent, and reflective level of life emerged when organisms stumbled on a way of passing on intricate, unpredictable information to others around them and fol-lowing them. It did so by inventing language and effectively binding together all human organisms, past, present, and future into a single mega-organism of potentially boundless achievement.

With that rhetorical but heartfelt flourish behind us, it is time to get down to sex. One of the most puzzling aspects of natural selection is the evolution of sexual reproduction. At first sight, sex looks like a good idea, in the sense that it endows a species with genetic flexibility and rapid response to changing conditions. However, there are problems.

Firstly, sex is unnecessary. Quite a few species manage to get along perfectly well without it. *Parthenogenesis* (virgin birth) is common among plants, where it is more properly called *parthenocarpy*. We have already mentioned parthenocarpic dandelions, but could add other common plants, such as blackberries (*Rubus*) and ladies mantle (*Alchemilla*). Some reptiles reproduce asexually, most notably New World lizards of the genus *Cnemidophorus* (family Teiidae), Old World lizards of the genus *Lacerta* (family Lacertidae), and the Brahminy blind snake (*Ramphotyphlops braminus*; family Typhlopidae). No mammal reproduces asexually, despite biblical assertions to the contrary.

Secondly, sex is unstable. Suppose a certain species reproduces sexually, producing a lot of offspring, of which half will be male and half female. For the population to remain approximately constant, all except about two of these offspring will die, leaving on average one male and one female. Now suppose a mutation occurs in one female and that she is able to reproduce asexually. Once again she will give rise to numerous offspring, of which about two will survive; however, those offspring, being clones of the mother, are both female. Both can reproduce parthenogenetically, giving rise to even more females. Provided the lone asexual female produces the same number of offspring as the sexual couple (a questionable assumption, of course, as fathers often have roles after copulation), after a few generations, the parthenogenetic female population will have swamped the initial population. There must be a counterbalancing advantage of sex that ensures stability.

Thirdly, sex is highly complex. Sexual reproduction depends on the intricate mechanism of *meiosis* in which, as we see in Chapter 2, the number of chromosomes in the germ-line cells (the gametes, the sperm and egg) is halved but brought back to its somatic (typical body cell) number on fertilization. What are the extraordinarily powerful selective pressures that lead to the development of this elaborate mechanism? There is nothing unusual about the development of complex mechanisms by jobbing together and modifying pre-existing anatomical and biochemical features—the numerous times that eyes have evolved independently is an example—but, as in the possession of an eye, there has to be a profoundly compelling pay-off, an offer the organism can't refuse.

The Oxford biologist William Hamilton (1936–2000), whom Richard Dawkins regarded as a candidate for the title of the most distinguished Darwinian since Darwin, thought he had identified the pay-off. Hamilton was deeply interested in parasites, and not long before he himself was ironically and tragic-ally struck down after being weakened by one (malaria), proposed that sex enabled an organism to stay one step ahead of the parasites it was prey to. The co-evolution of parasite and host, each providing a rapidly changing environ-ment for the evolution of the other, needs a rapid and special kind of response, which sex can provide. Careful analysis of the dynamics of co-existence, rather like the manoeuvring of nations during the Cold War, shows that sex provides an advantage because it provides a mechanism for storing genetic information that has become redundant but might be needed again when the parasite's genotype has reverted back to a previous incarnation. In other words, sex pro-vides a store for swords in the face of muskets, but muskets might run out of ammunition. Stored swords are useless, though, if muskets give way to nuclear weapons; that is, sex is useless if the parasite evolves a new strategy rather than reverting to a former one. This theory remains a speculation, for it is hard to confirm experimentally, and depends on a special evolutionary relationship between the parasite and the host.

It is easier to identify the mechanisms that sustain sex rather than the mech-anism by which the complex business originated. Firstly, sexually reproducing populations are more responsive to adjustments in the environment than parthenogenetic populations. Thus, advantageous mutations can occur in both parents separately, and confer reproductive advantage on their offspring; in parthenogenesis, one mutation must follow another. That is, mutation can occur in parallel in sexual populations but in series in asexual populations. Secondly, deleterious mutations are less likely to propagate in sexual popula-tions, because two afflicted parents can still produce a normal child (as becomes apparent from the viewpoint of Mendelian inheritance, Chapter 2), whereas an asexually reproducing organism can rid itself of a bad mutation only by under-going a back-mutation of the same gene, which is improbable. *Sexual dimor-phism* (the different appearance of males and females of the same species) is also relatively easy to explain—or at least to cook up plausible explanations for—particularly the extravagant displays that often mark out the male. For instance, in a corollary to his theory of the evolution of sex, Hamilton accounts for a male's flamboyant display as being a sign that he is healthy and free of parasites. The inspection of the male by the female—what we humans might term 'falling in love'—is then akin to a medical inspection.

Sex appears to confer advantages at a variety of levels, on populations, indi-viduals, and genes. Most evolutionary shifts give only a tiny advantage: to pay for sex, the advantage must be huge. Why should there be any advantage in mingling an unrelated stranger's genes with yours? The bottom line, though, is that the origin of sex, like the lengths to which organisms will go to achieve it, is still a mystery.

From the feeling that the Earth has moved, let's turn to the actually moving Earth. Nowhere have tectonic processes had a greater impact on our own existence than in the subtle changes that took place as the African crust rippled in reaction to the pressures it experienced as it wandered round the southern hemisphere.

About 20 million years ago the landscape of Africa was largely flat, covered across its width with tropical forest. Then the Earth moved. You would have started to note the difference about 15 million years ago, when a local uplift produced uplands of lava centred on what is now the region we call Kenya and Ethiopia. These uplands formed in a sensitive location, for the land beneath was moving apart. When the gap between them had widened, the uplands collapsed, to produce a deep, long fault, the Great Rift Valley, that now stretches from modern Mozambique, through Ethiopia, and on to the Red Sea and beyond, as far as Syria. The resulting newly elevated highlands cast a rain shadow on the eastern part of the continent, and tropical forest gradually decayed into open savannah. Now the landscape provided a rich variety of potential habitats—several *biomes*—with humid, hot, vegetation-rich regions in some locations and dry grassland in others. Not only were niches opened for exploration, but reproductive isolation was open for exploration and exploitation, because one variety of organism was unable to migrate across the natural barriers that had arisen. Organisms were trapped.

Organisms were trapped in physical space, but not in evolutionary space. One of the most important consequences of natural selection is the existence of *Homo sapiens*, the species regarded by *H. sapiens* itself as the apotheosis of evolution. Darwin was sensitive to the implications of his theory, that Man was descended from the apes rather than specially created. Those of a Christian disposition have also been troubled by the thought that the emergence of Man without there having been a Fall somewhat undermined the basis of the Christian Church, with the centrality to its dogma of the idea of redemption. Be that as it may, there is unequivocal evidence that you and I have descended from ape-like predecessors. The descent is so important for coming to an understanding of ourselves and our place in the biosphere that it is appropriate to spend some time considering it.

Natural selection needs to account for terrestriality (migration to life on the ground), bipedality (walking upright on two feet, leaving the hands available for manipulation), encephalization (expansion of the brain relative to the size of the body), and the emergence of culture. There is considerable debate about whether bipedality preceded terrestriality or followed it. One advantage of bipedalism is that it gives greater endurance for following herds, and by standing upright the animal can see further across the savannah to detect predators. Others argue that encephalization was the initial step, to some have thought that the adoption of culture—including the use of tools—was the springboard of our advance.

Primates are typically arboreal inhabitants of tropical and subtropical forest ecosystems. We recognize them by noting characteristic features of their hand and foot anatomy, how they move, their visual abilities, their dental architecture, and their intelligence. This last feature is central, for a principal characteristic of primates is the evolution of intelligence as a way of life. Dental features are important because they let us identify the type of diet, and in particular whether the animal was arboreal and living off soft fruits or terrestrial and living off harder seeds and grains. Primates fall into two principal groups, the *prosimians* and the *anthropoids*. The prosimians include lorises and bush babies; the anthropoids include monkeys, apes, and humans.

Figure 1.10 shows the human family tree, and it would be as well to keep an eye on it as remote human history is somewhat confusing and still being pieced together. Far back in the Palaeocene ('old recent') epoch of the Tertiary period (originally 'third' era) of the Cenozoic ('recent animal') era, the early prosimians separated into the modern prosimians, which will concern us no further, and the evolutionary branch that will lead to us. During the early Oligocene (ineffably, 'few recent'), the New World monkeys (the *platyrrhines*, which means simply 'flat nose' and includes marmosets, howler monkeys, and capuchin) set up camp in South America. Our branch of the tree separated later when the Old World monkeys (the *catarrhines*, meaning 'downward nose', and including the tree-dwelling colobus monkey and the terrestrial macaques and rhesus monkeys) emerged in Africa. The hominids first emerged about 30 million years ago, in the Miocene ('middle recent'), in the form of *Dryopithecus* (or 'oak dweller', reflecting their presumed tropical and subtropical forest habitat).[8] Early Miocene apes, such as *Proconsul*, were very similar to monkeys in posture: *Proconsul* itself had no tail (like an ape), but held its body parallel to the ground (like a

---

[8] The term *hominine* is now preferred by some for the group of animals resembling the human family, replacing the term *hominid*.

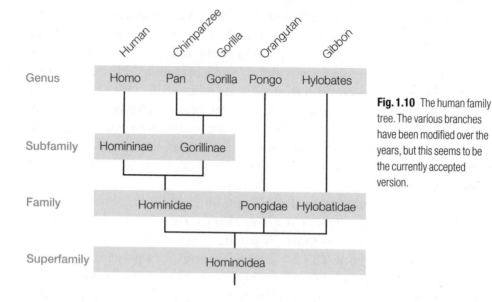

**Fig. 1.10** The human family tree. The various branches have been modified over the years, but this seems to be the currently accepted version.

monkey). *Proconsul* also had an opposable thumb, so there is a faint hint that it might have used simple tools. Hominid fossils are known throughout much of the Miocene in Africa and Eurasia, with the earliest specimens of a species of *Proconsul* dated at approximately 22 million years ago.

About halfway through the Miocene, between 10 and 15 million years ago, *Dryopithecus*, which has been found as far north as Spain and Hungary, diverged into several genera, which included *Sivapithecus* in India, Pakistan, and even possibly Turkey, and *Ramapithecus* in Africa. *Sivapithecus* is possibly ancestral to orangutan; *Ramapithecus* was long regarded as ancestral to humans, but is now thought to be just one of a number of ape-like species around at the time. The hominids living about 3 million years ago fall into two groups, one large-brained with small cheek teeth, the other small-brained with large cheek teeth. The former constitute the genus *Homo*; the latter the australopithecines of the genus *Austropithecus* ('southern ape'). The first of the latter, *A. africanus*, was discovered in 1924 during quarrying at Tang, near Johannesburg, and appeared to be an ancestor of *Homo*. Initially little notice was taken of this find, largely because the recent, sore memory of the Piltdown forgery had bred scepticism, and there was a view that Africa was an inappropriate location for the cradle of mankind; England was much preferred, the Home Counties preferably but the West Country would do. There is still uncertainty about where *A. africanus* fits into the phylogenetic tree.

In 1962, Louis Leakey, the doyen of hominid fossil hunters, digging in the

Olduvai Gorge in the Serengeti plain of Tanzania, came across the remains of a tool-using hominid which he identified as the new species *Homo habilis* ('handy man') from about 1.8 million years ago. Stone-tool making appears to have originated at roughly the same time as significant brain expansion, approximately 2.5 million years ago, and it is a matter of conjecture whether tool making drove brain size or vice versa. At the time, *H. habilis* was highly controversial, as some thought—on the grounds that there is much physical variation within species—that it was a big *A. africanus,* whereas others thought it was a little *H. erectus* (whom we meet in a moment). The current position appears to be that *H. habilis* is a true species, or at least a chronospecies en route to *H. erectus.* The current view also appears to be that the several skeletons that were all lumped together as *H. habilis* are actually two separate species, *H. habilis* and *Homo rudolfensis,* the latter (named after Lake Rudolf where it was found, now Lake Turkana in northern Kenya) having a slightly larger brain and a more modern brain structure. It is still unknown which of these species ultimately led to *H. sapiens,* for the 'obvious' answer, that it was the bigger-brained *H. rudolfensis,* is countered by the observation that this species appears to have other anatomical features that rule it out.

Now (about 2 million years ago), here (in Africa), *Homo ergaster* ('working man') staggers on to the scene under the burden of his stone toolkit. He is taller than his forerunners and has a bigger brain. The almost complete skeleton of 'Turkana Boy' (Fig. 1.11), discovered by Richard Leakey in 1984 on the west side of Lake Turkana, is a representative: the completeness of his body suggests that he may have drowned about 1.6 million years ago, for he was not ripped apart by carnivorous animals. *Homo ergaster* exploited wide open spaces and a temperate, drier climate; he employed tools that are classified as Acheulean. The name is taken from St Acheul in France, where a collection of relatively advanced bi-facial stone tools was found, such as a tear-shaped hand-axe. It is conjectured that *H. ergaster* was a hunter rather than a scavenger.

About 1.8 million years ago, *H. ergaster* spread into Asia, and there and in Africa evolved into *H. erectus* ('upright man') who was fully upright and bipedal, perhaps as tall as modern humans but with a heavier face, low forehead, and brow ridges (Fig. 1.12). The assumption is that *H. erectus* followed herds, which in turn followed the recession of forests as glaciation cooled the world, and pursued them through the tropical and subtropical zones of Saudi Arabia into south central China. From there, he crossed the landbridge into Java, where bits and pieces of 'Java man' were found in the River Solo by Eugene Dubois in 1891 in the course of his search for the 'missing link'.

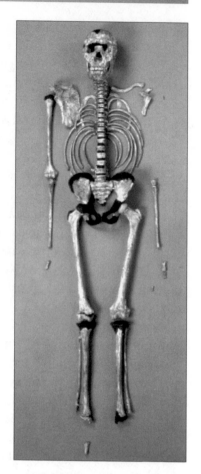

**Fig. 1.11** Misfortune for an individual can be a great prize for science. This is the almost complete skeleton of Turkana Boy (more formally, WT-15000), a specimen of *H. ergaster*, who went missing, presumed drowned, about 1.6 million years ago, and was discovered in 1984. Important inferences—some might consider them speculations—can be drawn from details of the skeleton. The relatively heavy bones (compared with modern humans) suggest routine physical exertion; the channel in the vertebrae through which the spinal cord runs has a smaller diameter than in modern humans, which suggests less nervous traffic, which in turn suggests less control over breathing and therefore, speculatively, the absence of a spoken language. The size of the pelvis, and its implication for females for the development of neonates, also suggest that infant helplessness, and the social organization that entails, were developed in the species.

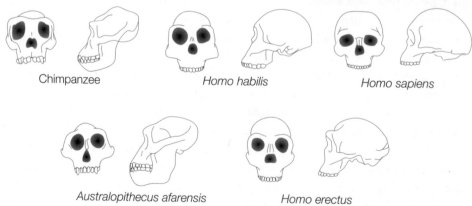

Chimpanzee            *Homo habilis*            *Homo sapiens*

*Australopithecus afarensis*            *Homo erectus*

**Fig. 1.12** The skulls of the chimpanzee and various *Homo* species. Note the disappearance of the brow ridge on moving through the sequence and the enlargement and rounding of the skull.

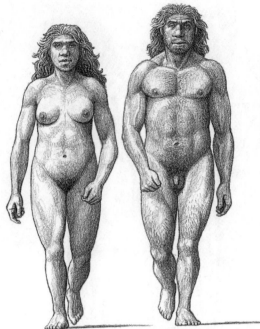

**Fig. 1.13** A reconstruction of the likely appearance of a Neanderthal couple; note the heavier features compared with modern humans. They had a heavy brow ridge and receding chin and forehead. It is unlikely that the Neanderthals interbred with *H. sapiens*; their line became extinct.

According to the 'out-of-Africa hypothesis', *H. sapiens* ('wise man') appears to have emerged in Africa about 150 thousand years ago, probably from *H. ergaster* but possibly from the not quite so bright (as judging from slender evidence about brain size) *H. erectus*. In due course, *H. sapiens* was to confront *H. neanderthalensis*, the paradigm 'caveman', whose remains were first discovered by quarry workers in 1856 in the Felderhofer Grotto above the Düssel River in the Neander Valley in Germany (Fig. 1.13).

The Neanderthals had larger bodies, were more muscular, were stockier, and were more barrel-chested than us, and although their brains were on average larger than modern humans their skulls have suggested to some commentators that they lacked frontal lobes. They appear to be adapted for a cold environment. Later, specimens were found in the Middle East, Asia, and northern Africa. Their stone technology is now classified as *Mousterian*, after a cave at Le Moustier in France. In due course, perhaps as a result of their contact with *H. sapiens* about 40 thousand years ago, they borrowed ideas from Acheulean technology and refined it to a version known as *Chatelperronian*, after a cave site of that name. The presence of ornaments made of antler and bone suggests that there was trade between the two species. A glimpse into Neanderthal home life is the discovery of a flute dating between 43 thousand and 67 thousand years

**Fig. 1.14** Cro-Magnon man was almost identical to modern man, shown here in a somewhat idealized form. This highly intelligent creature is in the process of bringing about a greater extinction of species than any of its competitors, including natural catastrophes. As the following pages show, however, despite its lack of self-control, this paradoxical animal is capable of exquisite understanding and, as others have shown, of exquisite artistic achievement.

ago and possibly of Neanderthal origin, with holes that match the seven-note diatonic scale of modern human music.

Apart from their musical evenings, the Neanderthals had a developed social organization, cooperated in hunting, probably communicated by language (although the structure of the larynx suggests that their language lacked much articulation), and appear to have buried their dead with what could be interpreted as tenderness (but might be rockfall). On the other hand, they may have been cannibals—which could also be construed as tenderness, the literal internalization of a loved one—if we were to discard our cultivated repugnance. If the one item of art, a single polished and carved baby mammoth tooth, is a single point from which reliable anthropological extrapolation can be made, then the Neanderthals were artistic.

Artistic they were, survivors they were not, for *H. neanderthalensis* vanishes about 30 thousand years ago, a dead-end of evolution, a dried leaf on the end of a twig on the bush of life. The extinction of the Neanderthals began about 40 thousand years ago with a wave of invasion that swept from east to west across Europe and ended about 27 thousand years ago. The incursors were the *Cro-Magnon* (Fig. 1.14), a variety of *H. sapiens* almost identical to modern humans and named after the site in central France (near Les Eyzies-de-Tayac in the

Dordogne) where the first specimens were found in 1868. Cro-Magnon was adapted to warmth, which supports the view that he (together with his she, of course) emerged from Africa. The Cro-Magnon swept all before them, perhaps on account of their greatly superior tools (that is, weapons) known as *Aurignacian*, which included bone and antler as well as refined stoneware, or because the Neanderthals had no defence against the diseases they carried, or because, lacking fully articulated speech, the Neanderthals could not communicate sufficiently effectively to coordinate their defence. The Cro-Magnon lived in tents and used animal traps, bows and arrows, hafts and handles for their knives, and bitumen to join them. War was now on an industrial footing; such was to be the lament of human history for the following millennia.

Modern humans, though, when not engaged in slaughter, were now equipped to reflect on their environment, on their own physical and psychological nature, and on the composition of the stuff that surrounded them and which they slowly learned to bend to their will. What follows is where this beginning led.

# T W O

# DNA

## THE RATIONALIZATION OF BIOLOGY

**THE GREAT IDEA**
*Inheritance is encoded in DNA*

*Almost all aspects of life are engineered at the molecular level and
without understanding molecules we can only have a very sketchy
understanding of life itself*

FRANCIS CRICK

EACH of us is about a hundred trillion selves. Each of our cells, and there are getting on for about a hundred trillion of them, most of them so small that about two hundred are needed to cover the dot on this i, contains a template for our complete body. In principle—always a dangerously suspicious word—your body shattered into its hundred trillion cells could spawn a hundred trillion yous, each of those new yous shattered again could become a hundred trillion more, and quickly you and your clones would dominate the universe utterly. Fortunately, there are physical and biological constraints that render this fantasy impossible. But even to contemplate the possibility suggests that we know about the cellular nature of life to an unprecedented degree.

We do. Darwin and his contemporaries, with the possible exception of one monk, knew nothing about the nature of heredity. Despite their grasp of the natural world and their profound insight into the consequences of competition, their understanding was hamstrung by their ignorance about the mechanism of inheritance. The most favoured mechanism at the time was *blending inheritance*, in which each parent poured their inheritable characteristics into the pot that was to be their child, and the child emerged from the mix. That such blending could not sustain natural selection, because novel adaptations would quickly be swamped, was used as a powerful argument against Darwin's views and delayed the general acceptance of his theory. Aristotle, though admirable in his pursuit of questions, had as usual got the answer wrong, demonstrating once again the failure of armchair speculation unsupported by experiment.[1] Noticing that blood washed through all organs of the body, Aristotle ascribed inheritance to the blood, a view that lingers even now as a metaphor. He considered semen to be purified blood which, on copulation, mingled with menstrual blood and brought forth the next generation.

The monk who held the key, of course, was Gregor Mendel (1822–84), born Johann to peasant stock on a farm in Heinzendorf (Hyncice), northern Moravia, a province in Austrian Silesia later incorporated into Czechoslovakia and now in the Czech Republic. Mendel's father, Anton, was a smallholder whose health and livelihood were wrecked by botany, in the form of a falling tree. Anton sold the farm to his son-in-law so that he could afford the fees for his son, whose life would be made by botany, at a school in Troppau and then at the university in Olmutz. Mendel's only way to achieve cheap education was to enter the Augustinian monastery of St Thomas at Brünn (now Brno), when he adopted the name Gregor, at the age of twenty-two, and was ordained to the priesthood in 1847. In a step that prepared his mind for the minor arithmetic of inheritance that he was to develop later, he was sent to Vienna to study science and mathematics with a view to becoming a teacher; but his studies there were feeble, especially in biology, and after two years he returned to his monastery, later to become its abbot (in 1868).

Mendel was incumbent of the prelature of the imperial and royal Austrian order of Emperor Franz Josef, meritorious director of the Moravian Mortgage Bank, founder of the Austrian Meteorological Association, member of the Royal and Imperial Moravian and Silesian Society for the Furtherance of Agriculture, Natural Sciences, and Knowledge of the Country, and—most importantly—a gardener. In the 1850s, at about the same time that Darwin was penning his thoughts, he began the studies for which posthumously he was to become famous. A number of questions have been raised—and vigorously rebutted—about the authenticity of his work or that of his assistants, for in 1936 the distinguished statistician and geneticist Ronald Aylmer Fisher (1890–1962)

---

[1]   Armchair speculation *allied with* experiment is, of course, extraordinarily powerful, being the core of the scientific method.

argued that the numbers Mendel had reported were suspicious. Further questions have also been raised about whether Mendel really knew what he was doing, and whether the myth that has grown up around his achievements is more to do with our hindsight than with his insight. Thus, the thrust of Mendel's work was to attempt to understand the rules of hybridization rather than the mechanism of inheritance. His motivation was the pursuit of the then prevailing view that new species arose from hybridization, with 'stable hybrids' being new species. His desperate aim was to create a new species: in that he resoundingly failed.

Mendel presented his findings—essentially a gloomy report of his failure—at the meetings of the Natural History Society of Brünn at sessions on 8 February and 8 March 1865 and published them as 'Experiments on plant hybrids' (*Versuche über Planzenhybriden*) in the transactions of the society in 1866. His results were entirely ignored, except for a misleading citation in W. O. Focke's *Die Pflanzen Mischlinge* (1881), and lay unremarked until 1900. They may have been ignored because in the view of the time they represented a failure to expose the rational basis of hybridization, and Mendel's drift into administration might also reflect his own disappointment at the dismal outcome of his lifetime's work. Then three botanists—Hugo de Vries in Holland, Carl Erich Correns in Germany, and Erich Tschermak von Seysenigg in Austria—discovered, they claimed, that they unknowingly, they claimed, had been replicating his work. There is a peculiar whiff of skullduggery in these reports, for it has been suggested that one of the authors (de Vries) delayed acknowledging Mendel's priority until it became clear that one of the others (Correns) was publishing similar work, so de Vries, having realized that he would have to cede priority anyway, announced Mendel's priority in an attempt to tarnish the lustre of Correns' claims. All manner of explanations have been offered for Mendel's neglect for thirty-five years, including the fact that he was an intrusive amateur, that he was too closely associated with a church from which nothing good could spring, that his deployment of mathematics—even of the simple arithmetic he required—was confusing to the biologists of the day. The truth may be simpler: until de Vries, Correns, and von Seysenigg revived his work and looked at it with more modern eyes, no one thought it relevant to the mechanism of inheritance.

Although Mendel did his work in the nineteenth century, its significance became clear only in the twentieth. Just as Planck quantized energy (see Chapter 7), we now realize that Mendel quantized heredity. We can now see that his achievement was to provide the evidence that led to the downfall of the then

**Fig. 2.1** Mendel's garden at his monastery. Mendel did his work on the common pea, which turned out to be a good choice, partly for reasons of economy but also because many of the characters of peas are genetically independent. The monastery garden is currently stocked with begonias.

prevalent blending theory of inheritance and its replacement in due course by a theory in which inherited information was carried in discrete units. For eight years the focus of his attention was the garden pea (*Pisum sativum*), which has a number of special features that make it ideal for the studies he carried out. For one thing, the structure of the flower itself is rather special, and makes it relatively easy to cross two plants or, as happens in the wild, to let them self-pollinate (to 'self'). Moreover, the plant shows a number of variable character-istics: for instance, its petals can be white or purple, its peas can be round or wrinkled, have green or yellow interiors, grow in pods that are green or yellow, and form plants that are dumpy or stringy. Moreover, and perhaps the true reasons, peas were cheaply available through a seed merchant, take up little room, and produce many offspring in a relatively short time. We can also sus-pect that pea soup made a tiresomely frequent appearance on the menu of the monastery of St Thomas. The one disadvantage of the garden pea is that it is not particularly photogenic in landscapes, and Mendel's experimental garden has been replanted, to please visitors, with more pulchritudinous begonias (Fig. 2.1).

Mendel was struck by the way that the hybridization of ornamental plants produced variants that recurred in later generations. He decided to look for the systematic pattern he thought might be buried in the observation. For the first two years he set about ensuring that he had true-breeding plants, so that green-pea plants bore green peas and yellow-pea plants bore yellow peas, and so on for the other characters. Next, he started a series of crosses and selfs. For instance, when he crossed green peas with yellow peas, all the peas of this first filial generation (the so-called $F_1$ hybrids) were yellow. However, when the hybrids were selfed, three-quarters of the peas in the next, $F_2$, generation were yellow and one-quarter were green. Mysteriously, the original green had reappeared. A similar pattern, with the same numerical ratios, emerged when he crossed and then selfed plants showing other characteristics. Clearly, a pattern was emerging, and patterns cry out for explanations.

Mendel built a hypothesis on the basis of his huge number of results. His first clue was the fact that his experiments led to variants in simple numerical ratios. To account for the discrete numbers that he had obtained for these ratios, he proposed that the difference between each characteristic (green and yellow peas, for instance) was due to the presence in the plant of different discrete units. Mendel used the term 'element' to denote the discrete heritable entities and referred to the different 'characters' when discussing the outward appearance, the phenotype, of his plants. Most of his reasoning was in terms of these observable characters, and it is only later interpreters who have directed attention to the role of the underlying 'elements'. These entities have received a number of different names since then, but are now universally known by the one suggested in 1909 by the Danish biologist Wilhelm Ludvig Johannsen, *genes*. More precisely, the different versions of genes that are responsible for a particular phenotype, such as that responsible for pea colour, are called *alleles*. So, green peas and yellow peas correspond to different alleles of the gene responsible for pea colour.

To account for the simple numerical ratios Mendel had identified, we can suppose that genes—we shall use the modern term—exist in pairs, with one pair corresponding to each character, and that each of the gametes (eggs and sperm in animals, ovules and pollen in plants) contains one of the genes. Then, when conception occurs (pollination in plants), the male and female gametes fuse randomly, so bringing individual genes back into pairs. Mendel had identified inheritable characteristics as *dominant* or *recessive*, and with hindsight we can see that this distinction also applies to genes. Therefore, if a dominant allele is paired with a recessive allele, then the phenotype will show the characteristics of the dominant allele. For instance, Mendel's experiments show that the yellow-pea allele is dominant over the green-pea allele, because when yellow true-breeding plants are crossed with green true-breeding plants, all the offspring are yellow.

We can illustrate these ideas symbolically. Let's denote the yellow pea allele as *Y* and the recessive green pea allele as *y* (that is the convention in elementary genetics: the dominant allele is denoted by a letter that indicates the trait and the recessive counterpart by the corresponding lower-case letter). The true-breeding yellow-pea and green-pea plants are *YY* and *yy*, respectively. The gametes of each plant are *Y* and *y*, respectively. When they cross, the offspring must be *Yy*, and all have yellow peas because yellow (*Y*) is dominant. Now we self these hybrids. Because the gametes of *Yy* plants are *Y* or *y* at random, the offspring of *Yy* plants will be *YY*, *Yy*, *yY*, and *yy*. Only the last, *yy*, corresponds to green peas (because *Y* is dominant in *Yy* and *yY*), so the plants are yellow and green in the ratio 3:1, just as Mendel observed. He was able to extend this simple scheme to the other characteristics and to combinations (green peas with a dwarf habit, for instance), and in every case found that the expected ratios were confirmed. (It is here that Fisher located his statistical attack, because the ratios were not exact, and the scatter in results—which could arise from bias, wishful thinking, in deciding whether a pea with a slightly crinkled surface was smooth or wrinkled, for instance—was suspicious.)

Not all inheritance is Mendelian in the sense that it obeys Mendelian laws and gives rise to simple statistics. Perhaps the worst advice in the history of expert advice was given by the German botanist Karl Wilhelm von Nägeli, of the University of Munich, who had not understood Mendel's arguments and suggested to him that he should divert his attention from peas and study hawk-weed (*Hieracium*) instead. But hawkweed reproduces by somatic parthogenesis (that is, asexually), and could hardly be less suited to demonstrate Mendelian heredity. Mendel must have felt somewhat gloomy as his experiments on hawk-weed led nowhere and certainly did not corroborate his ideas. He was also depressed by his results on the bean plant (*Phaseolus*), in which so many genes contribute to the characteristics he was assessing that the simple ratios he expected, and are so clear in *Pisum*, were concealed.

There are more subtle reasons why not all sexual heredity is Mendelian, for some genes are linked to others, and the inheritance of certain pairs of characteristics is not random. Moreover, many genes are *pleiotropic* in the sense that they govern more than one feature of the phenotype, and an organism is not a one-to-one mapping of traits on to genes. For example, a mutation in the fruit fly *Drosophila*, the hero of many genetic studies, results in lack of pigmentation in its compound eyes and in its kidneys (its Malpighian tubules); in another mutation, not only are the wings held out laterally, but the fly lacks certain hairs on its back. Even the statistics of straightforward Mendelian inheritance can be shrouded by secondary effects. For instance, the Manx cat has a gene, let's call it $t$, that interferes with normal spinal development in $Tt$ cats and results in the familiar tail-less phenotype; a double dose of the allele, to give a $tt$ cat, is un-survivable and $tt$ embryos die. Selfing $Tt$ cats will therefore give $TT$, $Tt$, and $tT$, a ratio of 1:2, in offspring that make it to birth, instead of the expected 1:3.[2]

There the matter rested for thirty-five years, until it was unearthed and reluctantly acknowledged under the possibly slightly murky circumstances we have mentioned. But while Mendel's observations slept, biology was travelling along another road that was destined to merge with his.

The quotable German biologist Ernst Haeckel (1834–1919) coined for us the term *phylogeny*, meaning the evolutionary history of a species, and proposed that 'phylogeny recapitulates ontology', where *ontology* is the development of an individual. He meant that the changes an embryo undergoes at it develops in the womb are a fast-forward version of the evolution of the species. He also proposed, with dire consequences twenty years after his death, that politics is applied biology. More relevantly to the current discussion, in 1868 he suggested that the nuclei of biological cells contain the information that governs inheritance. The German embryologist Walther Fleming took this suggestion further when in 1882 he discovered that nuclei of the cells of salamander larvae contain tiny rod-like structures that could be coloured by the absorption of certain dyes. On the basis of this observation, in 1889 Wilhelm von Waldeyer suggested the name *chromosome* ('coloured body').[3]

The numbers of chromosomes in the nuclei of cells are notoriously difficult to count because they are unravelled, entangled, and distributed through the nucleus until it starts to undergo division, whereupon they start to duplicate and divide. What we regard as lesser animals and by implication plants commonly have fewer chromosomes than us: we have twenty-three pairs, a house mouse has only twenty. The tomato, though, has twenty-two and, disconcertingly, the potato twenty-four. In fact, so difficult is the count that humans were long thought to have as many chromosomes as chimpanzees (twenty-four pairs); it was only by swallowing our pride and arguing that the number of chromosomes is unrelated to self-opinionated self-admiration that the correct number, twenty-three, could be accepted.[4]

2  The possession by Manx cats of different coloured eyes is unrelated to their tail-lessness.
3  Waldeyer was good at names, for he also coined the word 'neuron' in 1891.
4  At the time of writing, a well-known electronic encyclopaedia still puts the number at twenty-four.

At the turn of the century biologists had come to suspect that the chromosomes were the instruments of inheritance. Those chromosomes came into step with Mendelian inheritance in 1902 when Walter Sutton (1877–1916), a graduate student working in Columbia University, New York, having studied the sperm of grasshoppers (specifically, the plains lubber grasshopper, *Brachystola magna*, which is found throughout the plains of the western United States and Mexico, with its big cells and reasonably visible chromosomes), found that paired chromosomes did indeed separate, with one member of each pair going into a different cell. Sutton's discovery is commonly termed the *Sutton–Boveri theory*, because Theodor Boveri (1862–1915), a German biologist working on the ova of sea urchins, had claimed in 1904 that he had had the idea too and at much the same time as Sutton. Boveri had indeed contributed (along with others) some central ideas but—most importantly—had powerful friends.

At this stage we can infer that Mendel's genes are carried by Sutton's chromosomes. The world was ready for a new science, and in 1905 the term 'genetics' was suggested by the faintly odd William Bateson in a letter to the Cambridge zoologist Adam Sedgwick, and then in public in 1906 at the third international conference on hybridization. The ponderosity of his style, and perhaps the extent to which communication of science to the public has advanced in a hundred years, can be judged from his remark that the term

sufficiently indicates that our labours are devoted to the elucidation of the phenomena of heredity and variation: in other words, to the physiology of descent, with implied bearing on the theoretical problems of the evolutionist and systematist, and applications to the practical problems of breeders, whether of animals or plants.

Before taking a further step down into genetics and the underworld of inheritance, we need to be aware of what is involved in the two crucial processes of *mitosis*, the division of somatic cells (ordinary body cells), and *meiosis*, the formation of gametes (sperm and eggs, pollen and ovules) in the gonads (sexual organs) of animals and in the anthers and ovaries of plants. The intricacy of the latter process is one of the reasons why the evolution of sexual reproduction is so hard to understand and why there must be such an enormous evolutionary pay-off (Chapter 1). Still, Nature has risen to the task, and meiosis—a much more demanding task logistically than mitosis—takes place when and where it is needed. This is not a biology textbook, so I shall give only an outline of the two processes in so far as we need to understand them for what is to follow.

First, consider mitosis, the replication of somatic cells. A cell has a cyclic life, with less than about ten per cent of its time spent undergoing mitosis. The rest

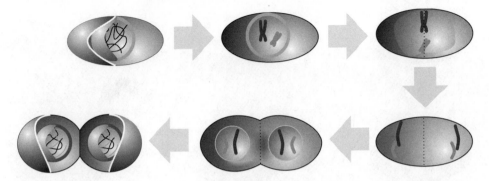

**Fig. 2.2** The process of mitosis, the division of a somatic cell into two replicas. Initially the chromosomes are spread out through the nucleus (depicted here as the inner sphere). When cell division commences, the chromosomes coil up, replicate, and form elongated X-shaped bodies (we show only two here; there are twenty-three such pairs in a human cell) consisting of two chromatids joined at the centromere. The chromosomes align on the central plane, the nuclear membrane dissolves, the chromatids separate, and are pulled apart into the cell cytoplasm. Once there, the nuclear membrane reforms and the cell membrane starts to close round each nucleus. Finally, the chromosomes unwind, and we have two identical diploid cells (cells with paired chromosomes) where originally we had only one.

of the time, though, is crucially important, as a lot of material is being prepared that will be used in the act of replication. For most of this apparently fallow but actually fecund time, our twenty-three pairs of chromosomes are extended and distributed in a tangled way throughout the cell nucleus. At the onset of mitosis (Fig. 2.2), the chromosomes contract by coiling up, in preparation for moving around more readily. At this stage, it also becomes apparent that each chromosome has already undergone replication, for each one consists of two identical bar-like units called *chromatids* joined together in a region called the *centromere* to look like an extended X. Next, the nuclear envelope dissolves and the nuclear components and the surrounding *cytoplasm*, the complex mixture of compounds and structures within the cell wall but outside the nucleus, merge into one. The chromatids are now pulled apart, a new cell membrane starts to form between the two troops of chromosomes (as we now regard the separated chromatids), a new nuclear membrane starts to form around each replication, the chromosomes uncoil, and we now have two identical cells in place of one.

Now consider meiosis, the formation of gametes. This process is rather more subtle than mitosis as the net outcome must be the formation of four cells, each with half the complement of chromosomes (twenty-three in humans). This process is a bit complicated, so follow the steps in Fig. 2.3, where we have focussed on one pair of chromosomes. Initially, the chromosomes are tangled together and fill the nucleus, but as meiosis begins they untangle and

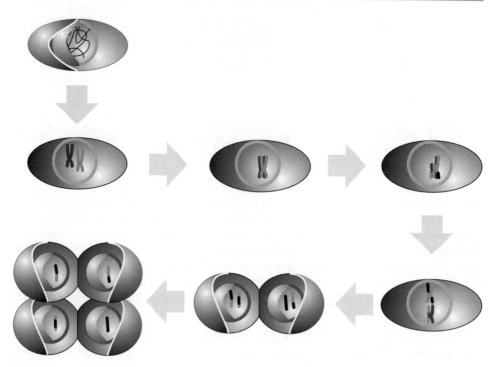

**Fig. 2.3** The process of meiosis, the formation of gametes. The strategy of meiosis is to turn a diploid cell into four haploid cells (cells with a single version of a chromosome) and to mix the genetic composition of the parental chromosomes. Once again, we show only a single pair of chromosomes in the parent cell. Initially, the two chromosomes are spread throughout the nucleus. However, as meiosis begins, they coil up and replicate to give two pairs of conjoined chromatids, just like in mitosis. However, corresponding pairs of double-chromatids migrate together and while lying side by side exchange genetic material. They then migrate to the central plane where a first mitosis-like division occurs (which we don't show in detail), to result in two cells, each containing two chromosomes. That is followed by a second mitotic division, in which the two chromosomes in each nucleus are separated again. We end up with four haploid cells, each containing a chromosome that is a genetic mixture of the two chromosomes in the parent cell. Reproduction is then, notionally but not mechanistically, the reversal of meiosis, in which the single chromosome in the gamete supplied by one parent combines with the single chromosome provided by the other parent.

contract. At this stage it becomes clear through a microscope that each chromosome has replicated and consists of two chromatids joined at a centromere to form the usual elongated X, just like in mitosis. Now, though, the chromatid pair from the father and that from the mother move together and form an elongated unit like the two sides of a zip. Each chromosome is attached to the nuclear envelope at its ends, which are called *telomeres* ('distant parts'); that anchoring probably helps one side of the zip to find its mate. While the two replicated chromosomes lie together, material in a chromatid representing the father's input is exchanged with material in the corresponding region of a chromatid supplied by the mother. This is the instant when genetic variation takes place in the organism.

After this literal crux in the organism's history, the process of *crossing-over*, the two pairs of hybrid chromatids are pulled into two regions, rather like in mitosis, to give two cells each of which contains a pair of chromatids. That is the 'first mitotic division' in the illustration. Then, in a 'second mitotic division' each of the pairs of chromatids is pulled apart into individual chromosomes, which now occupy individual cells. At this end point of the process, we have four cells where once we had one and the original genetic material from both parents has been distributed over all four cells. The chromosomes of one of those cells might contain the dominant *Y* allele of the gene for yellow peas; another might contain the recessive allele *y* for green peas. Mendel's arithmetic is about to emerge into his garden. Note, though, another facet of science: a great depth of complexity, in this case cell biology, may lie beneath a simple arithmetical observation.

Now it is time to unwrap the chromosome. What is the actual *stuff* of heredity? What is the physical embodiment of genetic information?

The idea that a chemical encodes inherited information had arisen during the nineteenth century, for after all, what else could it be? Once it was accepted, from about 1902 onwards, that proteins are long stringy molecules (usually coiled into a globule) built from a palette of about twenty amino acids in a definite sequence (more on this below), there was general enthusiasm for the idea that proteins encoded genetic information, with different sequences of amino acids conveying different messages from one generation to the next. Admittedly, there was the puzzling presence in the nuclei of cells of another type of molecule, called a 'nucleic acid' in recognition of its nuclear origin, which was composed of a string of another type of unit which we will come to later. These nucleic acids were regarded as boring and structurally too simple to be able to transmit the enormous amount of information carried in the chromosomes. It was widely supposed that they contributed merely to the structure of cells, rather like cellulose in plants.

That view was to change in 1944. The cornet player and biochemist Oswald Avery (1877–1955), born of British immigrants in Nova Scotia but who did all his seminal work in the United States, was investigating the different types of pneumococcus found in the mouths of pneumonia patients and in healthy people. It had been known since 1923 that pneumococci (the bacteria that cause pneumonia) came in several varieties: non-virulent forms looked rough to the eye whereas the virulent strains looked smooth. Frederick Griffith (1879–1941), working in the Ministry of Health in London on *Streptococcus pneumoniae*, showed that the rough and smooth strains could be converted into each other. Avery and his colleagues took up the work in 1930 and soon found that the transformation from one type of bacterium to another could be achieved with extracts from the cells and that a 'transforming principle', which appeared to be the effective agent, could be obtained. Avery then focussed on identifying the nature of the transforming principle. He found that proteases, which are enzymes that deactivate proteins, had no effect on its activity, so the principle wasn't a protein. He also found that lipases, which are enzymes that destroy lipids, the fatty substances that make up cell walls, also had no effect, so the principle wasn't a lipid. Having decided what the transforming principle wasn't, Avery went on to do a series of tests that showed conclusively that the principle was a boring old nucleic acid. The tables had been completely turned, and nucleic acids were on the way to being promoted, like Clark Kent into Superman, to being the most interesting and important molecules in the world.

Not everyone was convinced. Some were so attached to the protein theory of inheritance that they persisted in claiming that the transforming principle was perhaps an undetected protein associated with the nucleic acid. That view was to be rejected decisively in the next few years. In 1952, Alfred Hershey (1908–97) and his undergraduate assistant Martha Chase reported the results of their experiments on bacteriophages, which are viruses that infect bacteria. They noted that the element phosphorus occurs in nucleic acids but not in proteins, and that sulfur occurs in proteins but not in nucleic acids. Then, by tracing the whereabouts of each element by using radioactive versions of each one, they showed that during the process of infection, only the phage's nucleic acid, not any of its protein, entered the bacterial cell. That experiment convinced the world that a nucleic acid encoded hereditary information.

Meanwhile, progress had been made on the structure of one particular nucleic acid, *deoxyribonucleic acid* (DNA). This compound had been identified in 1868 by a Swiss doctor, Friedrich Miescher, in the cells obtained from discarded pus-soaked bandages of wounded soldiers in the German town of Tubingen. Pus is mostly an accumulation of white blood cells that accumulate to fight infection; although red blood cells of mammals have no nuclei, white blood cells do, and so are a source of nucleic acids.

To understand everything that follows, we need to know a little about the chemical composition of DNA. That is best achieved by picking apart its full name, deoxyribonucleic acid. The molecule is like a long thread, with units attached regularly along its length. The thread itself is built from alternating sugar molecules and phosphate groups. The sugar molecule is ribose, a close relative of glucose, from which one oxygen atom has been removed (hence the 'deoxy' and the 'ribo' parts of the name). As we can see from Fig. 2.4, ribose consists of a simple ring of four carbon atoms and one oxygen atom, with bits and pieces attached to the ring. The phosphate groups that link the deoxyribose rings together consist of a phosphorus atom (remember Hershey's experiment!) to which four oxygen atoms are attached. The backbone of DNA is just this alternation of phosphate and deoxyribose groups going on for hundreds of thousands of repetitions like a long fragile string of pearls.

That's the backbone. Attached to each deoxyribose ring is another molecule called a *nucleotide base*. The 'base' in this name has a technical origin, for in chemistry a base is a compound that reacts with an acid: in these compounds, the 'base' character refers to the presence of nitrogen atoms in the molecules, a

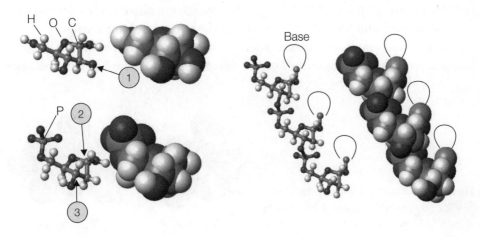

**Fig. 2.4** The structure of deoxyribonucleic acid (DNA). We can understand the structure of this complex molecule by seeing how it is built from simple components. At top left, we see the sugar ribose. This molecule consists of a ring of four carbon atoms (C) and one oxygen atom (O), with various other bits and pieces attached. Now imagine one oxygen atom, the one on the carbon atom at the south-east of the ring (arrow 1), removed, to give deoxyribose, and a phosphate group linked to the other end of the molecule. Now think of a molecular group—a nucleotide base (see Fig. 2.5, but represented by a blob here) attached to one of the ring carbon atoms (arrow 2), and the phosphate group as linked to another carbon atom of the ring (arrow 3) to give a long chain, as shown on the right. This chain is DNA.

common feature of bases in chemistry. Only four nucleotide bases occur in DNA, namely adenine (commonly denoted A), guanine (G), cytosine (C), and thymine (T). The structures of these molecules are all much the same and are shown in Fig. 2.5. As we can see from that illustration, the four bases fall into two pairs. Adenine and guanine have much the same shape, with two rings of carbon and nitrogen atoms stuck together. This structure is characteristic of a class of compounds that chemists call 'purines'. In contrast, cytosine and thymine have only one ring of carbon and nitrogen atoms. This structure is characteristic of compounds called 'pyrimidines'. To imagine a DNA molecule, think of one of these four bases as attached to each ribose group of the backbone, with a seemingly random choice of base at each location. You can perhaps begin to see why people thought that DNA was boring.

Once DNA had been identified as the genetic material, there was intense interest in its detailed structure. That structure began to emerge from the mists when the Austrian-American biochemist Erwin Chargaff (b. 1905), who had been born in Chernivtsi in the western Ukraine (then absorbed into Austria as Czernowitz) and had emigrated to the United States to work at Columbia University in New York, turned his attention to the problem. In 1950, using the new technique of 'paper chromatography', which enables closely relate species to be separated and identified by washing the mixture along a strip of paper, Chargaff found equal amounts of adenine and thymine, and equal amounts of guanine and cytosine, regardless of the tissue from which he had extracted the

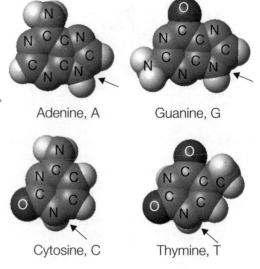

**Fig. 2.5** The four bases that form the letters of the genetic code. Adenine (A) and guanine (G) are purines, cytosine (C) and thymine (T) are pyrimidines. (The unlabelled small, light grey atoms are hydrogen.) The arrows show the nitrogen atom that forms the link to the ribose unit in DNA.

Adenine, A

Guanine, G

Cytosine, C

Thymine, T

DNA. That suggested that somehow an adenine was always associated with a thymine and that a guanine was always associated with a cytosine. He also found that the proportions of each pair of bases differ from species to species but are the same for different cells of the same animal. This observation indicated that there are not one but many DNAs and that the composition of each DNA is specific to the organism, just as you would expect for blueprints. Chargaff also found that whatever the species he used as a source of DNA, the total amount of purines (the double-ringed adenine and guanine) is the same as the total amount of pyrimidines (the single-ringed cytosine and thymine). All this information proved to be absolutely crucial to the recognition of DNA's structure, and with hindsight is almost enough to have led to the structure of the molecule.

The wind that finally blew away the remaining mist was the information emerging from X-ray diffraction studies being performed by the New Zealander Maurice Wilkins (b. 1916) and Rosalind Franklin (1920–58) at King's College, London, and the deployment of their results by Francis Crick (b. 1916 in Northampton) and James Watson (b. 1928 in Chicago) in Cambridge. As has been retold a thousand times, here is a story of skullduggery, rivalry, drive, application, animosity, tragedy, misogyny, deception, and above all imagination. That one of the most important discoveries of the twentieth century should have elicited most human emotions and attitudes is perhaps not at all surprising.

The tragic figure, of course, is Franklin, who died of ovarian cancer at thirty-seven, almost certainly induced by exposure to the X-rays that she had used in her work:[5] life did not surrender its secret without claiming life in return. Though tempting, it is inappropriate to elevate Franklin from tragic figure to tragic heroine and to cast her at the centre of the story. The facts about this very human story appear to be as follows. It has to be seen set against the background of mid-twentieth-century Britain, when, from today's viewpoint, men's attitudes to women were . . . undeveloped.

Wilkins was working on DNA at King's College, when the head of the laboratory, seeking to build up the X-ray unit, had invited Franklin to join the college and import her expertise in X-ray crystallography. She had acquired that expertise by working on the microstructure of coal at a laboratory in Paris and was keen to turn her attention to living rather than fossil life. It was not at all clear that she had succeeded in making the transition, for King's at that time

---

[5]  X-ray diffraction is a technique in which a beam of X-rays is passed through a crystal. The beam is scattered into different directions by the regular arrays of atoms and gives rise to a pattern of intensities that can be interpreted in terms of the locations of the atoms in the crystal.

excluded women from its common room.[6] Wilkins was absent when she arrived, and on his return appears to have been confused about her role. There was an immediate clash of temperaments, and each set up a laboratory to work on DNA. Both groups soon obtained quite good X-ray photographs of fibres formed from the molecule. Wilkins had met a young American biologist, James Watson, at a meeting in Naples, and showed him his pictures. That encouraged Watson to work on DNA's structure, and in September 1951 he moved to Cambridge to learn about X-ray diffraction in the laboratory then directed by Sir Laurence Bragg, one of the founders of X-ray crystallography. There he met Francis Crick, who was just finishing his doctorate.

In November 1951, the two streams of endeavour, one of careful measurement without the courage (or the impatience) to propose interpretations, the other of imaginative speculation without the resources (or the patience) to make measurements, came into collision. Watson went to London and listened to Franklin talk about her work. Hastening back to Cambridge, he and Crick built a model that they considered accounted for what Watson could remember of her data and invited the King's team to come and look at it. The construction of models—real physical models built from wire and sheets of metal—had proved to be a powerful technique for the elucidation of protein structure, and Crick and Watson were following the fashion of their time. The King's team arrived, and immediately dismissed the model as it did not accommodate their data. They also dismissed the potentially and, as it turned out, actually fruitful model-building approach. Moreover, Bragg ordered Crick and Watson to stop working on DNA and to leave it to the King's team, whose project it was. Attitudes to ownership in science as well as to women have changed since then: perhaps the next step marks the turning point of the former.

In 1952, Crick and Watson learned that Linus Pauling, who had been so successful with the structure of proteins and where Bragg's writ didn't run, was working on the problem. If Pauling was on the job, they argued, then the problem had already leaked from King's and they had as much right to work on it as anyone. Now something rather odd happens. At this point, Wilkins showed Watson one of Franklin's X-ray diffraction photographs without her knowledge (Fig. 2.6) and Max Perutz provided him and Crick with an unpublished report to the Medical Research Council in which Franklin summarized her most recent data. At last they had some definite numbers about the dimensions of the helical molecule, and could adjust their model to match. Within a few weeks, they were able to send Wilkins triumphantly their famous model, which

---

[6] I must not snigger: my own college common room at Oxford did not permit access of women until the 1970s.

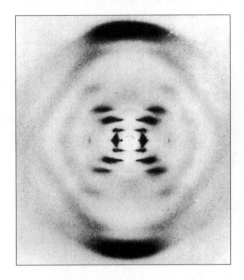

**Fig. 2.6** The crucial piece of evidence for the detailed structure of DNA was this X-like diffraction pattern obtained by Rosalind Franklin. It confirms that the molecule has the form of a double helix, and the details of the photograph can be used to determine the dimensions of the helices.

he accepted. A trio of publications, one from Crick and Watson, one from Wilkins's group, and one from Franklin's group (Franklin never knew that Wilkins had passed on her data), were published in *Nature* on 25 April 1953. The latter two provided the measurements in support of the speculations of the first. That date, 25 April 1953, is the birthday of modern biology.

The structure of DNA is the now famous and ubiquitously symbolic right-handed *double helix*, in which one long strand of nucleic acid is wrapped round another to form an entwined pair (Fig. 2.7), rather like, ironically, the stairways in the public entrance to the Vatican museum.[7] The key feature, though, is that the nucleotide bases of one strand match the nucleotide species of the other (Fig. 2.8), in the sense that adenine is always matched with thymine (which we denote A . . . T) and guanine is always matched with cytosine (G . . . C). That pairing accounts for Chargaff's observation that in all his samples the amount of adenine is the same as that of thymine, and the amount of guanine is the same as that of cytosine: the pairing ensures equal amounts. You should also notice that a relatively small purine (adenine and guanine) is always matched with a more bulky pyrimidine (thymine and cytosine), for in that way the double helix is uniform: two big purines would have resulted in a bulge and two small pyrimidines in a pinch. That pairing accounts for another of Chargaff's

[7] A photograph of this double helix can be found at http://www.planetware.com/photos/SCV/RVATMS3.HTM

**Fig. 2.7** The DNA double helix. The two strands of nucleic acid wind round each other to give an entwined double helix, with one narrow groove and one wide groove. The two strands are held together by hydrogen bonds between the bases, with a purine base (A, G), represented by the long rods, linked to a pyrimidine base (C, T), represented by the short rods. The pairing is always A … T and G … C.

observations, that the amount of purine (A + G) in a sample is the same as the amount of pyrimidine (T + C).

The adhesion between the two nucleic acid strands is due to a very special type of chemical bond known as a *hydrogen bond*. When I say special, I don't mean uncommon, for just about every water molecule in every ocean is linked to its neighbours by them, so there are about $10^{44}$ of them in the oceans alone, as well as many more in numerous other locations. A hydrogen bond is special in the sense that it forms in an unusual way, and between only a few certain types of atom, oxygen and nitrogen being among them. To form a hydrogen bond, a hydrogen atom (which is very small, so it can do this sort of thing) lies between two other atoms and acts as a kind of glue that binds them together. One of the keys to understanding the double helix is that—as we see in Fig. 2.8—thymine and adenine have just the right shape and arrangement of nitrogen, oxygen, and hydrogen atoms to form two hydrogen bonds very snugly. Similarly, cytosine and guanine also fit together snugly, but by forming three hydrogen bonds. That

**Fig. 2.8** The base pairing that holds the two strands of DNA together to give a double helix. The hydrogen bonds between the molecules are depicted by lines. Note that a purine pairs with a pyrimidine, and that the bulk of the two pairs is approximately the same.

hydrogen bonds are much weaker than ordinary chemical bonds holding atoms together to form stable molecules means that the two strands of the double helix can be pulled apart reasonably easily while leaving the nucleic acid strands themselves intact, just as water vaporizes without destruction of the individual water molecules.

We can now see why Watson and Crick were able to conclude their short but portentous paper with the coy remark that

It has not escaped our notice that the specific pairing we have postulated immediately suggests a possible copying mechanism for the genetic material.

Indeed, the fact that their model accounts so neatly for replication was the real reason why it was accepted so quickly even though a detailed, rigorous structure of the molecule did not become available until the late 1970s. To see the germ of the appealing and compelling idea, suppose the two strands have their nucleotide bases in the following sequence:

… ACCAGTAGGTCA …
… TGGTCATCCAGT …

with the first A in the upper strand linked by hydrogen bonds to the first T in the lower strand, C linked likewise to G, and so on. Then, suppose the strands separate into

… ACCAGTAGGTCA … and … TGGTCATCCAGT …

Now suppose that there is a supply of nucleotide bases in the cell. Then they will attach to the separated strands, each of which forms a template for the creation of a new strand, and form

… ACCAGTAGGTCA …    … TGGTCATCCAGT …
                  and
… TGGTCATCCAGT …    … ACCAGTAGGTCA …

Now we have two identical double helices where originally we had one. We have reproduction!

At this point it is relatively easy to make contact with the chromosome model of reproduction we encountered earlier in the chapter. All we have to do is to think of a chromosome as a strip of DNA. Then the process of mitosis is simply—a word, as always, we need to examine—the duplication of a double helix.

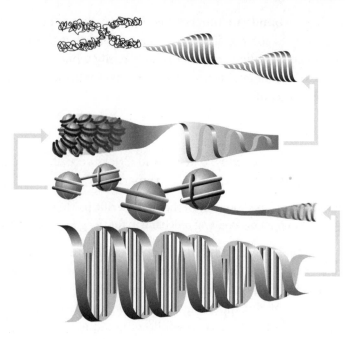

**Fig. 2.9** The double helix of DNA undergoes a great deal of coiling and supercoiling as it packs into a cell nucleus. This diagram represents the packing details. At the bottom we see the double helix of DNA itself. That molecule wraps round histone molecules, represented by the spheres, and the resulting coiled DNA coils to give the coiled-coil arrangement shown in the third rank of the diagram. That coiled-coil coils again, again, and again, to give a highly supercoiled pack of molecules, and the supercoiled molecule itself coils into the chromosome shown at the top.

Now, that word 'simply'. One of the problems that faces us is that a DNA molecule is very long: if the human DNA in one set of twenty-three chromosomes (with one DNA molecule in each chromosome) is stretched out and joined together, then it would be about 1 metre long, and all that stuff has to be confined to the tiny cell nucleus. Because the chromosomes are doubled, and there are about a hundred trillion cells in a human body, the total length of DNA inside each of us is enormous. Recall the two hundred cells needed to cover the dot of an i: those cells contain about 400 metres of DNA. To achieve this wonderful feat of packing, the double helix is wrapped round clusters of protein molecules called histones, which act like spindles. Then these spindles are wrapped round each other. That coil is itself coiled—it is *supercoiled*—around itself, and the tightness of the coil determines whether the chromosomes are bundled up, as they are during mitosis, or spread out through the nucleus, as they are for the rest of the cell's life (Fig. 2.9).

There are about 3 billion base pairs in human DNA, but only about five thousand in a small virus. We might feel proud to be so complex. The newt (*Triturus cristatus*), though, has 20 billion base pairs in its genome, which puts us into perspective. We can wriggle, newtlike, out of this embarrassment by arguing that a lot of DNA is redundant. Presumably that of the newt is particularly redundant, and may have arisen when at a recent stage in its evolution the species adopted a duplicate set of chromosomes in its cells (that is, became 'diploid', like us) after getting along with a single set (that is, had been 'haploid', like a gamete cell).

A DNA molecule is a store of information, essentially a message, handed down through the generations. That message contains all the information needed to construct and sustain the organism it inhabits. The obvious questions are what that information is, how it is encoded, and how it is interpreted.

The worker bees of the hive of cells that living organisms represent are the proteins. Proteins may be structural, as in muscle, cartilage, hoof, claw, and hair, or they may be functional, as in haemoglobin and the innumerable enzymes that control the processes that constitute 'being alive'. The specification of proteins is the central function of inheritance, so we can be confident that DNA is some kind of blueprint or recipe for our proteins. That is confirmed experimentally, for modification of DNA results in changes in proteins. Most often, that modification results in the malfunctioning of proteins that we call disease. Occasionally it is beneficial, in which case disease is promoted to the status of evolution.

As we have already mentioned, all proteins are strings of the small molecules called 'amino acids', which have the basic framework shown in Fig. 2.10. More formally, we say that a protein is a *polypeptide*, and typical proteins are polypeptides composed of around a hundred amino acid units (structural proteins might run to thousands). The entire panoply of about 30 thousand or so different proteins in the human body is constructed from just twenty different amino acids, so a DNA molecule must specify the sequence in which these twenty amino acids are linked together. Incidentally, there may be room for improvement here. Although organisms are built from these twenty components, there is an infinite number of other amino acids, and if Nature wanted to expand her repertoire (as perhaps she has already done on other planets), then she could scavenge the environment for other amino acids. Life on other planets might well be built from different amino acids, and we will have to be careful what we eat when we get there. Nature has in fact edged towards expansion on Earth, as a twenty-first amino acid, selenocysteine, in which a selenium atom replaces a sulfur atom, is occasionally needed for certain enzymes that help to protect cells against that most dangerous of elements, oxygen. If you happen to be reading this in northern central China, you might be in trouble,

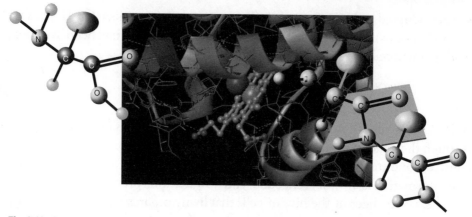

**Fig. 2.10** A protein is built up from amino acids, all of which have the general structure shown on the left in this illustration. The grey ellipse is different in each case, but all amino acids used in biology have this common layout. When two amino acids link together, the carbon atom of the –COOH group (on the right of the molecule) links to the nitrogen atom (on the left of the molecule). Many amino acids link together in this way, to give a long chain, as indicated in the structure on the right. This chain is called a *polypeptide* in general and, for two linked amino acids, a *dipeptide*. The –CONH– group indicated by the tinted plane on the chain is the *peptide bond*. We say that one peptide 'residue' (the rest of the amino acid molecule) is linked to another residue through a peptide bond. The long chain is typically twisted into helices, as can be seen from the fragment of haemoglobin shown in the background, where the helices shown there as ribbons are polypeptide chains.

because the soil there is unusually low in selenium and you are at risk from Kashin–Beck syndrome, which manifests as muscular problems.

Because a DNA molecule consists of a sequence of the nucleotides A, C, G, and T, it is natural to suppose that these are 'letters' that are combined together into 'words', *codons*, which specify the sequence in which amino acids must be linked. As there are only four letters but we need to specify twenty amino acids, together with indications of where to start and where to stop, the code obviously cannot be a one-letter or two-letter code. A one-letter code can specify only four amino acids, and a two-letter code can specify only sixteen. A three-letter code, with ACG standing for one amino acid and CAT for another, and so on, can specify up to $4^3 = 64$ amino acids and punctuation marks, which is more than enough. Suspecting Nature's natural parsimony (that is, the unconscious but efficient use of scarce resources and the unconscious but effective avoidance of the unnecessary deployment of energy), we can suspect that the genetic code is a *triplet code*, a code based on three-letter codons. There is no *a priori* reason for dismissing a variable code, in which two bases specify some amino acids, three specify others, and so on; but Nature did not adopt that inelegant solution, and mercifully the early workers who set about breaking the genetic code appear not to have explored that blind alley. One advantage of a triplet code is that it allows Nature to expand her repertoire by using some of the redundancy in the code to encode for new amino acids. There is already a hint of how this extension could evolve. We have just seen that a twenty-first amino acid, selenocysteine, is sometimes incorporated: the triplet code for that amino acid is TGA, which is also used as a stop signal, and changes its function depending on whether selenium is available. If selenium is available, then TGA says 'use selenocysteine'; if it isn't, then TGA says 'grind to a halt, stop building this protein'.

The code-breakers did explore blind alleys, some of great elegance, but were doing it in an Aristotelian manner from armchairs; once again, experiment came to the rescue and showed that Nature had not adopted the most elegant, economical schemes that humans might have selected if they had been in charge. The genetic code was a code-breakers dream, for the symbols of the code were so few (four) and the output was not an order of battle but any one of only about twenty choices. At the time, 1953, there was almost no data, for no one knew any of the nucleotide sequences in DNA and the known sequences of amino acids in proteins were very sketchy: Frederick Sanger (b. 1918) had almost completed his sequencing of the protein insulin (which he did in 1955), but that was about all. There was plenty of scope for unconstrained imagination.

The Russian physicist George Gamow (1904–68) certainly had an unconstrained imagination, for he had initiated the Big Bang theory of the origin of the universe and had devised a theory of the origin of the elements. He was interested in everything, and it was natural that his attention should turn to the hottest problem of the 1950s, the genetic code. Gamow came up with a brilliant idea: proteins grew on the outside of the double helix in the diamond-shaped cavities in the grooves of the helix. These cavities were formed by four nucleotide bases, the two at the top and bottom of the diamond on one strand and the two other corners a base from the same strand and its partner from the other strand. Ingeniously, this is a triplet code even though it involves four nucleotides, because the last two (a complementary base pair like A . . . T) count as one (because if one base is A the other must be T). He then envisaged the amino acids as settling into their appropriate niche and an enzyme coming along to join them all up. He went on to suppose that diamonds that were related by flipping horizontally or vertically coded for the same amino acid, with the result that only twenty distinct codons remained, which is just the number he thought he needed. Cleverness tripped there, however, because there is then no redundancy and no room for codons for starting and stopping. Gamow thought, with the optimism that comes from enthusiasm, that there would probably be a way round that problem.

Gamow's *diamond code* had another special property: it is an *overlapping code* in the sense that each nucleotide base contributes simultaneously to three codons. Thus, the sequence AGTCTTG consists of the codons **AGT**CTTG, A**GTC**TTG, AG**TCT**TG, AGT**CTT**G, AGTC**TTG**. An overlapping code is very efficient and compact, which would seem to make it an attractive candidate for Nature to adopt. Nature had other ideas. One problem with overlapping codes is that many amino acid sequences are ruled out. For instance, suppose we wanted to code for a dipeptide, a minuscule protein consisting of two amino acids. An example is the sweetening agent aspartame (which is sold as Nutra-Sweet), a combination of slightly modified forms of the amino acids aspartic acid and phenylalanine. Because there are twenty naturally occurring amino acids, there are $20 \times 20 = 400$ possible dipeptides. To code for two amino acids with an overlapping code, we need four bases, such as CCGA to get CCGA for the amino acid proline (as it turns out) and CCGA for arginine. But there are only $4 \times 4 \times 4 \times 4 = 256$ possible combinations of four nucleotide bases, so many dipeptides cannot be encoded (aspartame being one of them). However, these forbidden combinations started to turn up, which showed that Nature didn't use the elegance of an overlapping code: she demanded more flexibility for her activity in the never-ending demanding game of evolution. Sidney Brenner (b. 1927) did the definitive analysis of this problem: he showed that all possible overlapping codes were ruled out by the known amino acid sequences. Another more notional nail in this now firmly hammered coffin is that a change in one letter can affect the composition of a protein by up to three amino acids. Thus, if AGTCTTG were to mutate to AGGCTTG, then it would consist of the codons **AGG**CTTG, A**GGC**TTG, AG**GCT**TG, and so on, with possible dire consequences for the protein, and the organism, which often cannot survive a change in even one base.

There was one further blind alley lined with economical and elegant ideas that speculative physicists favoured but Nature turned out to disdain. One problem was punctuation. How do we know where to start? Even in a non-overlapping code, ... AGTCTTG ... could be read as ... (AGT)(CTT)(G ..., ... A)(GTC)(TTG)( ..., ... AG)(TCT)(TG ..., and so on. The different choices represented by these examples are called *frame-shifted readings* of the code. Crick suggested that machinery existed in the cell only for certain codons and that the code had to be such that all frame-shifted readings were nonsense. In this example, let's suppose that the correct reading is ... (AGT)(CTT)(G ..., then AGT and

CTT would be valid codes, but the frame-shifted readings GTC and TCT would be nonsense. A code of this kind is termed *comma-free*, because it can be read unambiguously without punctuation. When the sixty-four candidate codons are investigated with this restriction in mind, it turns out that twenty can be legitimate, exactly the number it was presumed are required. For instance, TTT is ruled out because the combination . . . TTTTTT . . . contains reading-frame ambiguities such as . . . (TTT)(TTT) . . . and . . . T)(TTT)(TT . . . Because the code appeared to provide exactly the number of codons required and to avoid the problem of frame-shifted readings, it was immediately and universally accepted.

But not by Nature. She put her foot down on this kind of unconstrained speculation in 1961, and stopped fertile imaginations wasting even more time. The putting down of her foot was identified by Marshall Nirenberg and Heinrich Matthaei, who showed that TTT was, after all, a valid codon and that it meant phenylalanine.[8] So, the elegant, restrictive, comma-free code bit the dust.

It turns out that Nature had been bluffing in her typical, unconscious, unwittingly witty way. She had evolved the simplest code of all, not caring about redundancy and not paying special attention to the reading-frame problem in the code itself. The actual genetic code, which was gradually pieced together during the 1960s, is highly redundant, with up to six codons referring to the same amino acid and three meaning stop (Fig. 2.11). This redundancy is very clever, as we can see by hindsight, because it makes 'mistakes' in replication less likely to have fatal consequences. For instance, CCT, CCC, CCA, CCG all code for proline, so a mistake in copying the last letter is unimportant. Even where a single letter change is significant, the outcome is often the replacement of one amino acid by a similar one. For instance, the change from TTT to TAT results in the replacement of phenylalanine by its cousin tyrosine. The code is almost optimal in this respect. Finally, because all sixty-four codons are viable, Nature has scope for variation and experimentation, as we have already mentioned.

The interpretation of the code by the apparatus within the cell was the third hurdle that had to be surmounted. The basic problem is that DNA is confined to the cell nucleus whereas protein synthesis takes place in the surrounding

---

8  They dealt with RNA, which we haven't mentioned yet but will do shortly: in RNA thymine is replaced by uracil, U, and they actually showed that UUU was the code for phenylalanine.

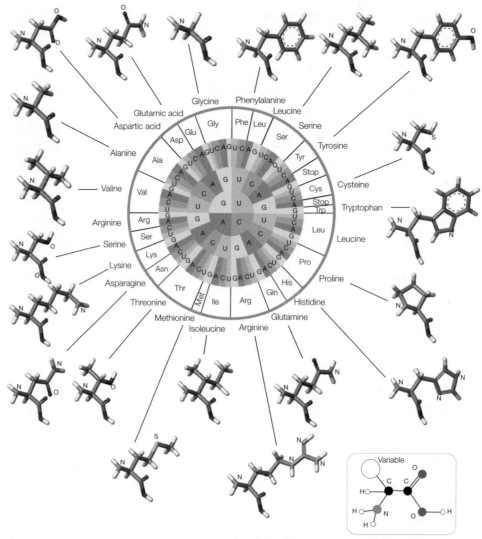

**Fig. 2.11** The genetic code and the structures of the amino acids the three-letter codons denote. For instance, reading out from the centre, the codon UAC codes for tyrosine (Tyr). Note that U stands for uracil (Fig. 2.12). All the amino acids have the layout shown in the inset. Note that some amino acids occur in more than one location and that the code is highly redundant, especially in its third letter. For instance, ACG, ACU, ACT, and ACA all code for threonine (Thr).

cytoplasm. A DNA molecule is far too big to get out into the cytoplasm through the nuclear membrane, so how does the information it carries get to where it is used?

Enter *ribonucleic acid* (RNA), a more primitive version of DNA. Ribonucleic

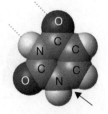

Uracil, U

**Fig. 2.12** The base uracil, U, which appears in place of thymine in the RNA molecule. Uracil differs from thymine by the loss of a methyl group ($CH_3$) at the north-east corner of the latter molecule. The arrow indicates the point of attachment to ribose and the dotted lines mark the location of the hydrogen bonds the molecule forms to adenine.

acids have the same general structure as DNA, consisting of a sugar–phosphate backbone with nucleotide bases hanging from it. However, the sugar is ribose rather than deoxyribose (hence the R of RNA in place of the D of DNA), in which the original additional oxygen atom of ribose has not been lost. Secondly, in place of thymine RNA has the subtly different but very similar pyrimidine uracil (U, Fig. 2.12). It is not entirely clear why U occurs rather than T or why ribose rather than deoxyribose is used in the spine: it is probably due to the slight difference in strengths of the hydrogen bonds that the molecule can form. One major difference is that RNA consists of a single strand. It is presumed that RNA was the original encoding substance but its function was taken over by the more stable DNA at an early stage in evolution. Some support for this view comes from the observation that RNA can also act as an enzyme. That function resolves one problem about the origin of life: which came first, the chicken (the enzymes needed to make use of genetic material) or the egg (the genetic material needed to specify enzymes).

There are two main types of RNA, namely *messenger-RNA* (mRNA) and *transfer-RNA* (tRNA). First we concentrate on mRNA, for it carries the information encoded in DNA out into the cytoplasm. To pick up the message, mRNA is synthesized rather like DNA replicates, with one strand of DNA exposing itself and an enzyme, *RNA polymerase*, using that strand as a template to produce mRNA. Only one strand of DNA is used, but it is not necessarily the same strand that is used throughout the chromosome, and copying always occurs in the same direction along the strand (so we don't get the equivalent of Beethoven played backwards). The copying proceeds almost at machine-gun rates: vertebrate RNA polymerase duplicates about thirty bases a second, and it takes about seven hours to replicate a cell's full complement of DNA. About one base in a million is copied wrongly, but proof-reading enzymes keep a watchful eye and most errors are corrected, leaving about one error in about 10 billion bases. When copying reaches a 'stop' codon, mRNA stops forming and is transported away from the DNA and out through pores in the nuclear membrane into the cytoplasm, carrying its precious information.

**Fig. 2.13** A ribosome consists of two components of unequal size which come together to form a single unit (right) when transcription takes place. Each unit is a small factory. The large unit typically consists of two ribosomal RNA (rRNA) molecules about 2900 and 120 bases long, respectively, and about thirty-two different proteins, in most cases in single copies. In the small unit there is one rRNA molecule about 1540 bases long and one copy each of twenty-one different proteins.

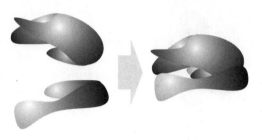

Lying in wait are the *ribosomes* (Fig. 2.13). These cunning little organelles (specialized components of the cell with specific functions) are aggregations of protein and RNA that lurk as two separate blobs, then combine into a single functional unit when they attach to the mRNA that emerges from the cell nucleus into the chemically perilous world of the cytoplasm. The other component of the cytoplasm that we need to notice at this stage is the transfer-RNA, the nucleic acid that does the actual construction of the protein. Figure 2.14 shows, in a variety of ways, the shape of a tRNA molecule. There are two important parts. The *anticodon* loop is the bit that recognizes the codon in mRNA. For instance, if the codon is CGU, coding for arginine, then the anticodon will be the comple- mentary sequence GCA, which can find the codon CGU by a matching of hydrogen bonds and sticking to it like Velcro. The other important part is the amino acid *attachment site* at the end of the nucleic acid chain. This is another

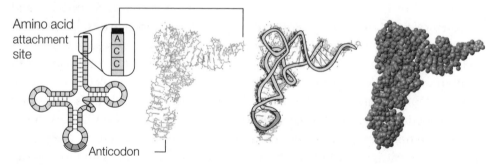

Amino acid attachment site

A
C
C

Anticodon

**Fig. 2.14** A transfer-RNA (tRNA) molecule. Biological molecules are so complex that various representations are used to depict them, depending on the features it is intended to emphasize. On the left we see a schematic layout of the locations of the bases (the squares) and the general shape of the molecule. The anticodon is the part that is used to identify a codon sequence in messenger-RNA and the appropriate amino acid is attached to the site shown. The second illustration shows the individual bonds in an actual tRNA molecule (this is yeast phenylalanine RNA). To help the eye, the third structure depicts the backbone of the structure overlaid on the line structure. Finally, the fourth structure shows all the atoms, and gives an idea of the actual volume-filling shape of the molecule, but details are hard to identify (except by other molecules).

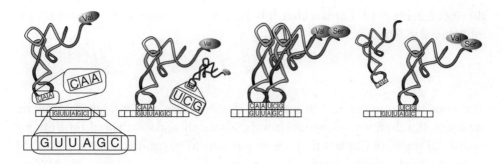

**Fig. 2.15** Protein synthesis under the guidance of messenger-RNA (mRNA, the strip of letter on the tape) and the action of transfer-RNA (tRNA). The action takes place inside a ribosome, which is not shown here. A tRNA with the anticodon CAA and loaded with a valine molecule lands on the codon sequence GUU. Soon, another tRNA, this one with the anticodon UCG and loaded with a serine molecule, wanders by and settles on to its codon, the sequence AGC. Enzymes then join the valine and serine molecule, to give the dipeptide Val–Ser, the discharged CAA tRNA floats away, to encounter another valine molecule somewhere, and the ribosome moves on to the next codon and awaits the arrival of the appropriate tRNA molecule and its amino acid. In this way, the polypeptide chain is built up in the order specified by the mRNA.

Velcro-like part of the molecule, and has a sequence of nucleotides that can stick to one and only one amino acid, in this case arginine.

Now we can visualize what happens in the cytoplasm once a ribosome clamps round a piece of mRNA. The ribosome pauses over the first codon, various tRNA molecules try their luck, but have the wrong anticodon to stick (Fig. 2.15). Then along comes a tRNA molecule with an anticodon for GUU with valine stuck on its attachment site. This one sticks, and in doing so enables the ribosome to ratchet along to the next codon, which may be AGC. In due course, a tRNA with an anticodon for AGC comes along, carrying the serine molecule which it has bumped into and captured elsewhere in the cytoplasm. The anticodon attaches to the codon, bringing its serine molecule close to the valine molecule; an enzyme shifts the valine off its tRNA and attaches it to the serine molecule, so forming the dipeptide valine–serine, and the discharged original tRNA ambles off into the solution in unconscious search for another valine. Now the ribosome ratchets along to the next codon, and the process is repeated. Gradually, the protein chain grows and the information originally in the nuclear DNA is converted into functional protein. Life is under way.

We can summarize the story so far as follows. The *central dogma* of genetics is that the flow of information and activity is DNA → RNA → protein. Only very rarely does the information flow from RNA to DNA (as we see later). The presumed inability of a protein to influence a DNA molecule is consistent with the incorrectness of Lamarckian inheritance of acquired characteristics (as introduced in Chapter 1).

The enormous importance of knowing the structure of DNA should now be apparent, but there are numerous loose ends we should touch on. Loose ends, I might call them, but in fact they are mountains of current activity and a world without end in current research.

First, there is the link to evolution, the molecular basis of the subjects we discussed in Chapter 1. Replication and transcription are never perfect: even nucleotides and amino acids make mistakes as they grope around blindly, responding to shape and electric charge, fitting as best they can, but sometimes getting stuck in the wrong position and not able to back out of their error before they are clamped into position by the next arrival or the formation of a bond. The DNA may replicate falsely when it forms the next generation, an mRNA molecule may make a mistake when it reads the DNA, a tRNA molecule may attach to the wrong codon, and even a tRNA correctly attached might bring along the wrong amino acid. All except the first, though, are transients: they will affect a cell but not the body as a whole. Only the first, which is called *somatic mutation*, affects the entire body, for a wrong step early in the organism's development will be replicated and replicated and fill the entire body. When meiosis occurs and the gametes are formed, mutated DNA there enters the germ line and is ready for handing on to that cleverly contrived extension of the body, the next generation. This type of mutation is called *germinal mutation*.

Replication is clearly a dangerous business, as so much can go wrong. We can be confident that it is a reasonably stable process, with mutations arising infrequently, for otherwise we wouldn't be here. One day, of course, we (our species) won't be. One reason for DNA's longevity is that each cell has a sophisticated policing and repair system which can identify mutations and correct them. Another reason is that DNA consists of a lot of junk, regions called *introns* that just come along for the ride and don't code for anything (aren't 'expressed'). The serious parts of DNA, the actively coding regions, are called *exons*. If mutations occur in the introns there are no consequences for the organism because that genetic material is not expressed as protein. A great deal of our DNA is intron junk, for in Nature's so-called elegant and economical but actually squalid way, she doesn't bother to clear out junk if it falls into disuse, but just drags it along through the generations. That's rather odd, because it means that a lot of that most precious resource, energy, goes into the propagation of uselessness. Perhaps the junk has a function we haven't yet recognized. Perhaps it is the perfect way of ensuring the propagation of information through the generations, never exposing itself to the hazards that accompany overt activity. Junk DNA may be pure, eternal, unexpressed information, with no other purpose than purposeless existence. This purposeless DNA is highly successful, for about 98 per cent of the DNA we lug around is junk, with only 2 per cent useful in the sense of coding for proteins.

It is easy to imagine a variety of mutations of DNA. A *base substitution* is the replacement of one base by another. Some base substitutions are silent in the sense that the mutant codon codes for the same base as the original, so the protein that results is unaffected. Other base substitutions, though, may change the message, and the severity of the effect depends on the degree to which the resulting substituted amino acid differs from the original in the protein. *Addition mutations* or *deletion mutations* are additions or deletions of complete base pairs: they may upset the interpretation of the DNA because instead of . . . ATGGTCT . . . being read as . . . (ATG)(GTC)(T . . . the deletion of the second T results in the message being read as . . . (ATG)(GCT)( . . . and from that point on the protein may be entirely different and rendered non-functional. On the other hand, it might strengthen the jaws of the cheetah or the olfactory sensitivity of a deer.

Mutations may be spontaneous or induced. Spontaneous mutations occur at a steady rate and constitute a *molecular genetic clock* that is steadily ticking away within the biosphere. The rate of mutation in a given gene is approximately constant, so by noting the number of amino acid differences between two species we can infer the time at which they diverged from a common ancestor. This is the kind of information we had in mind in Chapter 1, where we remarked that evolution is predictive, for in no case has this kind of information ever been in conflict with information about the sequence of species. The molecular clock also renders cladistic diagrams of descent (like the fragment in Fig. 1.2) quantitative, by attaching to them a scale of time. Mutations may also be induced by environmental factors, such as exposure to nuclear and ultraviolet radiation, ingested chemicals, and oxidation by virulent oxygen-containing species such as a superoxide radical (an oxygen molecule with an additional electron), that run out of control: that is the price we pay for using oxygen and striving for longevity.

Although the central dogma identifies the flow of information from DNA through RNA to protein, we did remark that there are exceptions. A *retrovirus* contains a single-stranded RNA molecule that makes use of a double-stranded DNA molecule of a host for its own replication. *Human immunodeficiency virus* (HIV), the virus that causes *acquired immune deficiency syndrome* (AIDS), is a retrovirus: it attacks the immune system and opens the body to uncontrollable infection. This virus was isolated in 1983 by Luc Montagnier at the Pasteur Institute in Paris, Robert Gallo at the National Cancer Institute in the USA, and Jay Levy at the University of California at San Francisco. The HIV virus attaches to the T-lymphocytes, a type of white blood cell, transferring its RNA and the enzyme *reverse transcriptase*. These molecules arrive at the vicinity of the DNA molecule in a chromosome and the enzyme synthesizes a DNA version of the viral RNA and a copy of that newly made DNA. At this stage there is a double strand DNA version of the viral RNA. This DNA is incorporated into the DNA of the host and then viral mRNA is synthesized from that DNA by using the replication mechanism of the cell. Now the viral mRNA is interpreted to make the proteins needed to build more virus particles. These particles then bud from the cell, taking some of the cell wall as their protective membrane. This process ablates the surface of the lymphocyte and in due course kills it, so reducing the organism's immunity to attack by other infections. A retrovirus is also believed to be the causative agent of a variety of cancers, including some found in humans.

A *restriction enzyme* is an enzyme produced by various species of bacteria and which can recognize a particular sequence of nucleotide bases in a DNA molecule and cut the DNA at that location. The fragments of DNA generated in this way can be joined together by other enzymes called *ligases*. Pieces of DNA that can replicate independently of the DNA in the host cell in which they are grown are called *vectors*; they include *plasmids*, the circular DNA molecules found in bacteria. Vector molecules that have inserted DNA sections are called *recombinant DNA*. These vectors construct multiple copies of a particular piece of DNA, amplifying the original material and producing large amounts of a *DNA clone*. The members of the colony so formed may be the desired product, as it is in the genetically engineered production of insulin, or—in gene therapy —it may be reinserted back into the original organism.

More recent methods of modifying DNA include the direct technique of *microinjection* in which genetic material containing the new gene is injected into the recipient cell by using a glass needle with a fine tip. The cell looks after its own (or, at least, another organism's own) and provides a mechanism whereby obligingly it brings the injected genes into the nucleus of the host cell and incorporates them into it. Genes can also be incorporated by creating pores in the cell membrane and allowing the incoming genes to work their way into the cell. In *chemical poration* the cells are bathed in solutions of special chemicals; in *electroporation* the cells are subjected to a weak electric current. If you think these techniques are too refined, then you can resort to *bioballistics*, in which small slivers of metal are coated with genetic material and then simply shot into the cell. I am reminded of a scene in one of the Indiana Jones movies in which after his opponent goes through a marvellously elaborate traditional piece of sword-play, Jones casually shoots him.

Speaking of shooting, another major consequence of understanding DNA is its forensic application in the form of *DNA profiling* or, more informally, *DNA fingerprinting*. Actual fingerprints, the patterns of ridged skin on the distal finger phalanges, were suggested as a means of identifying suspects in 1880 by Henry Faulds, a Scottish physician working in Tokyo, and shortly thereafter used to eliminate an innocent suspect and identify the perpetrator in a burglary there. A hundred years later, the identification of a person has moved from the tips of his fingers to every cell in his body, with the invention of 'DNA fingerprinting' by Alec Jeffries in 1984 at the University of Leicester. We need to understand two features of the technique: one is the amplification of tiny quantities of DNA; the second is the actual fingerprinting. DNA profiling is such an important technique in forensics, paternity testing, and evolutionary studies, that it has undergone tremendous development over the past twenty years, with a variety of flavours used in different circumstances. Here we sketch a typical approach.

Kary Mullis (b. 1944), the inventor of the *polymerase chain reaction* (PCR), claims that the idea came to him in 1983 during a moonlight drive in the California mountains, which must be one of the nicest ways to win a Nobel Prize. Polymerase, recall, is the enzyme that helps replicate a DNA strand by using it as a template; the same enzyme can be used *in vitro*. For it to work, the enzyme needs an abundant supply of the nucleotide bases and two *primers*, which are short sequences of about twenty nucleotides, to get the reaction going. First, the DNA strands are separated (the DNA is 'melted') by heating the mixture, then the solution is cooled, which enables the primers to stick to the appropriate parts of the DNA strand—the primer molecules jostle around until they find exactly their complements, and then bind—and act as boundaries to the region of the molecule that is to be duplicated. Finally, the temperature is raised again to the value where the polymerase can function efficiently, and a complementary strand grows on the template. Because the enzyme must withstand the high temperature of the melting phase, it is extracted from a bacterium, such as *Thermus aquaticus*, that is found in hot springs. The whole cycle takes about three minutes. It is then repeated over and over, around thirty or forty times, with the gradual production of tens of millions of copies of the strip of the original DNA lying between the primer markers (Fig. 2.16). That means that the target region of even a microscopic sample of DNA can be amplified and made ready for inspection.

The profiling technique itself makes use of the *polymorphism* of our genes, the fact that the DNA molecules may have significant differences between individuals. For instance, the junk DNA of our introns may contain long sequences of gibberish DNA, which has accumulated during meiosis. Here we focus on *variable number of tandem repeats* (VNTR), such as varying numbers of . . . CGATCGATCGATCGAT . . . in the same region of the DNA of different

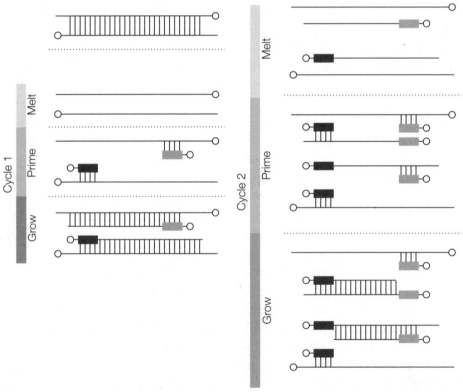

**Fig. 2.16** This sequence of diagrams shows how a PCR reaction takes place. At the top left, we see a representation of the target DNA double helix. In the first step (lower left), the strands are separated and the primers are attached to each one. Then enzymes grow complementary strands on the template provided by each strand. The double strands are melted again, and primers are attached to each one. Enzymes now build a complementary version of the strands, as before, but now replicas of the DNA lying between the two primers, the target sequence, appear in the mixture, and after a number of cycles come to dominate.

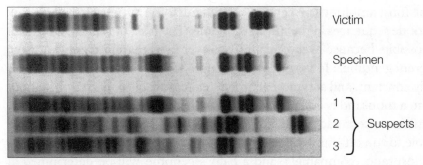

**Fig. 2.17** DNA fingerprints of a victim, a specimen, and three suspects. The profiles point clearly to Suspect 1 as the guilty party.

individuals and which have accumulated. As these tandem repeats lie in an intron region, they are not expressed and the individual and any onlooker is completely unaware of their existence, unlike variations in exons, such as those for brown or blue eyes (the latter being the result of the absence of brown pigment).

Now suppose we have used PCR to amplify a part of a DNA molecule that is highly polymorphic between individuals. The action of a restriction enzyme such as *AluI*, which jostles around until it finds the sequence AGCT, latches on to it, and then snips the molecule apart, or *Eco*RI, which attaches when it stumbles on to GAATTC and then snips at that point, will fragment the amplified DNA regions into a number of fragments of different sizes that depend on the numbers of tandem repeats in the individual. The sample is then dragged through a gel by applying an electric current, a process called *electrophoresis*. Because the tiny fragments can wriggle through the forest of cross-links in the gel more easily than the big fragments, the sample is separated into a series of bands that look a bit like a bar-code (Fig. 2.17). The pattern of bands is a portrayal of the spectrum of tandem repeats in the sample and therefore characteristic of the individual.

In this way, or sophistications of it, rapists have been brought to book, the innocent have been cleared, czars have been identified, pseudo-Anastasias revealed, evolutionary relationships identified, robbers trapped by a single hair, children reunited with their families (not least in Argentina, when whole families were brutally redistributed), and denying fathers found despite their protestations of innocence. Few molecular advances—the development of penicillin and the contraceptive pill are of similar rank—have had such a direct impact on society.

One of the most ambitious projects of the twentieth century was to identify all the nucleotide sequences in the human genome. The task, of course, is intrinsically impossible, because everyone who has ever lived (except identical twins) has a different genome. However, the differences of the composition of exons is reasonably constant, and a 'typical genome' is a reasonable concept: only about one base in a thousand is different between individuals, so individuals differ by about 3 million letters, a lot of them inconsequential. One day, perhaps, we will each be able to run off our own genome, and take it to our physicians (and perhaps our insurance companies), and a baby's genome will be determined at birth: the information in it would just fit on to a DVD and would last a lifetime.

The magnitude of the task can be appreciated by thinking about the size of the human genome. There are about 3 billion nucleotide bases in your genome. This book contains about 1 million letters, so your genome is equivalent to a library of about 3000 volumes. Suppose you regarded yourself as a really clever chemist, able to determine the order of bases at the rate of one an hour by carrying out a series of reactions and identifications of the products using conventional laboratory techniques. Three billion hours is 34,000 years of work. To achieve your aim in a decade rather than that ludicrous length of time, you would have to work 3400 times faster, and sequence DNA at the rate of one base a second, twenty-four hours a day, seven days a week. To make sure of the sequence, you ought to repeat your work several times. Ten times would give you confidence in your sequence, meaning sequencing at the rate of ten bases a second.

Amazingly, that was effectively achieved. Like the two previous pivotal steps in genetics, Mendel's original quantization of heredity and the Watson–Crick model of DNA, the Human Genome Project was fraught with clashes of precedence and propriety. This is not the place to give details of the genome wars, which centred largely on the morality of retaining information about the human genome for that most invigorating of elixirs, private profit, for it has been covered in full elsewhere by the major protagonists, the driving Craig Venter (b. 1946) and the humane John Sulston (b. 1942), not to mention the other principals, Frank Collins and Eric Lander. The squabble marred a moment in human history that should have been the pinnacle of achievement; but life, and in particular its genome, is like that. In a few years the animosities will have become as forgotten as the Franco-Prussian War, and we shall remember the achievement if not the means.

The crucial procedure was to determine the sequence of every nucleotide base in every strand of DNA in every chromosome. The procedure was based on Frederick Sanger's work in which, after successfully sequencing a protein, he turned his attention to DNA and sequenced all 5375 bases of the virus fX174 in 1977. His procedure was as follows. First, Sanger synthesized a new strand of DNA complementary to a single-strand template in such a way that the final letter carries a radioactive label (a molecule with one atom replaced by a radioactive isotope). To achieve this, he included in the usual mix of enzymes and nucleotides one modified version of a nucleotide called a dideoxynucleotide. When that modified nucleotide is incorporated it stops replication and gives a strip of DNA that terminates at that base. Then he repeated the procedure with dideoxynucleotides for the other three letters of the alphabet. Because the strands terminated at different points on the template molecule, each run gave a DNA molecule of different length. When the mixture was dragged through the tangled undergrowth of molecules making up a gel, the different length molecules are spread out, and show up as different patches on an X-ray film. The modification used in automatic sequencing machines is to use labels that fluoresce in different colours when illuminated by laser light, so A shows up as red, C as green, and so on. Then the sequence can be identified electronically.

The second crucial component is the implementation of this procedure on a production-line basis that can identify thousands of bases an hour. There are two principal approaches. One is to work in sequence through known strips of DNA. The other, the 'shotgun approach', is to shatter DNA into myriad pieces, and then to sequence this jumble. The challenge of the latter is to reassemble the DNA sequence from these fragments. It is at that point that supercomputers play a central role in the reassembly. Broadly speaking, the orderly approach is more accurate but the shotgun approach is faster. In practice, each draws on the other.

The first draft of the human genome was released in 2001, about fifty years after the structure of DNA was determined and about a hundred years after Mendel's work was recognized and genetics began. The consequences of knowing the human genome are incalculable, both in terms of likely good and likely bad. Like all great science, knowledge has the potential for pleasing devils as well as angels. At least, we can now paste our recipe on to the spacecraft we send into interstellar space, so that there is a fleeting chance that humans can be recreated even though all physical manifestations are lost. At best, on our immediate Earth, we can come to realize our consanguinity and not waste our aspirations in petty squabbles that stem from the difference of a few letters in our genes.

# ENERGY

## THE UNIVERSALIZATION OF ACCOUNTANCY

**THE GREAT IDEA**
*Energy is conserved*

*Energy is eternal delight*
WILLIAM BLAKE

**N**EITHER the pulsating biosphere that emerged from inorganic Earth, nor the molecular activity that sustains and propagates it now, could have done so without an influx of energy from the Sun. But what is this thing called energy? The word might spring from everyone's lips, and a scientist might see it as binding the universe into a comprehensible, living whole; but what is it really?

Poets, in their inimitable way, mastered the concept of energy well before it came to the attention of scientists. Thus, Sir Philip Sidney, writing in 1581 in *The Defence of Poesie*, drew attention to 'that same forcibleness or *Energie* (as the Greeks call it) of the writer'. He had in mind vigour of expression rather than an aspect of the motion of the musket ball that later killed him. The Greeks actually called it ἐνέργεια, which translates literally into 'in work', and we can sense the etymological trail that leads to literary forcefulness. In our own time, the general public has taken energy to heart and has convinced itself that it knows exactly what it is, finds it costly, senses its essential contribution to the modern world, and is fearful of the prospect of its unavailability.

Energy is still an aspect of literary discourse, but it has taken on a new, rich, and precisely circumscribed life within science. That was not always so. The scientific use of the term can be traced back to 1807 when Thomas Young (1773–1829), who worked as a professor of natural philosophy at the scientific firmament that was the Royal Institution of Great Britain and later, in the admirable polymathic ways of the time, contributed to the deciphering of the Rosetta stone, seized the term for science when he wrote that 'the term energy may be applied, with great propriety, to the product of the mass or weight of a body into the square of the number expressing its velocity.'[1] Like many pioneers, Young's claimed 'great propriety' was but half-baked, and we will have to do some work to complete its baking. In doing so, we shall come to understand the modern interpretation of energy and see the significance and importance of its conservation.

To grasp the nature of energy, we need to understand two very important features concerning events and processes in the world. One concerns the characteristics of the motion of bodies through space; the other the nature of heat. The description of motion through space was essentially complete by the end of the seventeenth century. It then took a surprisingly long time to wrestle with and finally conquer the nature of heat. That was not achieved until the middle of the nineteenth century. Once motion and heat were understood, scientists had effectively mastered the nature of events, or so they thought at the time.

The Greeks touched, rather uselessly, on the motion of bodies and confused the world for two thousand years: their armchair style of questioning was far better suited to mathematics and ethics than to physics. Thus, Aristotle (384–322 BCE) speculated that an arrow was kept in flight by the action of vortices in the air behind it and concluded, therefore, that an arrow must quickly come to rest in a vacuum. As so often, science clarifies by inversion of accepted opinion, and we now know that exactly the opposite is true: an arrow is slowed by the resistance of the air, not pushed along by it. The evidence in those cumbrous times for the necessity of a sustaining force was plentiful, for oxen needed to strain at creaking wooden carts to maintain them in motion: how absurd it was to think otherwise, for then farmers would have been led to harness the oxen behind a moving cart to stay its natural motion! Arrows and stones in flight were more problematic, for there were no obvious oxen involved. Aristotle's ever fertile mind saw vortices in the air that urged the arrow forward, and thereby saved his theory.

Aristotle also had more general delusions about the cause of events and the

---

[1]  His lectures while professor of natural philosophy (1801–3) at the Royal Institution, London, were published as *A course of lectures on natural philosophy and the mechanical arts* (1807).

motion of objects.[2] As rules of thumb his delusions were quite reasonable, and he is to be admired for ceaselessly searching for explanations and pressing Nature for answers. However, besides being utterly wrong, his opinions lacked what we now regard as explanatory power and were totally incapable of being rendered quantitative. He envisaged, for instance, a series of concentric spheres, with the spherical Earth at the centre surrounded in succession by the sphere of water, the sphere of air, and the sphere of fire, the whole being encased in the crystal spheres of the heavens. In his model, matter sought its natural place, so earthy objects fell Earthwards after being hurled upwards initially, and fiery flames flew upwards, striving for their natural abode. It is easy to pick holes in such a model from our current point of view, but it held sway in people's minds for two millennia, perhaps because they were in the grip of acceding to authoritative teaching rather than relying on their own observations, or perhaps because they lacked encouragement into the exercise of the inquisitiveness that was needed to set observation against authority.

Galileo's principal contribution to this particular story was his discarding of the blindfold of authority and, with his eyes opened to observation, his *experimental* demonstration that Aristotle's version of events was wrong. Galileo asserted that a body maintained its state of motion without there being an impressed force. He arrived at this conclusion by considering a ball rolling down an inclined plane and then up a corresponding plane and noticed that whatever the angle of the second plane, the ball rose to the same height. He concluded that if the second plane were made horizontal, then the ball would roll for ever, for it would never attain its initial height. The introduction of an inclined plane was itself a stroke of genius, for it slowed processes—the falling of bodies—to the point that they could be studied quantitatively and precisely, and thus impression gave way to observation.

Galileo's conclusion was a major turning point in science, for it emphasized the power of abstraction and idealization that I mentioned in the Prologue, the latter being the disregarding of interfering factors that cloud the essentials of an experiment. Of course, Galileo never *explicitly* demonstrated that the ball would roll on and on for ever, and in any experiment of this kind a real ball would in fact come to a stop sooner or later in an apparently decidedly Aristotelian way. However, Galileo realized that there are *essential* components of behaviour on the one hand and extraneous influences on the other. The latter include friction and air resistance: by reducing them (by polishing the ball and surfaces

---

[2]  No doubt readers of this book in two thousand years' time will find our delusions similarly quaint; but at least they are more potent than Aristotle's.

**Fig. 3.1** Newton, and modern physics, were born in this room early on Christmas Day, 1642. The furniture is not original.

of the planes, for instance), he could edge towards ideality and the exposure of essential behaviour. In Aristotle's experience of the world, with oxen tramping through mud dragging heavy carts, the extraneous influences completely overwhelmed the essential behaviour of the cart.

Galileo's torch passed to Newton. According to the old-style calendar, Isaac Newton (1642–1727)[3] was born the year that Galileo died (Fig. 3.1), so romantically inclined transmigrationist semiophiles can see the passing on of a special spirit. Unlike Galileo, Newton was by all accounts a most disagreeable and petulant man, but one of the greatest of all scientists. Almost single-handedly, he brought mathematics to bear on physics, and so opened the way to modern quantitative physical science. He did more: he invented the mathematics he needed to pursue his programme, and his *Principia*,[4] published in 1687, is a monument to the power of the human intellect applied to the rationalization of observation.

Euclid's five axioms for the formulation of geometry, which we explore in Chapter 9, summarize the structure of space, so through them we know where we are. Newton's three laws summarize motion in that space, so through them we know where we are going. In a slightly simplified form they are as follows:

---

[3] For all the information you could ever want about Newton, go to http://www.newton.cam.ac.uk/newton.html

[4] More fully, *Philosophiae naturalis principia mathematica*, or *Mathematical principles of natural philosophy*.

1. *A body continues in its state of uniform motion in a straight line unless it is subjected to a force.*

2. *The acceleration of the body is proportional to the force applied.*

3. *To every action there is always opposed an equal reaction.*

From these three simple statements sprang the whole edifice of *classical mechanics*, as the description of motion based on Newton's laws is called, and the prediction and understanding of the motion of particles, balls, planets, and—these days—satellites and spaceships.

Newton's first law is just a reaffirmation of Galileo's anti-Aristotelian observation and is sometimes called the *law of inertia*. His second law is commonly regarded as the richest of the three laws, for it lets us calculate the path of a particle through a region where a force is acting. Where a force pushes from behind, we go faster in the same direction; when it pushes from ahead, we slow down. If the force pushes from the side, then we veer in the direction it impels us towards. The law itself is written in the form

$$Force = mass \times acceleration$$

with the mass (more specifically the *inertial mass*) a measure of the particle's responsiveness to the force. For a given force, the acceleration is large when the mass is small but the acceleration is small when the mass is large. In other words, high inertial mass indicates low responsiveness, and vice versa. A sharp eye will detect a tautology in this law, for it defines mass in terms of force and force in terms of mass.

Because acceleration is the rate at which speed changes, you can probably appreciate that buried inside Newton's second law is a way of predicting the path of a particle that is subjected to a given force, a force that might vary from place to place and take on different values at different times. 'Buried' is a good term here, because the calculation of paths can be a very tricky exercise, more akin to exhumation than algebra. Still, it can be done in a number of simple cases, and even complex fields of force, such as those near double stars or even around our own Sun when the interactions between the planets are taken into account, can be tackled by using computers (Fig. 3.2). In short, we can interpret the second law as meaning that, provided we know where a particle—or even a collection of particles—lies at a given time, then in principle we can predict where to find it and where it will be going at any later time. The prediction of these precise trajectories was one of the glories of classical mechanics.

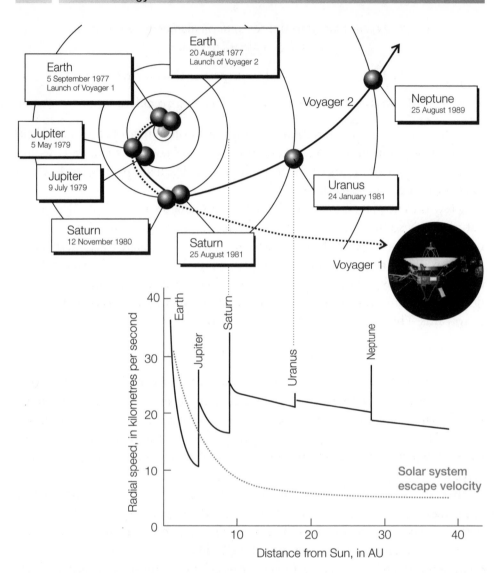

**Fig. 3.2** The orbital paths of spacecraft are calculated using Newtonian mechanics. The computations are complex, as the spacecraft are subject to the influences of the planets. The upper diagram shows the paths of Voyager 1 and Voyager 2, which began their journeys in 1977 and have been functioning ever since. Voyager 1, now the most distant human-made object in the universe, is leaving the solar system at a speed of about 3.6 AU per year (1 AU, one astronomical unit, is the mean radius of the Earth's orbit round the Sun, and corresponds to about 150 million kilometres), 35 degrees out of the plane of the planetary orbits. Voyager 2 is also escaping from the solar system at a speed of about 3.3 AU per year, 48 degrees out of the plane in the opposite direction. The lower graph shows the boosts to the speed of the spacecraft as they swung by each of the planets. These gravity-assisted boosts ensured that they have enough speed to reach their targets and then to leave the solar system.

Newton's third law is deeper than it looks. At first sight, all it seems to imply is that if a bat exerts a force on a ball, then a ball exerts an equal and opposite force on the bat. We can indeed feel the force exerted on a ball when we strike it with a bat or kick it with our foot. The real significance of the third law, though, is that it implies a 'conservation' law. Now, conservation is what this chapter is all about, so we are starting to home in on our quarry. However, we have to do some unpacking of the concepts involved.

A *conservation law* is a statement saying that something doesn't change. That might seem to be the most boring kind of comment possible in science. In fact, it is commonly the deepest and most significant type of scientific law because it gives insight into the symmetry—essentially the shape—of systems and even into the symmetries of space and time. The particular conservation law implied by Newton's third law is the *conservation of linear momentum*. In classical mechanics, the *linear momentum* of a particle is simply the product of its mass and its velocity:

$$Linear\ momentum = mass \times velocity$$

This definition means that a cannon ball moving rapidly has a high momentum but a ping-pong ball moving slowly has a low momentum. The linear momentum is an indication of the force of the impact of the moving body when it strikes an object, a difference illustrated by the impact of a cannon ball compared with a table-tennis ball. The law of conservation of linear momentum states that the total linear momentum of a collection of particles doesn't change provided they are free of any externally applied force. So, for example, when two billiard balls collide, their total linear momentum is the same after the collision as it was before. We have to unpack the full significance of 'linear momentum' before we can comprehend this statement.

Momentum is a directed quantity in the sense that two particles of the same mass moving at the same speed but in different directions have different momenta. Two billiard balls rolling towards each other at the same speed have equal but opposite linear momenta and their total linear momentum is zero. When they collide head-on, they come to a standstill, so the linear momentum of each one suddenly becomes zero and the total after the collision is again zero. We see, in this instance, that although the momentum of the individual particles changes, the total linear momentum is conserved. This conclusion is entirely general: whatever the individual linear momenta of the particles initially, the sum of those momenta (allowing for the different directions as well as the magnitudes of the momenta) will be the same after the particles have interacted

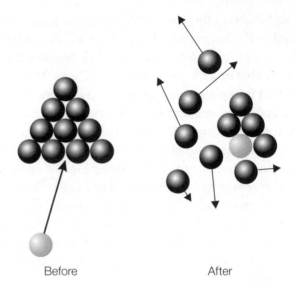

**Fig. 3.3** Collisions, and interactions in general, conserve linear momentum, with the result that the total linear momentum after a collision is the same as the total linear momentum initially. Here we see a collision of a ball with a group of balls. The linear momentum of the cue ball is indicated by the length and direction of the arrow on the left. The linear momentum is transferred to six of the 'red' balls, and their individual momenta are given by the lengths and directions of the arrows on the right. If you put those arrows head to tail without changing their orientations, you will find that they add up to the length and direction of the original arrow.

Before          After

as it was before (Fig. 3.3). Billiards itself is a game based almost entirely on the principle of the conservation of linear momentum: every collision between the balls or of the balls with the cushions conforms to the law and leads to different trajectories over the table, depending on the original angle of approach.

Now we can take a giant but controlled leap from the billiard hall to the universe. The funny thing is, because linear momentum is conserved in any process, there must be a fixed amount of linear momentum in the universe. So, when you drive off in your car, even though you pick up momentum as you accelerate, and change the direction of your momentum as you drive round a corner, something somewhere takes up the momentum so that the total in the universe doesn't change. You actually push the Earth a little bit in the opposite direction as you drive off: you accelerate the Earth in its orbit if you drive off in one direction, and decelerate it if you drive off in the opposite direction. The mass of the Earth is so great compared with the mass of your car, though, that the effect is wholly undetectable however much rubber you burn. But it's there.

I have said that a conservation law is a consequence of—a window on to—the symmetry of something or other. The something or other in this case is space itself, so the symmetry of space is ultimately responsible for the conservation of linear momentum. The symmetry of *space*, the *shape* of space: what can that mean? In this instance, all it means is that space isn't lumpy. As you move through empty space in a straight line, space stays exactly the same: everywhere it is smooth and unvarying. The conservation of momentum is just a sign that space isn't lumpy, and Newton's third law is a 'high level' way of saying the same thing.

There is another consequence of Newton's third law, another conservation law, and another insight into the shape of space. We have been discussing *linear* momentum, the momentum of a particle travelling in a straight line. There is also the property of *angular* momentum, the momentum of a particle travelling in a circle. A rapidly rotating heavy flywheel has a very large angular momentum; a slowly rotating bicycle wheel has a low angular momentum.

Angular momentum can be transferred from one object to another if the first object exercises a *torque*, a twisting force, on the second, and the responsiveness of the second body to the torque depends not just on its mass but how that mass is distributed. For instance, it's harder to accelerate a wheel when its mass is concentrated on its rim than when the same mass is concentrated near the axle. That's why flywheels have their steel concentrated near the rim (Fig. 3.4), as that distribution is good at damping out variations in angular speed: metal near the axis is less effective and therefore wasteful.

Angular momentum is conserved, provided the system is free of externally applied torques. Suppose two spinning billiard balls collide in a glancing blow; then angular momentum may be transferred from one to the other and the spin of one may be transferred in part to the other. Nevertheless, the sum of the

**Fig. 3.4** A flywheel has a lot of mass concentrated at a large distance from its axis. Such a wheel requires a large torque (twisting force) to change its angular momentum. In the model steam-driven traction engine shown in the illustration, the flywheel (the uppermost of the wheels shown) helps to preserve the piston's steady motion.

angular momenta after the collision is the same as it was initially: angular momentum is conserved. The same is true in general: the total angular momentum of a collection of interacting particles can be neither created nor destroyed. Even if the spinning billiard ball slows down by friction, the angular momentum is not lost: it is transferred to the Earth. As a result, the Earth spins a little bit faster (if the billiard ball was initially rotating in the same direction as the Earth) or a little bit slower (if the ball was rotating in the opposite direction). As you drive in a screw in the northern hemisphere, you speed up the Earth's rotation, but slow it down again when you pause or stop; when you do it in the southern hemisphere, you slow it down then speed it up when you stop. The universe as a whole appears to have zero angular momentum, as there is no rotation of the universe as a whole. That's the way it will always remain, because we cannot generate angular momentum; we can only transfer it from one bit of the universe to another.

So what does the conservation of angular momentum tell us about the shape of space? Because angular momentum is about rotational motion, we can suspect that its conservation tells us about the shape of space as we turn round. In fact, the conservation of angular momentum reveals that if we travel in a circle around a particular point, then we don't encounter any lumps in space. The conservation of *linear* momentum stems from the uniformity in space as we travel in a straight line; the conservation of *angular* momentum stems from the uniformity of space as we travel in a circle. More technically, the conservation of linear momentum tells us that empty space is homogeneous and the conservation of angular momentum tells us that it is isotropic. Newton's third law is telling us what we might think is obvious: that space is uniform wherever we go (so long as we stay away from externally applied forces and torques). However, the fact that the law has measurable consequences means that our armchair speculation about the nature of space is open to experimental verification, which is a wonderful thing.

You may have noticed that *energy* has not yet played a role in the discussion. Newton didn't use the term and died a century before Young proposed its adoption. His formulation of mechanics, for all its originality and elegance, was essentially the physics of the farmyard (or, more closely, that of the ice rink), using the almost literally tangible concept of force. You and I know, we think, exactly what a force is, for we know when we exert or experience one. Its adoption by Newton as the central aspect of his mechanics is a sign that physics had barely moved from the farmyard. As we saw with Galileo, the march of progress in science has commonly been accompanied by a transition from the tangible to the abstract, for thereby the grasp of the subject becomes larger. There are many suits of clothes, but essentially only one human skeleton: once we understand the skeleton, we understand so much more than we would understand by watching the flapping of the clothes. The introduction of energy marks the emergence of the abstract in physics and the extraordinary illumination that spreads across the world in its light.

That illumination took half a century to spread across the world. At the start of the nineteenth century, energy was still a literary term; by the middle of the century, energy had been captured by physics. Its final acceptance can be dated with some precision, for in 1846 William Thomson (1824–1907, from 1892 Lord Kelvin) was still able to write that 'physics is the science of force', but in 1851 he was proclaiming that 'energy is the primary principle'. This transition took place in two stages: first in studies of the motion of individual particles (including the particles we refer to as planets), and then in the action of the elaborate collections of particles we call steam engines.

Dawn rose on particles in a series of bursts of illumination during the opening years of the nineteenth century. First, as we have seen, Thomas Young suggested that the term *energy* be applied to the quantity obtained by multiplying the mass of a particle by the square of its speed. This energy of motion was perceived as a measure of *vis viva*, or living force, and regarded as a sensible measure of the vigour of the events taking place in a collection of particles. Paradoxically, the greater the living force of a cannon ball, the more death and destruction it could achieve.

Young's identification of energy with *mass × speed²* was not quite right. He had arrived at his suggestion by considering the force that a moving object exerts when it collides with something, and the somewhat subtle recognition that the force exerted by a given body increases four-fold if its speed is doubled. That is true, but the numerical factor in Young's expression is wrong. His error was recognized in about 1820 when it was realized that the concept of *work* (which we discuss below) can be combined with Newton's second law to deduce that the energy arising from motion is better expressed as one-half of this quantity. For some time, the resulting quantity was called *actual energy*, but shortly the name shifted to *kinetic energy*, and that term is now used universally. We write

$$\textit{Kinetic energy} = \tfrac{1}{2} \times \textit{mass} \times \textit{speed}^2$$

Thus, a heavy body moving rapidly has a high kinetic energy, whereas a light body moving slowly has a low kinetic energy. A falling ball acquires kinetic energy as it accelerates. Unlike linear momentum, the kinetic energy is the same whatever the direction of the moving particle: a ball moving horizontally at a given speed has the same kinetic energy regardless of its direction, but its linear momentum is different for each direction of travel.

The 'work' to which we have referred is a crucial concept in the study of energy and deserves a moment's explanation. We have to understand what a scientist means by work, which is not quite the same as its everyday meaning. In science, *work* is done whenever an object is moved against an opposing force. The further we move the object, the greater the work we have to do. The greater the opposing force, the greater the work we have to do. Raising a heavy object against the pull of gravity (the opposing force, for it resists upward motion of the weight) involves doing a lot of work. Lifting a piece of paper off a table also involves work, but not very much. Raising the same object through the same distance on the Moon, with its weaker gravity, involves doing less work than on Earth.

Raising a block of metal against the pull of gravity is more interesting than you might think. First, let's imagine just pushing it over a slippery, frictionless surface, a chuck on an ice rink. The block accelerates for as long as we go on

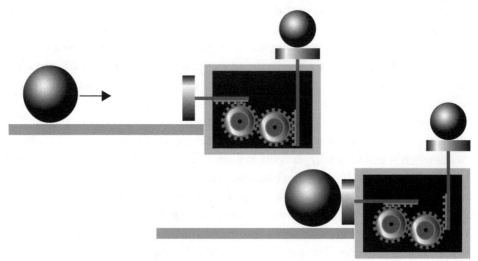

**Fig. 3.5** The motion of a body can be used to do work, so it corresponds to a form of energy, and is known as kinetic energy. In this device, the ball crashes into the piston, and the motion of the piston is converted, through chain of gears, into the raising of a weight represented by the other ball. The work done in raising the second ball (which is proportional to its weight and the height through which it is raised) is equal to the kinetic energy of the rolling ball.

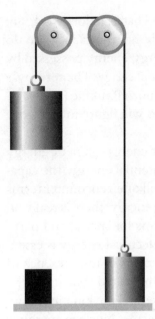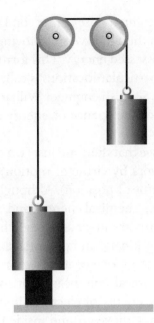

**Fig. 3.6** Even though an object is stationary, it may still possess energy by virtue of its position: this form of energy is known as potential energy. On the left, the heavy weight is ready to be lowered. On the right, the heavy weight has been lowered to the platform, and in the process has raised the light weight. The heavy weight has therefore done work, and therefore must have possessed energy initially. That energy was its original potential energy.

pushing. As a result, its kinetic energy increases from zero initially to whatever value we choose or the point at which we lie exhausted and cease exerting a force, with the block sliding across the ice at a constant speed away from us. The work we have done has been converted into kinetic energy, the energy of motion. (The factor of ½ in the expression for kinetic energy was introduced to make sure that these two quantities, the work done and the kinetic energy achieved, are equal.) Now, we can turn this comment round, and let the block, moving steadily across our Galilean, frictionless table, collide with some kind of contraption that can convert its motion into the raising of a weight (Fig. 3.5). All the kinetic energy is converted into work, the same quantity of work that we put into accelerating it initially.

This observation motivates the following definition: *energy is the capacity to do work*. That, in fact, is all that energy really is. Whenever you see the term energy used in a technical rather than a literary sense, all it means is the capacity to do work. A lot of stored energy (a heavy mass moving rapidly) can in principle do a lot of work—raise a heavy weight through a great height. An object that possesses only a little bit of energy (a light mass moving slowly) can do only a small amount of work—raise a light weight through only a small height. Doubling the speed of the object quadruples the work it can be harnessed to do.

Now we take the next step. Suppose we raise a weight to a certain height and attach it to a series of pulleys that can raise another weight (Fig. 3.6). When we release the first weight, it raises the second weight. That is, it does work. So, the first weight, even though it wasn't moving initially, has the capacity to do work. It therefore possessed energy. This form of energy, energy possessed by virtue of being at a particular location, is called *potential energy*. The term was coined in 1853 by the Scottish engineer William Macquorn Rankine (1820–72), one of the founders of the science of energy and who will figure in this story again.[5]

At this stage, we see that there are just two forms of energy—kinetic energy (the capacity to do work by virtue of motion) and potential energy (the capacity to do work by virtue of position). Although you will often encounter terms like 'electrical energy', 'chemical energy', and 'nuclear energy' there is really no such thing: these terms are just handy shorthand terms for special and particular combinations of kinetic and potential energy. Electrical energy is essentially the potential energy of negatively charged electrons in the presence of positive charges. Chemical energy is a bit more complicated, but it can be traced to the potential energy of electrons in molecules and the kinetic energy of their motion as they move around inside the molecule. Nuclear energy is analogous, but arises from the interactions and motion of subatomic particles inside atomic nuclei. The exception to the universality of the terms kinetic and potential energy is the energy of electromagnetic radiation (for instance, the energy of light, such as that carried from the Sun to the Earth and used to warm us or drive photosynthesis and the production of food). As far as energy stored in matter is concerned, it is entirely composed of kinetic and potential energy. So, at this point, we really do understand all there is to know about energy.

Well, not quite. We don't know everything, as you can judge from the pages that remain in this chapter and the fact that other chapters also elaborate the concept of energy. Energy deserves all this space because it is so central to the universe and all the structures and events in it. In fact, the two great foundations of science are causality, the influence of one event on a subsequent event, and energy. Causality is essentially the coherence and consistency of the chain of commands that keeps the universe moving and which we disentangle to achieve understanding; energy is the ever watchful guardian of propriety, ensuring

---

[5]  Two of the founders of the science of energy, and specifically thermodynamics, have temperature scales named after them. Thomson (Lord Kelvin) is commemorated in the *Kelvin scale*, on which absolute zero, the lowest attainable temperature, lies at −273 °C, and Rankine is commemorated in the *Rankine scale*, in which absolute zero lies at −460 °F.

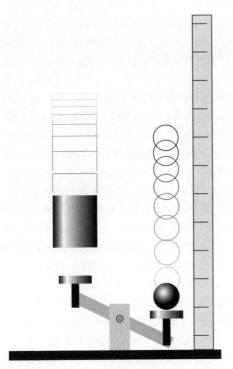

**Fig. 3.7** In this abstract form of a fairground 'test your strength' machine, the kinetic energy of the falling weight on the left impels the ball on the right upwards. The kinetic energy of the falling weight (perhaps a mallet) is converted into the work of raising the ball.

that causality causes only legitimate actions. As we shall see, energy is truly the currency of cosmic accountancy.

Let's start to unpack the concept of energy. Potential energy is potential because it can be converted into *vis viva*, actual energy, kinetic energy. Suppose we cut the cord holding the weight on high. It plunges down (we are doing this experiment on Earth, in the Earth's gravitational field) and accelerates as it falls. The instant before it strikes the ground, it has acquired a lot of kinetic energy and has lost all its potential energy.[6] It still has the capacity to do work. With a suitably designed contraption, we could capture the kinetic energy by letting the falling weight hit a lever that impels another weight upwards, rather like the old-fashioned strong-man sideshow in a fairground, where striking a lever with a hammer drives a weight upwards towards a bell (Fig. 3.7). Indeed, such a sideshow epitomizes perfectly the central content of this chapter. We have to conclude that potential energy and kinetic energy are freely interconvertible.

The experiment we have done also implies that the *total energy*, the sum of the kinetic and potential energies of the first weight, is constant. Thus, we arrive at the *conservation of energy*, the observation that energy can neither be

---

[6]  By convention, for events taking place close to the surface of the Earth, particles on the surface itself are ascribed zero gravitational potential energy.

created nor destroyed, that the total energy is constant. This conclusion can be proved formally by using Newton's second law, so in a sense that law is a statement of the conservation of energy just as the third law is covertly a statement about the conservation of momentum.

Both the other conservation laws we have encountered (those of linear and angular momentum) have been associated with a symmetry, and have told us something about the shape of space. The obvious question that now comes to mind is whether the conservation of energy is a consequence of symmetry, and if so of what. In Chapter 9 we shall see that we should not think of space alone, but of *spacetime*, and that time should be treated on an equal footing with space. We should be able to sense that whereas the conservation of momentum stems from the shape of space, the conservation of energy stems from the shape of time. This is indeed the case, and the fact that energy is conserved stems from the fact that time is not lumpy: it spreads smoothly from the past into the future with no squashed bits or stretched bits. So deep is the relationship between conservation laws and the symmetry of spacetime that the conservation laws survive even when Newton's laws of motion fail, for the conservation of momentum and energy survive even in relativity and quantum mechanics.

Because Newton's second law is also effectively a statement of the conservation of energy, we can see that the law is a direct consequence of the smoothness of time just as the third law is a direct consequence of the smoothness of space. Such an explanation is now thought by most scientists to be more convincing than the spur to the fervently religious Thomson's and many of his contemporaries' enthusiasm for the conservation of energy, which they considered to be a consequence of God's bounty. God, they argued, had endowed the world with a gift of energy, and that energy could neither be increased by human intervention nor, being divine, be destroyed by any of our activities.

This analysis of the behaviour of particles in terms of kinetic energy, potential energy, and the conservation of energy had become established as the currency of physics by 1867 and the publication of Thomson and Tait's magisterial *Treatise on natural philosophy*. By then, there was a realization that the concept of energy helped to unify whole swathes of physics. Thus, in 1847 the polymath Herman von Helmholtz (1821–94) used the concept to show the underlying unity of mechanics, light, electricity, and magnetism. Yet, despite this success, there was a nagging problem that threatened the whole edifice, the problem of heat.

Heat had long been a mysterious phenomenon, yet with the development of the steam engine and the dependence of national economies, and hence success in war and trade, on their efficient operation, it had moved to the centre of scientific attention. The problem, though, was not only that the nature of heat was unknown, but it appeared to lie outside the reach of contemporary physics.

Heat had for long been thought by many to be a fluid called *caloric* (taking its name from the Latin for 'heat', *calor*), one of the 'imponderable', weightless, fluids so beloved of early investigators. Not only was caloric imponderable (and hence conveniently undetectable by weighing), it was also 'subtle' in the sense that it could penetrate everywhere, even between particles that were stacked closely together. We might snigger at these misconceptions, but not everyone today can explain what is meant by 'heat' and, moreover, the language of caloric still pervades everyday language, for we speak of heat as 'flowing' as though a fluid from a hot to a cold body.

Caloric was eliminated from science in 1798, by the scientist, inventor, politician, womanizer, soldier, hypocrite, benefactor, statesman, reformer, and spy Benjamin Thompson, Count Rumford (1753–1814). Thompson was born in Massachusetts, fled to England in 1776, established the Royal Institution in 1799, and travelled on to Bavaria, where he was appointed minister of war, minister of police, court chamberlain, state councillor, and count of the Holy Roman Empire. He chose his title from the name of the town Rumford (later Concord), New Hampshire, where the first of his wives was born.[7] Caloric was eliminated as a result of Rumford's observations on the boring of cannon that he was supervising in the Munich arsenal. He recorded:

18.77 lb of water in an oak box. Initially 60°F; after two horses had turned the lathe for 2½ hours, the water boiled.

His conclusion from this and related experiments was that heat could be produced continuously and was inexhaustible. That being so, it had to be generated by friction and was therefore to be regarded as the motion of particles making up the metal of the cannon rather than a fluid secreted in the metal.

There was still a long way to go before heat had been incorporated quantitatively into science, its true atomic nature determined, and finally accommodated in the law of the conservation of energy. The impetus to understand heat arose, as we have indicated, from the central importance of the steam engine in industry, and it is not surprising that most of the developments that led to our

---

[7] He later lost his heart to Madame Lavoisier, after her husband Antoine (Chapter 5) had lost his head. The marriage was not a success.

current understanding of heat were made by groups of north British scientists centred on Glasgow and Manchester and with close links to manufacturing industry.

One theme that will recur throughout this book is that a sign of the advancement of science is the elimination of fundamental constants. This is our first glimpse of what is involved and the clarification that ensues. In the nineteenth century (and, it must be admitted, in parts of the world in the twenty-first), work was measured in one set of units (ergs, as it happened, but the details are unimportant) and heat was measured in another (calories). That different units were used to measure these two quantities concealed the fact that the quantities are essentially the same. Much effort was expended during the nineteenth century in trying to measure the 'mechanical equivalent of heat', the work that could be obtained from a given quantity of heat, and effectively finding a conversion factor from calories to ergs. This effort was an essential part of the progress of science and a part of the experimental foundation of the law of the conservation of energy. However, from our current viewpoint, it was a waste of time. Don't mistake me: it was an *essential* waste of time. It was essential because it helped to show that heat was a form of energy, that no more work was produced than heat was absorbed, and that no more heat was produced than work was done. It was a waste of time only because now that we recognize work and heat as two aspects of a single entity, energy, we measure them in the same units and no longer need to convert from one unit to another.

He who deserves most credit for wasting his time in such an exceptionally fruitful manner is James Joule (1818–89). Joule, born in Manchester the son of a wealthy brewer, had sufficient of his own funds to pursue research until the money ran out in about 1875. In a celebrated experiment, Joule used vigorously rotating paddle wheels driven by a falling weight to stir water, and measured the rise in temperature of the water (Fig. 3.8). Thus, he was able to demonstrate that work could be transformed into heat. By comparing the work needed to raise the temperature of water to the quantity of heat needed to achieve the same effect, he was able to measure the mechanical equivalent of heat. Although he managed to measure this now useless quantity, he deserves unbounded praise for establishing the equivalence of heat and work and thus showing that the quantity he had spent so much time trying to measure was of no importance. In a fitting commemoration of his contribution, the units in which both work and heat are measured, and energy too, of course, is the joule.[8] A joule (J) is quite a small unit of energy: each beat of the human heart does about 1 J of

---

[8]   One joule (1 J) is the work required to move through 1 metre against a force of 1 newton (1 N).

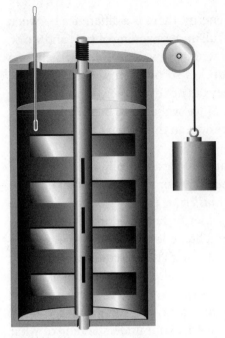

**Fig. 3.8** An idealization of Joule's apparatus for measuring the mechanical equivalent of heat. The falling weight drives the paddles through the water inside the insulated container. The work done can be calculated by the distance through which the weight falls. The temperature of the water is monitored, and the rise in temperature is then used to calculate the heat required to achieve the same effect.

work. Each day, corresponding to about a hundred thousand beats, your heart does about a hundred thousand joules of work driving blood through your body, so you need to consume enough food to supply that quantity of energy just to keep ticking over. (Thinking about it takes a lot more.)

The work of Joule and his contemporaries established without doubt that heat and work are forms of energy, and that by taking them into account, the balance sheet of energy remained intact. Even with lumbering machines that lived off heat and snorted steam, not just the much simpler collection of particles that make up the bodies treated in Newtonian dynamics, energy was proved to be conserved.

The apparently universal validity of the law of the conservation of energy eliminates the possibility of a perpetual motion machine ever being produced. A *perpetual motion machine* is a device that produces work without consuming fuel. That is, it creates energy. The energies of fraudsters, however, do seem perpetual, and all manner of weird machines are still being exhibited and invariably, when analysed or simply dismantled, shown to be fraudulent. We are so confident that energy is conserved that scientists (and patent offices) no longer take claims of its overthrow seriously, and the search for perpetual motion is now regarded as the occupation of cranks.

Although heat and work are two faces of energy, there is a difference between them, just as common sense suggests. The full understanding of heat and work, and how they were manifestations of energy, had to await the development of a molecular understanding of the distinction. As so often in science, with the understanding came the realization that they did not exist: there is no such *thing* as heat, and there is no such *thing* as work! Since we seem to be surrounded by both in our everyday world, there must be more in this remark than meets the eye. Let's burrow into it.

First, what do I mean when I say, apparently paradoxically and against the thrust of all that has gone before, that neither heat nor work is a form of energy? The crucial point is that both are ways of *transferring* energy from one location to another. Work is one way of transferring energy; heat is another. There is no such thing as 'work' stored in an engine and being let out as we drive along a road or lift a load. In exactly the same way (although it runs counter to the way we use the term in casual conversation), there is also no such thing as 'heat' stored in an object even though we might think of that object as being hot. Heat is a way of *transferring* energy: it is energy in transit, not energy possessed by anything. Perhaps you can see that if I am to clarify your understanding of what is meant by heat, you have to discard all your preconceptions based on the colloquial and imprecise use of the term in everyday conversation. To coin a term, scientists often take a familiar word, strip the flesh and fat from it, and use the bone that lies beneath. As so often, scientists refine language not to be excluding and cold, or even to undermine the livelihood of poets, but so that they *really* know what they are talking about.

Work is energy transferred in such a way that, in principle at least, that energy can be used to raise a weight (or, more generally, move an object against an opposing force). There was no work stored in the engine before the event; there is none stored in the moved object after the event. There was this abstract entity *energy* stored in the engine before the event; the moved object has a higher energy after the event—its kinetic energy may be higher or, if it is a raised weight, its potential energy may be higher. Energy has been transferred from engine to object through the agency of work: work is the agent of transfer, not the entity transferred. The weasel words 'in principle' will not have gone unnoticed. They mean, in this instance, that the energy escaping from the engine (or whatever device we are considering) could have been used to raise a weight even if in fact it didn't. For example, the work might have been used to drive a generator that drove an electric current through an electric heater. The end product was hot water, rather than a raised weight. However, we could have used the energy to raise a weight, so it was released as work.

Heat is energy transferred as a result of a temperature difference, with energy flowing from hot (high temperature) to cold (low temperature). There is no heat stored in the source before the event; there is none stored in the receiving object after the event. There was energy stored in the source before the event; the heated object has a higher energy after the event—some water, for instance, might have evaporated or some ice melted. Energy has been transferred from source to object through the agency of heat: heat is the agent of transfer, not the entity transferred.

Everything becomes clear when we consider events on a molecular scale. Suppose we could look at the motion of atoms just outside the engine. To be definite, let's look closely, really closely, at the piston that is being pushed out by expanding gas (in a car engine) or the influx of steam (in a steam engine). If we could see the atoms of the piston, we would see them all moving in the same direction as the piston moved out (Fig. 3.9). After all, macroscopic, observable motion is the uniform motion of innumerable atoms. There is no piston in a steam turbine; instead the force of the steam drives the turbine blades round and we can use this motion to do work. If we could see the atoms of the blades, we would see them all moving in the same circular direction as the blades rotated. When a wire is connected to the poles of an electric battery, the electrons that make up the electric current—a stream of electrons—move through it. If we could see the electrons in the wire, we would see them all moving in the same direction. That electric current can be used to do work, for instance, by including an electric motor in the circuit. In every case, work is associated with the uniform motion of atoms (or electrons). That is what work is: it is the transfer of energy that stimulates *uniform* motion of atoms in the surroundings.

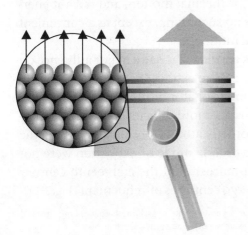

**Fig. 3.9** When work is done, energy is transferred in such a way that atoms are moved in a uniform, directed way. In the magnification of this piston that is moving upwards, we see how the atoms are all moving in step. They transfer this motion to an object resting on or connected to the piston, and bring about, for instance, the raising of a weight.

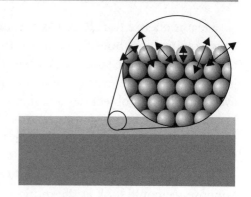

**Fig. 3.10** When energy is transferred as heat, the motion of the atoms is disorganized. We can imagine the atoms of the hot object and its conducting wall (the horizontal slabs) as oscillating vigorously about their locations, and jostling each other. That jostling transmits energy into the surroundings, where the atoms pick up this thermal motion.

What about heat? Once again, we look through an imaginary microscope of such power that we can see the motion of atoms. Now there is no piston or turbine blade to move, no movable part of the hot object. Instead, energy seeps out through a conducting wall. Now there is no net motion of the surrounding atoms, but we see them jiggling about at random (Fig. 3.10). As energy leaves the object and enters the surroundings, the atoms in the surroundings jiggle ever more vigorously, and pass the energy of their jiggling to their neighbours, which in turn hand it on to theirs. In short, the transfer of energy as heat is the transfer of energy that stimulates *random* motion of atoms in the surroundings.

The random jiggling motion of atoms is called *thermal motion*. It is not heat. Heat is the mode of transfer of energy. We should never say 'heat is transferred', except in so far as we understand that that is a convenient way of saying that energy is being transferred as heat or by heating. Heat, in fact, is better regarded as a verb than a noun. Heat is not heat energy. There is no such thing, even though the term is widely used (there is only kinetic and potential energy, each of which contributes to the energy of thermal motion, and radiant energy). Heat is not thermal energy. There is no such thing, except as a convenient way of referring to the energy of thermal motion.[9]

This atomic distinction between work and heat has had a major influence on the development of civilization. It is quite easy to extract energy as heat: the energy just has to tumble out in a random jumble of atomic motion. As such, early humans were soon able to achieve it. It is far more difficult to extract energy as work, for the energy has to emerge as orderly atomic motion. Other than the bodies of animals, devices to achieve this orderly mode of extraction were not constructed (except in a few sporadic instances) until the eighteenth century and to achieve efficiency have had to undergo centuries of refinement (Fig. 3.11).

---

[9]  I am being pedantic, of course. I should really admit that all nouns—cats, dogs, heat, thermal energy, chemical energy—are but convenient ways of referring to things. But I want to purge and purify your thoughts.

**Fig. 3.11** Here is a junkpile of some of the sophisticated devices that are needed to extract energy as work. The ability to extract energy in this way, rather than just as heat, was a relatively late development of civilization.

Now we can see how heat has been brought into the fold and how energy is truly conserved. That is, now that we realize that energy can be transferred as heat or work, we can conclude that energy is conserved both in the domain of *dynamics*, the motion of individual bodies and the interconversion of kinetic and potential energy, and also in *thermodynamics*, the interconversion of heat and work. Energy is truly the universal currency of cosmic accountancy, for no event takes place in which energy is either created or destroyed. Thus, energy is a type of constraint on the events that are possible in the universe, for no event can occur that corresponds to a change in the total amount of energy in the universe. That conclusion would have pleased Thomson and Clerk Maxwell, who had become enthusiastic about the conservation of energy largely through their belief that God had endowed the universe with an infinitely wisely chosen fixed amount of energy at the Creation and that mankind had to make do with what an infinitely thoughtful God had thought appropriate.

The question that perhaps occurred to Thomson and Maxwell is how much energy there is in the universe, for that would be a quantitative gauge of God's munificence: they probably presumed that the amount was infinite, for anything less would indicate a bound to God's generosity and thus an unacceptable hint of divine meanness. Because energy is conserved, if we could assess the total energy present now, that would be the same as the original bountiful endowment. So, how much energy is there now? The honest answer is that we don't know. However, there is a clue which points to the total.

First, we have to overcome, as always in science, our prejudices. There certainly seems to be a lot of energy: you have only to think about volcanoes and hurricanes on Earth, and the brilliance of the stars, to conclude that the universe is endowed with a colossal fund of energy. In fact, there is more than meets the eye, because (as we shall see in more detail in Chapter 9) mass is equivalent to energy, so all matter is a form of energy (through $E = mc^2$). If we were to add together the masses of all the stars in all the galaxies of the visible universe we would get a huge total mass and therefore a huge total energy. However, in science as in life, we have to be circumspect. There is another contribution to the energy, the gravitational attraction between matter. Attraction lowers the energy of the interacting bodies, so the more there is of it, the *lower* the energy. One way to think of that is to ascribe the energy of gravitational attraction a negative value, so the greater the attraction, the greater the reduction of the total energy.[10] Because of its negative contribution, as we add in all the gravitational interactions between the stars in galaxies and between the galaxies, our original huge total energy gets whittled away.

Does it get whittled away completely? It's beginning to look like it. We can judge the net total energy of the universe by examining its rate of expansion (this topic is taken up in more detail in Chapter 8). If the negative gravitational interaction overwhelms the positive contribution of mass, then the long-term future of the universe will be for its expansion to slow, then reverse, and finally to collapse in on itself in the Big Crunch. That is just like throwing a ball up into the air with too little kinetic energy for it to escape: eventually, the pull of gravity brings it back to Earth again (Fig. 3.12). That future is increasingly thought to be unlikely. On the other hand, if the gravitational attraction is weak, then the universe will expand for ever. That is like throwing a ball up with such a colossal amount of kinetic energy that it can escape the pull of gravity and speed off into intergalactic space and still be moving as it approaches infinity. That remains a possible future: observation hasn't ruled it out.

---

[10]  The energy of attraction between the Sun and the Earth contributes a whopping $-5.3 \times 10^{33}$ J to the total, so gravitational potential energy is far from negligible even though gravity itself is weak.

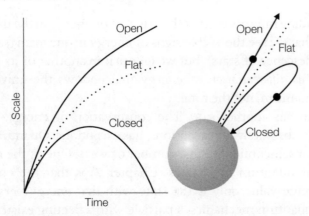

**Fig. 3.12** The paths from the sphere indicate what happens when we throw a ball up on the surface of the Earth. If we throw it up relatively gently (with less than its escape velocity), it falls back down. If we throw it up vigorously (with more than its escape velocity), it escapes to infinity, and is still moving as it approaches infinity. The dotted path indicates what happens when we throw it with exactly the escape velocity: it just escapes, but slows to a standstill as it approaches infinity. The dotted line is the dividing line between escape and capture. The graph indicates how this idea applies to the universe as a whole. If gravity is strong (because there is a lot of matter in the universe), then the universe will collapse at some time in the future (like a ball thrown up and falling back). If gravity is very weak (because there is not much matter in the universe), then the scale of the universe will increase for ever (like a ball thrown up and speeding off for ever). If gravity and the outward motion are exactly in balance, then the universe will expand for ever and glide towards a standstill (like the ball thrown with the escape velocity).

If the positive and negative contributions to the energy are exactly equal, the universe will also expand for ever, but its expansion will become slower and slower as it gets bigger and bigger, and in the far distant future we can think of the universe as hovering between continued expansion and collapse. That is like throwing up a ball with exactly the right escape velocity so that it has *just* enough kinetic energy to escape, but as it approaches infinity, it has slowed to a standstill.[11] Because such a ball is not moving, it has zero kinetic energy, and because it is infinitely far from the Earth and out of reach of its gravity, it has zero potential energy, so it has zero total energy. Because energy is conserved, although it had changing amounts of kinetic and potential energy, the total energy of the ball must have been zero all along. There are complicating factors related to possible additional effects leading to the acceleration of the universe as it expands (see Chapter 8), but it looks as though the total energy of the universe is in fact very close to zero. In fact, it may be exactly zero. If that proves to be the case, then it really does look as though God was somewhat parsimonious in His provision of energy at the Creation.

---

[11]    The escape velocity on Earth is 11 km s$^{-1}$, the same for an object of any mass.

The misleading impression that there is a lot of energy in the universe stems from the fact that we see the visible signs of energy in one form (such as matter and the incandescence of stars), but we ignore it in another of its forms (gravitation). It is this *differentiation* of energy that endows the universe with its spectacular dynamism, not the total.

Every coin has another side. The conservation of energy, the law that appears to have absolutely no exceptions, has exceptions. Quantum mechanics undermines our self-confidence in a number of ways. One of the many bizarre implications of quantum mechanics (Chapter 7) is that the energy can be ascribed a definite value only if the state with that energy persists for ever. According to quantum mechanics, a particle with a fleeting existence does not have a definite energy, and for brief instants of time the energy of the universe cannot be ascribed a definite value and therefore its energy need not be conserved. Perhaps very short-lived perpetual motion machines can be built after all!

FOUR

# ENTROPY

## THE SPRING OF CHANGE

**THE GREAT IDEA**

*All change is the consequence of the purposeless collapse of energy and matter into disorder*

*Not knowing the Second Law of thermodynamics is like never having read a work of Shakespeare*[1]

C. P. SNOW

A QUESTION anyone might forget to ask is why anything happens at all. Deep questions are often thought wrongly to be naive questions; yet deep, seemingly naive questions properly investigated can expose the heart of the universe. That is certainly true of this particular question, for we shall see that by pursuing the answer we are led into a full understanding of the driving force of all change in the world. We shall come to understand the simple events of everyday life, such as the cooling of hot coffee, and we shall see at least the ankle of the explanation of the most complex events of everyday life, such as birth, growth, and death.

The answer to our question about the origin of change lies in the field of science known as *thermodynamics*, the study of the transformations of energy, particularly of heat into work. Thermodynamics does not have a reputation for light-hearted frivolity, for the perception of it is encumbered by its origins, the examination of the efficiencies of steam engines. A steam engine, it is easy to think, is the epitome of lumbering, and surely can have nothing to do with the exquisite delicacy of the opening of a leaf, let alone the formation of an amusing opinion. Steam engines symbolize the heaviness of industry and by extension the oppression and social burden arising from industrialization (Fig. 4.1).

---

[1] *The two cultures.*

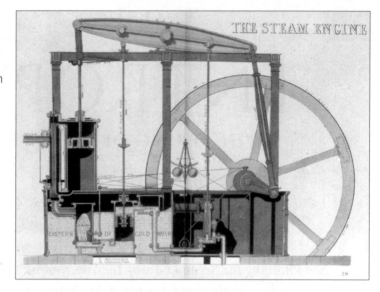

**Fig. 4.1** A steam engine might look like a cumbersome brute, but in fact it epitomizes the workings of the universe. As this chapter unfolds, we shall see that, when expressed in a suitably abstract way, all the events of the world, both inside us and outside us, are driven by steam engines.

They represent filth rather than purity, urban rather than rural, and ponderousness rather than delicacy. How can such clanking, wheezing, steaming, whistling, leaking behemoths have anything to do with understanding the delicate network of events that surrounds us, enriches us, and pervades every aspect of this wonderful world?

We have begun to see that science illuminates by embracing ever greater abstraction. So it is in this connection too. When we strip away the iron to leave the *abstraction* of a steam engine, we obtain a representation of the spring of all change. That is, provided we look at the *essence* of a steam engine, its abstract heart, and ignore the details of its realization—the steam, the leaky pipes, the drips of oil and grease, the rattles, bangs, and rivets—we find a concept that applies across the range of all events. Science is like that: science distils from reality its essence, its great ideas, then finds the same phantom spirit elsewhere in nature. The identification of the same spirit inhabiting different events means that we acquire a common understanding of a whole swathe of the world. Through a poet's eyes, we see the superficialities of events, which does not mean that the events are never emotionally or spiritually stirring. But through a scientist's eyes, we penetrate the surface and see the spirit within. In this chapter, we peel back the skin of events and find the spirit of the steam engine within.

The realization that the steam engine epitomized all change emerged through the nineteenth century and reached its fulfilment early in the twentieth. That is another problem with thermodynamics: its aura is so Victorian. Like that era, thermodynamics might be thought to be a subject of the past and, except for engineers, to have little relevance to our understanding of the modern world. But the roots of thermodynamics lie deep and their ramifications spread throughout the structure of the modern world. Interpreted in a modern manner, thermodynamics is among the most relevant of subjects.

To set the scene, I will stir the pond of nineteenth-century history and bring the minds of four dead scientists to the surface. These four—Sadi Carnot, William Thomson (Lord Kelvin), Rudolph Clausius, and Ludwig Boltzmann—made major contributions to the distillation of the spirit of the steam engine. We shall trace the emergence of the great idea of 'entropy', a concept that lies at the heart of this discussion, through their eyes before revisiting it from a more modern perspective.

In the early nineteenth century, the steam engine epitomized wealth; later we shall see that it actually epitomizes change, but we will settle for wealth for the moment. With her engines pounding away across the country, pumping mines to render them viable and efficient, driving looms and thereby powering the economy, soon to be put on wheels and within hulls to enhance and facilitate mobility, trade, and that euphemism for aggression, defence, England was a driven nation. The steam engine was pervading and transforming the social and economic structure of the nation as the computer was to do a century or so later. The anxious and envious eyes of the French watched from across the Channel, hamstrung as they seemed to be by their lack of accessible coal. The pressing aim of engineers was to enhance the efficiency of the steam engine, to get more work for less coal. Was water the best medium, or could they use air? Was high pressure better than low? What of temperature? Could the analytical French use reason to beat the pragmatic English at what had become the latter's own game?

One bright light shone through the foggy clouds that obscured the answers to questions like these. Sadi Carnot was finally born in 1796. I say finally, because his determined parents had two earlier attempts at generating Sadis, but both—in succession named Sadi—died young. Their third and finally successful Sadi survived a little longer, until cholera struck him down at thirty-six. Although his life was brief, his less than four decades were enough to grant him the immortality that notable contributions to science can bring.

Carnot was fundamentally wrong about his perception of a steam engine, but the strength of the *essence* of the steam engine that he identified is such that it shone through even his fundamental misconception. Carnot followed (at least in the original formulation of his ideas, although he did change his mind much later) the then prevailing view that we encountered in Chapter 3, that heat is a fluid—caloric—that flows from a hot reservoir to a cold sink and in the process may turn an engine, much as a water wheel is turned by the flow of water. He also considered, once again in the prevailing fashion of the time, that heat, being a fluid, was neither created nor destroyed as it flowed from source to sink. On the basis of this false model he was able to prove the astonishing result that the efficiency of an idealized steam engine—one that ignores the effects of friction, leaks, and so on—is determined only by the temperatures of the hot source and cold sink, and that it is independent of both the pressure and the identity of the working substance.[2] Thus, to achieve the greatest efficiency, the hot reservoir should be as hot as possible and the cold sink should be as cold as possible. All the other variables were fundamentally irrelevant.

These counterintuitive conclusions were thought too absurd by the engineers of the day, and Carnot's little book, *Réflexions sur la puissance motrice du feu* (1824), languished largely unread and forgotten. But not quite. The slender threads that keep the pulse of crucial ideas alive through history brought Carnot's book to the attention of William Thomson (1824–1907), later to be Lord Kelvin of Largs. As we saw in Chapter 3, Kelvin, as it will be convenient to call him, in collaboration with James Joule, had already contributed to the overthrow of the caloric theory and had identified heat as a form of energy. The contemporary world had come to recognize that energy, not heat, is conserved, and that heat and work, being two aspects of energy, can be transformed into each other. The concept of the flow of caloric through an engine gave way to one in which the flow was of energy, and the engine itself was perceived as less like a mill and more as a device for converting some of that energy from heat into work. That remains the principle of all so-called *heat engines*, devices for converting heat into work, which include steam engines, steam turbines, jet engines, and internal combustion engines.

Kelvin was stimulated by Carnot's *Réflexions* to reflect himself on the efficiency of steam engines and to render Carnot's work more quantitative. Carnot had developed his ideas by using simple arithmetic and had not drawn

[2] The efficiency of a perfect engine, the ratio of work done to the heat supplied, working between the temperatures $T_{hot}$ and $T_{cold}$, where these temperatures are on the thermodynamic (absolute, Kelvin) temperature scale, is equal to $1 - T_{cold}/T_{hot}$.

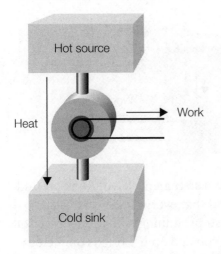

**Fig. 4.2** This is the sort of diagram that we shall use to represent a steam engine—more generally, a heat engine. There is a hot source from which energy is withdrawn at a high temperature, a device for converting heat into work (in an actual steam engine that would be the piston in its cylinder), and a cold sink into which 'waste' heat is discarded.

on the rigour of mathematics to express his ideas in a more modern and compelling idiom.

To appreciate Kelvin's contribution, we can imagine standing in front of a typical nineteenth-century steam engine. On casual inspection of the engine we would probably conclude that the piston in its cylinder is the essential component, for it is the device that taps into the flow of energy and bleeds some off as motion and therefore work (Fig. 4.2). Alternatively, we might conclude that the hot reservoir is the crucial component, for it is the source of the energy that is to be converted into work. Kelvin, however, formed the seemingly bizarre view that although these two components are obviously important and involved much careful design and construction, the *essential* component of a steam engine is the cold sink—the surroundings into which waste heat is discarded. In this view, the crucial part of the engine didn't seem to be there, and didn't have to be designed or constructed: it was simply the surroundings of the parts that were built. Science often proceeds in this way: springing forward by inverting common sense, by illuminating an old problem with light from a new direction. The Hungarian biochemist Albert Szent-Györgyi (1893–1986) expressed this aspect of science particularly well, when he said that scientific research consists of seeing what everyone else has seen but thinking what no one else has thought.

Kelvin's conceptual somersault led him to promote his recognition of the central role of the cold sink to a universal principle of nature: *all viable engines have a cold sink* (Fig. 4.3). Kelvin didn't express his principle in quite those

---

[3] More precisely, he said: No cyclic process is possible in which the sole result is the absorption of heat from a reservoir and its complete conversion into work.

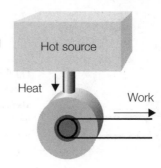

**Fig. 4.3** The Kelvin statement of the Second Law asserts that this engine won't work. Every viable heat engine has a cold sink into which some heat must be 'wasted'.

words,[3] but they are the essence of his formal statement. If you look around, and examine any steam engine, you will find that every single one has a cold sink. Take away the cold sink and the engine stops running even though you still have plenty of energy stored in the hot reservoir and even though connected to that reservoir is a well-lubricated and greased piston in a cylinder. The cold sink is essential. Remove it, and the engine stops. In fact, the principle applies to any other kind of engine that converts heat into work, including the internal combustion engines that impel our cars and the jet engines that impel our planes. The cold sink is harder to identify in these more sophisticated devices, but careful analysis of the flow of energy shows that it is there. In an internal combustion engine, for instance, we can think of the exhaust valves and manifold as the sink for the deposit of waste heat. Here is the first glimmer of the recognition that there is a steam engine *notionally* inside every kind of heat engine, for the essential component, the cold sink, and the essential act, the discarding of waste heat, is present in every one. Could it be that living organisms, which are even more sophisticated than internal combustion engines, are also built on the same abstract principles? The leviathan of thermodynamics is starting to stir.

*All viable engines have a cold sink* is one statement of the Second Law of thermodynamics. The law is not normally expressed so succinctly, but this form of words captures its essence. At this stage it has the typical form of an empirical law, one that is a direct summary of experience: there is the potential for abstraction, but in this form the law could have been stated by any keen observer. In this form, too, the universality of the law looks somewhat restricted. It is a summary of the structure of heat engines on Earth, and probably, if they have them, of heat engines built by extraterrestrials everywhere else in the universe. But the law does not appear to have the broad sweep that embraces life, the universe, and everything. Be patient: let the story unfold.

At about the same time, in 1850, the German physicist Rudolph Clausius (1822–88) was also beavering away on what was then the hot topic of the day,

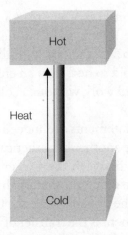

**Fig. 4.4** The Clausius statement of the Second Law asserts that processes like this are never observed. Provided there is no external intervention, energy is never observed to flow as heat from a cold body to a hot body.

namely heat, and published his reflections in a paper *Über die bewegende Kraft der Wärme* (On the motive force of heat). He too had noticed a common feature of nature and had the stature of a respected scientist to publish what others might think a simpleton's observation: *heat does not flow from a cooler to a hotter body* (Fig. 4.4).[4] Clausius was, of course, far from being a simpleton, and in this and subsequent papers he developed this notion into a quantitative principle of great power. For the moment, though, we will stick with the empirical form of the law and see that it does indeed conform to everyday experience. To do so we need to note that the law does not prohibit the transfer of heat from cold to hot: that, after all, is what we achieve in a refrigerator, which pumps heat out of the contents of the refrigerator and deposits it in the warmer surroundings. The point is that to achieve refrigeration, we have to do work: the refrigerator must be connected to an electricity supply that drives its mechanism. Clausius's remark applies to a process that is not interfered with in any way, a process that can occur without us having to drive it. In short, Clausius's statement refers to 'natural' or 'spontaneous' changes, which are changes that occur without having to be driven by an external agency. Thus, cooling to the temperature of the surroundings is spontaneous; but heating to above the temperature of the surroundings is not spontaneous, as it has to be driven (for instance, by forcing an electric current through a heater in contact with the object). In science, 'spontaneous' has no connotations of fast: the slow flow of thick tar from an upturned barrel is spontaneous, even though it may be exceedingly slow. Spontaneous in science has the connotation only of 'natural', not of 'rapid'.

[4]  As before, we are paraphrasing the truth. A closer statement of Clausius's observation is as follows: 'No cyclic process is possible in which the sole result is the transfer of energy from a cooler to a hotter body.'

Thermodynamics is an Amazon of concepts. Like the Amazon, thermodynamics is a confluence of many streams. The Kelvin and Clausius tributaries turn out to be part of the same river of ideas. In fact, they are logically equivalent, for if heat could flow spontaneously from cold to hot, then an engine could work without a cold sink; and if an engine could work without a cold sink, then heat could flow spontaneously from cold to hot.

To see that Kelvin's and Clausius's statements are indeed equivalent, let's use a hypothetical sinkless engine to drive another hypothetical sinkless engine in reverse (Fig. 4.5). The only difference between the two engines is that the temperatures of their energy sources are different, that of the driving engine being arranged to be lower than that of the driven. As we see from the illustration, the net effect of the overall arrangement is the transfer of energy from the cooler source to the hotter, which is contrary to Clausius's statement of the Second Law. Therefore, if Kelvin's statement is false, then so too is Clausius's. Now we show the opposite, that if Clausius's statement is false, then so too is Kelvin's. For this demonstration, we let an engine run, discarding waste heat into a sink. Then we allow all that waste heat to run back into the hot source, which is contrary to Clausius's view of what can happen naturally (Fig. 4.6). The net effect of the overall arrangement is the conversion of heat from the hot source into work, with no heat discarded into the cold sink, which therefore needn't be there. That conclusion is contrary to Kelvin's statement. We conclude that, because the failure of each statement implies the failure of the other,

**Fig. 4.5** This arrangement (top) shows that if Kelvin's statement of the Second Law is false, then Clausius's statement is also false. The engine on the left is arranged to drive the engine on the right in reverse, and therefore to convert work into heat that is deposited in the 'hotter' reservoir. The net effect (bottom) is the transfer of heat from the 'hot' to the 'hotter' reservoir, which is contrary to Clausius's statement.

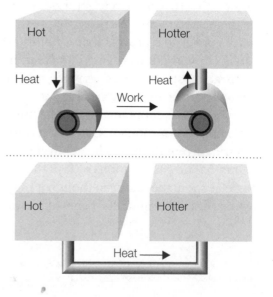

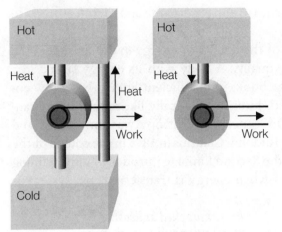

**Fig. 4.6** This arrangement shows that if Clausius's statement of the Second Law is false, then Kelvin's is also false. The engine (left) produces work and deposits some heat in the cold sink. However, there is also a device that transfers that discarded heat into the hot source. The net effect (right) is the elimination of the need for the cold sink, which is contrary to Kelvin's statement.

then the two statements are actually logically equivalent: they are equivalent statements of the Second Law.

Having two statements of the Second Law is just a little bit uneconomical. We should suspect that the Kelvin and Clausius statements are also different aspects of a single more abstract concept, a single more abstract statement of the law. In uncovering this more abstract, hidden statement we take the first step towards the recognition of the universality of the steam engine. As we have seen several times before, and as I have already emphasized in this chapter, journeying into abstraction is the essence of the power of science, for it increases science's reach and strengthens its grip on phenomena.

We saw in Chapter 3 how the concept of energy emerged to become the supreme currency of physics. There we were concerned with the quantity of energy, and saw that the phenomena of physics were rationalized once the conservation of energy had been identified. The First Law of thermodynamics recognizes this conservation by asserting that the energy of the universe is constant. We have no quarrel with that law in this chapter. However, just as two libraries may contain the same number of books, one in order and the other as a random pile, and therefore differ in the quality of the service they can provide, so energy has a qualitative face that affects its efficacy. The quality of stored energy is measured by that famously elusive property, the *entropy*. I say elusive, but we shall see before long that entropy is a concept much easier to grasp than energy; it is merely because energy hangs from everyone's lips in daily discourse but entropy barely dares to speak its name that we consider the former an old friend and the latter a dragon. One purpose of this chapter is to dispel the difficulty that inappropriately attaches to entropy's name and to shift entropy to its rightful place in daily discourse.

Entropy, loosely, is a measure of the quality of energy in the sense that the lower the entropy the higher the quality. A body with its energy stored in a refined, carefully ordered way, like books in an efficient library, has a low entropy. A body with its energy stored clumsily, chaotically, like the books in a random pile, has a high entropy. The concept of entropy was introduced and rendered quantitatively precise by Rudolph Clausius in 1856 in the course of the development of his statement of the Second Law. He introduced it by defining the change in entropy that occurs when energy is transferred to a system as heat.[5] Specifically he wrote

$$\textit{Change in entropy} = \frac{\textit{energy supplied as heat}}{\textit{temperature at which the transfer occurs}}$$

Thus, if a certain energy is supplied as heat to a body at room temperature, then there is an increase in entropy that we can calculate from this formula (note that the temperature to use in the denominator is on the absolute scale). As you sit there reading this sentence, you are generating heat that is spreading into your surroundings, so you are increasing their entropy.[6] If the same amount of energy is supplied as heat to the same body at a lower temperature, then the change in entropy is greater. If energy leaves a body as heat, then 'energy supplied as heat' is negative, so the change in entropy is negative. That is, the entropy of the body decreases as it loses energy as heat, like a cooling cup of coffee. Note that the change in entropy is given by the energy transferred as *heat*, not by any energy transferred as work. Work itself does not generate or reduce entropy.

Before I pull back the curtain and show you what entropy really is, let's see whether the concept does indeed unite the laws proposed by Kelvin and Clausius. In fact, Clausius proposed that both statements can be housed under one roof by the statement that *entropy never decreases*.[7] Consider first the Kelvin statement, which is equivalent to saying 'your engine will work only if you waste some energy', expressed in terms of changes in entropy. Suppose we claim to have invented an engine that uses all the heat and doesn't need a cold sink. Clausius would say the following:

[5]  We saw in Chapter 3 that heat is a mode of transfer of energy that makes use of a difference in temperature. Heating stimulates chaotic, random thermal motion.

[6]  You are equivalent to a 100-watt lamp, so you release energy (by consuming food) at about 100 joules per second. If your surroundings are at 20°C (corresponding to 293 K), you are generating entropy at the rate of about 0.3 joules per kelvin per second.

[7]  More formally: The entropy of an isolated system increases in any spontaneous change.

You have removed heat from the hot source, so the entropy of that reservoir has gone down. All that heat is converted into work by the machinery, so the energy enters the environment as work. But work doesn't change the entropy, so the net effect is the decrease in entropy of the hot source. According to my statement, entropy never decreases. Therefore, your engine can't work, just as Kelvin asserted.

Now consider Clausius's original statement, the one about heat not flowing from cold to hot. Suppose we claim to have observed heat flowing in the wrong direction, such as by finding ice forming in a glass of water put in an oven. Clausius would now say the following:

Energy left the cool object (the water in the glass) as heat, so its entropy went down. Because the temperature is low, and the temperature occurs in the denominator of my expression for the change in entropy, that decrease in entropy is large. The same energy enters the hot region (the interior of the oven), so the entropy of that region increases. However, because its temperature is high, that increase in entropy is small. The net effect is the sum of a small increase and a large decrease, giving a decrease overall. According to my statement, entropy never decreases, so heat cannot flow spontaneously from cold to hot, just as I asserted before.

We see that the degree of abstraction represented by Clausius's introduction of entropy neatly captures the two empirical laws that seemingly portrayed two different aspects of the world: the statement of the Second Law in terms of entropy is like a single cube that rotates to appear as a square, representing Kelvin's statement, or a hexagon, representing Clausius's statement. Clausius's statement that *entropy never decreases* is a succinct summary of experience and is the more sophisticated, more abstract statement of the Second Law. Clausius himself summarized the thermodynamic condition of the world in his famous pair of statements that jointly summarize the First and Second Laws:

*Der Energie der Welt ist konstant; die Entropy der Welt strebt einem Maximum zu.*

That is, the energy of the world is constant; the entropy strives towards a maximum.

There was considerable opposition when the Second Law was first expressed in terms of entropy, for it offended the sensibilities of the age: whereas it was easy to accept that the energy of the universe is constant (for energy was perceived initially as a divine gift that no amount of human fiddling could augment or reduce), how can something increase in abundance? Where does it come from? Who or what is pouring entropy into the universe and thereby lubricating the wheels of spontaneous change? Such was the alien spirit of the law that considerable efforts went into searching out counterexamples; but to no avail. There has never been an exception to the Second Law, wherever it is applied. It is applied to predict the spontaneity of simple physical processes, such as cooling of hot objects to the temperature of their surroundings (and to eliminate the reverse as unnatural) and the spontaneous expansion of gases into the available volume (and the elimination of the reverse). The Law is also used to predict whether a chemical reaction will run in one direction or another, such as to judge whether carbon can be used to reduce an ore (as can be used for iron) or whether electrolysis must be used instead (as for aluminium). It applies to the intricate network of biochemical reactions that make up that complex property of matter we call life. There is nothing that it cannot touch, and nowhere that it has ever failed; it is now regarded as a rock of absolute stability that has universal and perpetual validity.

But what does it mean? What is this thing called entropy, and what does its inability to decrease actually mean? What is the physical significance of entropy? How can we internalize the concept and come to think of it as a friend? The Second Law summarizes succinctly the workings of the world as embodied in the Kelvin and Clausius statements, and provides a means of assessing quantitatively whether or not a process is spontaneous. However, it is more a doorway to understanding than a final elucidation. We must press on the door, open it, and see physically what drives the universe in one direction rather than another. In other words, what lies behind entropy and what is the deep substructure of the Second Law?

The door we now press on opens into the molecular basis of matter. When we step into this world, we see solids composed of rank upon rank of atoms, molecules, or ions (charged atoms), each one wobbling slightly around its mean location. We see liquids composed of molecules jostling past one another, not only as the liquid flows but also as it lies apparently dormant and lifeless in a pool. We see gases composed of molecules in flight, colliding, bouncing apart, and travelling far and fast seemingly at random. This is the world where the interpretation of entropy lies and where we can start to visualize how change is accompanied by its increase.

The short-sighted Austrian physicist Ludwig Boltzmann (1844–1906) saw further into the nature of matter than any of his contemporaries until, that is, he hanged himself in the face of their incomprehension and rejection of his ideas. Entropy, he showed, is a measure of disorder: the greater the disorder, the greater the entropy. A solid, with its neatly packed rows of molecules, is more ordered than a liquid, with its closely packed but reasonably mobile molecules, and a solid has a lower entropy than the liquid into which it melts. A gas, with its freely flying molecules, is more disordered than a liquid, and a gas has a higher entropy than the liquid from which it evaporates.

Changes of entropy accompany heating as well as changes of physical state. Thus, when we heat a solid, before it melts its molecules jiggle around more violently as its temperature rises, and we conclude that because this disorderly thermal motion increases, then so too does its entropy. The same is true when we heat a liquid, for as we raise its temperature its molecules move more vigorously and the whole collection of tumbling, migrating molecules becomes more disordered. When we heat a gas, the molecules move with a wider spread of speeds and therefore the molecules have a greater disorder in their thermal motion: once again, raising the temperature of the gas brings about an increase in entropy. When a gas expands to fill an enlarged volume, its disorder and therefore its entropy increases even though we keep its temperature the same, because although its molecules have the same spread of speeds, we become less confident that a molecule will be found in a given small region of the container. When energy escapes from a hot object as heat, the thermal motion of the surrounding molecules increases as the energy spreads over them, and the entropy of the surroundings increases. In short, entropy increases as the *thermal disorder* of a substance becomes more vigorous, with the increasing thermal motion of its atoms. Entropy also increases as the *positional disorder* increases, the range of available positions of its atoms.

Wherever we encounter increasing disorder, we encounter increasing entropy (Fig. 4.7). That is why entropy is such a simple concept: all we have to keep in mind is that it is a measure of disorder. In most simple cases we can judge with a moment's thought whether the entropy increases or decreases when a change occurs. The only tricky point—it is not actually tricky, just a reflection of the precision with which we have to think in thermodynamics—is that to apply Clausius's dictum about entropy as a signpost of change, we have to think in terms of the *total* entropy change, which means the total entropy change of the object of interest *and* the rest of the universe. The latter is easier than it seems, because the entropy of the rest of the universe increases if energy escapes into it as heat, and it decreases if energy flows from it as heat into the object of interest. That's all we have to keep in mind.

A final preliminary point is that it should now be clear that entropy is not increased by something physical being added to the universe. An increase in

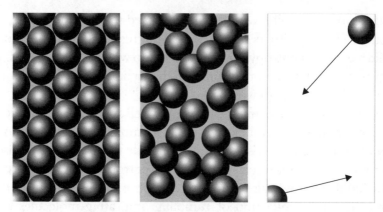

**Fig. 4.7** The entropy of the samples in these three panels gets progressively greater from left to right. The panel on the left represents the orderly array of molecules in a solid: this sample has a low entropy. The middle panel represents the less orderly arrangement of molecules in a liquid: this sample has a higher entropy. The panel on the right represents the highly chaotic structure of a gas (the words 'gas' and 'chaos' come from the same root), with molecules dashing about randomly: this sample has the highest entropy.

entropy reflects the increasing disorder of the world, the reduction in quality of its constant amount of energy. There is no external cosmic *source* of entropy: the increase in entropy is merely the rise of disorder of the energy and matter that we already have. As such, the concept of entropy is much easier to grasp than that of energy. It is very hard to give a concrete definition of energy. We could mumble to the effect that it is the capacity to do work, or (as will become apparent in Chapter 9) that it is an aspect of curved spacetime, or even that it is the curvature itself; but to be honest, none of these definitions seems quite concrete enough to grasp. Entropy, on the other hand, is a breeze. All we have to do is to think of the disorder of the distribution of energy and matter, and we have a complete qualitative mastery of the concept. Boltzmann, alas, was driven to his death by the inability of scientists of the time to come to terms with this profoundly simple insight (Fig. 4.8).

This molecular interpretation of entropy may seem far removed from Clausius's definition of an entropy change in terms of the heat supplied and the temperature at which it is supplied. However, we can bring the two into line, and thereby see how disorder is the underpinning of Clausius's definition. The analogy I like to use to show the connection is that of sneezing in a busy street or in a quiet library. A sneeze is like a disorderly input of energy, very much like energy transferred as heat. It should be easy to accept that the bigger the sneeze, the greater the disorder introduced in the street or in the library. That is the fundamental reason why the 'energy supplied as heat' appears in the

**Fig. 4.8** Boltzmann's tombstone has as his epitaph one of the central equations linking the concepts of thermodynamics to the behaviour of atoms and molecules. The form of the equation is *entropy = constant ×  logarithm of the number of possible atomic arrangements*. So, as the number of atomic arrangements increases (like going from solid to liquid and then gas), the entropy increases. This formula renders the qualitative ideas we have described into a precise, numerical, quantitative form.

numerator of Clausius's expression, for the greater the energy supplied as heat, the greater the increase in disorder and therefore the greater the increase in entropy. The presence of the temperature in the denominator fits with this analogy too, with its implication that for a given supply of heat, the entropy increases more if the temperature is low than if it is high. A cool object, in which there is little thermal motion, corresponds to a quiet library. A sudden sneeze will introduce a lot of disturbance, corresponding to a big rise in entropy. A hot object, in which there is a lot of thermal motion already present, corresponds to a busy street. Now a sneeze of the same size as in the library has relatively little effect, and the increase in entropy is small.

We can now begin to see what the Second Law is struggling to express. The statement that entropy never decreases in any natural change is the same as saying that *molecular order never increases on its own accord*. Molecules distributed at random—as in a cloud of dust—will not form themselves spontaneously into the Statue of Liberty. A gas will not collect spontaneously in one corner of a container. Energy widely dispersed—as in the warmth of a table top, with in- numerable atoms vibrating randomly—will not spontaneously flood into a small region: an egg does not spontaneously cook when left on a cool table.

We can turn this understanding round and see why the signpost of spon- taneous change points in the direction of increasing entropy. The key thought is that localized, orderly matter and energy tend to disperse. Atoms in their random jigglings tend to migrate into new niches; the energy of random jigglings is passed on as atoms jostle their neighbours. The natural direction of change is towards ever greater disorder, be it the disorder of location of matter or the disorder of location of energy, positional disorder or thermal disorder. Order naturally decays into disorder; energy degrades and disperses. Like it or not, the world is getting worse.

That the world is getting worse, that it is sinking purposelessly into corruption, the corruption of the quality of energy, is the single great idea embodied in the Second Law of thermodynamics. It is an extraordinary insight to know that all the changes going on around us are manifestations of this degradation. The spring of the universe that the Second Law has revealed is the unstoppable degradation of the universe as energy and matter spread in disorder.

Now, you might consider that there are certain difficulties with this bleak world view. If the direction of the universe is towards degradation, what room is there in it for the emergence of exquisite structures, of people, and of noble thoughts and deeds? Such a view certainly caused some consternation among the Victorians, who viewed the seemingly unstoppable improvement of Man, particularly pale versions of Man in the less torrid regions of what they regarded as the upper hemisphere of the globe, as a source of pride and a motivation. How could righteous Empires justifiably impose ennobling Civilization if both masters and mastered were drifting inexorably into hopeless Degradation? How could the increasing mastery of matter be compatible with a future of the universe that was drifting inexorably towards a Hogarthian gutter? Surely, although the Second Law might summarize the steam engine rather neatly, it did not summarize the actions of Man—or even those of a cockroach.

To resolve this paradox, the crucial point to grasp is that no change is an island of activity: change is a network of interconnected events. Although drift into degradation might take place in one location, the consequence of that drift might be to ratchet up a structure somewhere else. I am reminded of a medieval clock, such as the astronomical clock in Prague (Fig. 4.9), where the falling of a weight drives an elaborate parade of events. Overall, there is a dispersal of energy, an increase in entropy, as the weight falls and friction dissipates its energy as heat into the surroundings. However, because the downward motion of the weight is linked through an intricate train of gears to model

**Fig. 4.9** Detail of the mechanical clock in Prague. This clock is an allegory on the Second Law, because although there appear to be systematic events taking place, they are driven by the generation of greater disorder elsewhere as the driving weights fall. Food is like the weight, enzymes are like the gear wheels, and our own actions are like the motions of the figures. That is not to say that there is no free will, but the argument that allows for free will is longer than this margin will allow.

moons, suns, stars, and apostles, its falling generates the sense of organized behaviour and complex—almost purposeful—display. If we wilfully ignored the clockwork, we might conclude that organized events, the acts of the apostles, were occurring naturally. But we who have an inner knowledge know that there is clockwork driven by a naturally falling weight.

The clock in Prague is an allegory on the workings of the Second Law. Although elaborate events may occur in the world around us, such as the opening of a leaf, the growth of a tree, the formation of an opinion, and disorder thereby apparently recedes, such events never occur without somehow being driven. That driving results in an even greater production of disorder elsewhere. The net effect, the sum of the entropy change arising from the reduction of disorder at the constructive event and the entropy change arising from the increase in disorder of the driving, dissipative event, is the net increase in entropy, an overall production of net disorder. So, wherever we see order emerging, we must lift the curtain and see greater disorder being produced elsewhere. We, indeed all structures, are local abatements of chaos.

Indeed, here we forge another link to the development of life we explored in Chapter 1, for the fact that men have nipples, for instance, is a direct consequence of the Second Law of thermodynamics. The ceaseless decline in the quality of energy expressed by the Second Law is a spring that has driven the emergence of all the components of the current biosphere. In a very direct sense, all the kingdoms of creation have been hoisted out of inorganic matter as the universe has sunk ever more into chaos. The spring of change is aimless, purposeless corruption, yet the consequences of interconnected change are the amazingly delightful and intricate efflorescences of matter we call grass, slugs, and people. That men have nipples is a consequence of the common origin of animals and the fact that the Second Law drives Nature forward, making do with what is available, without foresight, always blindly, and sometimes with an inappropriate long-term effect.

Elsewhere, the location of the greater rise in disorder that drives an increase in order may be very local or very far away. It might even be inside us. The clock-work inside us is biochemical, with cogs fashioned from protein, not iron; but nevertheless, it works in much the same way. It also models the workings of a steam engine. So, let's return to the steam engine with eyes opened to the concept of entropy. We shall see what the engine truly is, its most abstract anatomy, and in particular see why the cold sink is so crucial to its operation.

We can think of a steam engine, or any heat engine, as involving two steps (Fig. 4.10). The first step in the operation of the engine is the drawing down of energy as heat from the hot reservoir. The loss of energy from the reservoir lowers its entropy, as its atoms now possess less thermal motion than before. The energy we have extracted flows through the mechanism for converting heat into work (the piston and the cylinder in an actual steam engine) and enters the cold sink. If all the energy we have extracted from the hot source enters the cold reservoir, the entropy of that reservoir increases. However,

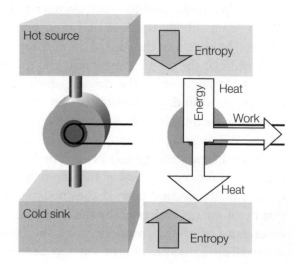

**Fig. 4.10** The thermodynamic analysis of the operation of a steam engine (or any heat engine). Energy leaves the hot source as heat and thereby reduces its entropy. Some of that energy is converted into work, which has no effect on the entropy. The rest of the energy is deposited into the cold sink, so generating a lot of entropy. Provided the temperature of the cold sink is lower than that of the hot source, the overall entropy will increase even if the energy discarded as heat is less than that extracted from the hot source. The difference in energy extracted and discarded can be extracted as work.

because the temperature of the sink is lower than that of the source, the increase in entropy is larger than the original decrease (remember the parable of the quiet library). Overall, the entropy of the device will increase, because the decrease in entropy of the source is overcome by the larger increase in entropy of the sink. So, the flow of heat from source to sink is spontaneous.

Now for the crucial point. So far, our engine has produced no work, and we would have obtained the same outcome if we had simply put the hot source in direct contact with the cold sink. However, the transfer of energy out of the hot source remains spontaneous even if we convert some of it—not all of it—into work and deposit the rest into the cold sink. It is certainly the case that if we withdraw energy as heat from the hot source, then we get a reduction in entropy, as before. However, we can get a compensating increase in entropy by releasing a smaller amount of heat into the cold reservoir. For instance, if the temperature of the cold sink is half that of the hot source (using absolute temperatures), then we can get a compensating increase in entropy by allowing only half the extracted energy to enter the cold sink, leaving the other half for us to use to do useful work. The engine operates spontaneously—that is, it is a useful, viable device—because there is an overall increase in entropy even though we are extracting some energy as work.

We can now see that the cold sink is essential. Only if the cold sink is present and some energy is released into it is there any hope of the entropy increasing overall. The extraction of energy from the hot source corresponds to a decrease in entropy. The transfer of energy to the surroundings as work leaves the entropy unchanged, so at this stage of the story there is an overall decrease in entropy. For the engine to act spontaneously (and engines that do not act spontaneously, in the sense that they have to be driven, are worse than useless), it is essential to produce some entropy somewhere to ensure that overall there is an increase in entropy. That is the role of the cold sink: it acts as the quiet library, being the location of a large increase in entropy even though only a small amount of energy is discarded into it. However, note its importance: there must be 'waste' and a vessel for that waste, if the engine is to be viable. The cold source is indeed the source of the engine's viability, for without it there can be no increase in entropy.[8]

The steam engine represents the fact that, to get work—a constructive effort—out of any process, it is essential that there should also be a dissipation

---

[8] I leave it as a challenge for you to show that the energy that must be discarded to make the engine viable depends on the ratio of the temperatures of the hot and cold reservoirs, and that in this way it is possible to derive Carnot's expression for the efficiency quoted in footnote 2.

of energy. Just withdrawing energy from the source doesn't work: you have to throw away some heat to stir up the cold sink (which may be simply the surroundings, not necessarily a constructed part of the engine) in order to make the engine work. Wherever we find construction, we find linked to it at least as great destruction.

Let's look at some of the changes going on in the world and see how, though they are constructive, they take their life from destruction elsewhere. First, the outside world. Any act of building, as in the building of a wall, needs work to be done as the masonry is lifted into position. To achieve that work, an engine must be used (including the food-powered muscle-engines of animal bodies), and for the engine to be viable, entropy must be generated by dispersing energy into the environment. Thus, the engine of a hoist, a heat engine of some kind, acts by dispersing energy into its surroundings. That is true even if the hoist is electrical, with the dispersal of energy taking place at some distant power station. All the artificial structures of the world, from the great pyramids to the merest hovel have been built on the back of the dispersal of energy.

We can look more closely at the manner in which the dispersal of energy is achieved by considering the chemical reactions that are used to raise the temperature of the hot source. I will concentrate on a conventional steam engine for this discussion, because although the principle of an internal combustion engine is the same as far as this process is concerned, it is realized in a technologically more intricate way and I don't want to distract you with details. A steam engine is an *external* combustion engine, with fire heating water outside the piston, so the sequence of events is easier to follow.

Let's suppose that the fuel is oil, a mixture of hydrocarbons (compounds built from carbon and hydrogen only), like the sixteen-carbon-atom-long chain shown in Fig. 4.11. This is the molecule typical of fuel oil and diesel fuel; it is also closely related to the molecules of fat that are present in meat and which help to lubricate the muscle fibres as well as acting as an insulating layer and a reserve of fuel. That we eat foodstuffs closely related to diesel fuel, some more than others, is no accident, but the thought is a little sobering.

When the oil burns, molecules like the one in the illustration are attacked by oxygen molecules of the air. Under the onslaught of this attack, the carbon chain breaks up and the hydrogen atoms are stripped from it. The carbon atoms are carried away as carbon dioxide molecules and the hydrogen atoms are carried away as water molecules. A great deal of heat is produced because the newly formed bonds between the atoms are stronger than the original bonds in the fuel and in the oxygen, so energy is released when the weak old bonds are

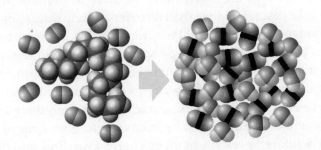

**Fig. 4.11** A hexadecane molecule (in the centre of the cluster of molecules shown on the left) is representative of the hydrocarbon molecules found in fuels and the fats of foods. It consists of a chain of sixteen carbon atoms (the dark spheres) to which are attached thirty-four hydrogen atoms (the small pale spheres). Think of the molecule as writhing and wriggling as it tumbles past its writhing and wriggling neighbours. When this molecule burns, oxygen molecules attack it, the carbon atoms are carried off as sixteen separate carbon dioxide molecules and the hydrogen atoms are carried off as seventeen separate water molecules (right). There is a large increase in positional disorder. Moreover, heat is released into the surroundings as stronger bonds are formed between the atoms than in the original materials. As a result, combustion is accompanied by a large increase in entropy.

replaced by strong new ones and the atoms settle into energetically more favourable arrangements. And why does the hydrocarbon burn? Because in doing so there is a huge increase in disorder and therefore in entropy. There are two principal contributions to this increase in entropy. One is the release of energy, which disperses into the surroundings and raises their entropy. The other is the dispersion of matter, as long, orderly chains of atoms are broken up and the individual atoms spread away from the site of combustion as little gaseous molecules. The combustion is portraying the content of the Second Law.

Let's for a moment suppose that the energy released in the combustion is confined to the flame. This hot region of burning fuel is in contact, through metal walls, with the water we want to heat. The violent jostling of the atoms in the flame corresponds to a high temperature. The gentle jostlings of the water corresponds to a low temperature. We have already seen that the entropy of the world increases as heat flows from a hot to a cold body, so the flow of energy from our combustion into the water is an entropy-increasing, spontaneous process.

Now the water is hot, and in principle its temperature may rise to being as hot as the flame itself. However, as the temperature of the water rises, there comes a point at which the water boils. Why does it do that? Because, of course, the formation of steam becomes a spontaneous process once the temperature has reached a certain value, the 'boiling point' of the water.

To understand why the water boils, we have to examine the changes in entropy that take place. Here we shall discover a funny feature of boiling, a different thermodynamic viewpoint. First, we note that there are two conflicting contributions to the change in entropy when water turns into steam. There is a big increase in entropy when the liquid becomes a gas. This increase suggests that there is always a tendency for water to evaporate. However, the evaporation of water needs energy, because attractions between the water molecules that hold the liquid together must be overcome to give a gas of independent molecules. Therefore, as water vaporizes, energy must flow into the liquid. This inward flow of energy lowers the entropy of the surroundings because it corresponds to an efflux of energy from them. At low temperatures, the decrease in entropy of the surroundings due to this efflux of energy is large (the quiet library again), and even though there is an increase in entropy of the water as it evaporates, overall there is a decrease in entropy. Therefore, at low temperatures, evaporation is not spontaneous. When, however, we raise the temperature of the surroundings, the decrease in their entropy becomes smaller (the busy street), and at a sufficiently high temperature the overall change in entropy of the water and the surroundings becomes positive. Now the water has a spontaneous tendency to evaporate, and it boils. Here is the funny feature that we mentioned would emerge. We see that the effect of increasing the temperature is to reduce the change of entropy of the surroundings to the point at which the overall change in entropy is positive. It is as though to achieve evaporation, we have to soothe the restraining effect of the surroundings by raising their temperature.

At this point in the story, the Second Law has appeared three times: in governing the combustion, in governing the flow of heat from the flame into the water, and in the evaporation of the water. It now enters a fourth time: the flow of energy through the engine and the conversion of some of it to work. We have dealt with that stage earlier, and there is no need to go over it again. The point this discussion has sought to make, though, is that each stage of the operation of the engine, from the combustion of the fuel to the effecting of external change, is brought about by the natural tendency of matter and energy to disperse. The world is driven forward by this universal tendency to collapse into disorder. We, and all our artefacts, all our achievements, are ultimately the outcome of this purposeless, natural spreading into ever greater disorder.

That—the achievement of achievements—of course, is why we have to eat. We have to ingest a supply of energy which we can allow, through the elaborate metabolic processes that pervade our bodies, to spread into the environment. As it does so, it generates enough disorder there for the world to grow a little more disordered while we construct, for instance, these words or our self. Eating is a more complex operation than refuelling because, unlike the fuel that powers our cars, we use much of the ingested matter for repair and growth. However, in so far as food is a source of energy, it is fuel to stoke the hot reservoir of the steam engine inside us, and it drives us and our actions forward by virtue of the dissipation of some of the ingested energy into waste.

The steam engine inside us—or at least its abstract essence—is distributed through all our cells and takes on thousands of different forms. We will look at just one realization of the biological steam engine. One molecule that appears abundantly in every cell is adenosine triphosphate (ATP, Fig. 4.12). As we can see from the illustration, this molecule consists of a hefty organic part and a short tail of phosphate groups (phosphorus atoms surrounded by oxygen atoms). The business end, as far as we are concerned, is the phosphate tail. This molecule is like the hot reservoir of a steam engine. When it springs into action at the request of enzymes in the cell, it discards the terminal phosphate group, becoming adenosine diphosphate, ADP, as a result. The energy released is used to power constructive events within the cell, such as the construction of a protein or the preparation of a neuron for the transmission of a signal. The forward momentum of the reaction comes from the dissipation of matter (the release of the phosphate group) and of energy, which can stir up thermal disorder. Thus, the construction of the protein or the formation of an opinion can be traced back to this tiny analogue of a steam engine.

For the cell, and us, to go on living, a phosphate group—not necessarily the same one—must be reattached to ADP to reform ATP. That reconstruction can be achieved by coupling the reaction that effects the reattachment to a more powerful steam engine, another metabolic reaction that dissipates matter and

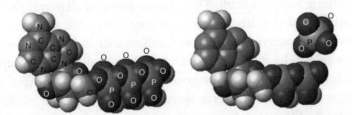

**Fig. 4.12** The molecules adenosine triphosphate (ATP, left) and adenosine diphosphate (ADP, right) act as the hot and cold reservoirs of the notional steam engines inside us. To rebuild ATP from ADP by reattaching the phosphate group, we have to couple the engine to an even more powerful engine (in the sense of generating more entropy). That is, we have to eat. And to grow the food we eat, the Sun has to burn (in its nuclear way).

energy even more potently. That is why we have to eat. We ingest material that acts as the fuel for the steam engine that drives the formation of ATP from ADP, which in turn drives our growth and activities.

That food itself must be forged by coupling the reactions that form it to ever more powerful notional steam engines, engines that dissipate ever more effectively. The ultimate steam engine is the Sun, for the energy it dissipates into its surroundings drives the reactions that constitute photosynthesis, the formation of carbohydrates from carbon dioxide and water. So, ultimately, our activities and aspirations are driven by the energy released when nuclei fuse together in the Sun. The ancients were perhaps right to worship the Sun as the giver of life; but they did not appreciate that it is the driving force of universal corruption.

The same type of illumination of the molecular basis of life that we met in Chapter 2 also springs from the Second Law. Life is a process in which molecules bumble around, just happening to have the right shape to fit in a niche, and able to release their burden. The molecular basis of reproduction illustrates the unconscious activities of molecules and energy. Life got under way because this bumbling around gave opportunities for natural selection, opportunities for molecules to use blind, unconscious, undirected bumbling to build the great web of activities we call being alive. Life, at root, is molecular bumbling.

The question that probably springs to mind at this point is whether the dissipation of matter and energy can continue for ever. Or will the universe become so infinitely disorderly that entropy can no longer increase and events come to an end?

The speculative termination of the natural tide of events by the topping out of entropy is called the *heat death* of the universe. Then, because things can't get any worse, nothing happens at all. There is one point to clarify: if the universe were to succumb to heat death, it would not mean that time would come to an end. Events would continue—atom would collide with atom—but there would be no net change. All the steam engines, notional as well as actual, would be at rest, for entropy could no longer be generated. Others take a more sanguine view, and argue that if the universe were to start to contract, then entropy would decrease as the space for energy and matter were to become more confining. Thus, they muse, events would reverse, with anti-Kelvins and anti-Clausiuses ruling the day, perhaps to bounce out again with entropy increasing once again in a newly revitalized universe.

Let's try to sort out the issues involved. First, let's accept the prevailing view, which we explore in more detail in Chapter 8, that the universe will not turn in on itself and contract into a Big Crunch. So in practice there is no need to fret about the possibility of time in some sense reversing, with the unnatural becoming natural, when the universe starts to fall in on itself. But scientists like to explore the boundaries of thought, and we should be able to decouple the question of the thermodynamic future of the universe from its cosmological future. In other words, suppose we (that is, cosmologists) are wrong about the long-term future of the universe and that it will in fact collapse. What then? Will the natural become unnatural, the non-spontaneous spontaneous?

The ever-imaginative and highly visual British mathematician Roger Penrose (b. 1931) has looked the collapsing universe in the face and has suggested that there might be a gravitational contribution to entropy. In other words, disorder can stem from the very structure of spacetime rather than just from an untidy disposition of the things that inhabit it. He accepts the singularity of the initial moment, the Big Bang, but considers the possibility that the singularity of the final moment, the Big Crunch, could be a point with a far more complex structure (Fig. 4.13). Thus, although in their final days all the matter and energy in the visible un iverse might be compressed back into a single point, and therefore have extraordinarily low entropy, the structure of the spacetime that they inhabit is so complex that the disorder is greater than at the initial moment of creation. So, entropy will go on increasing from now until eternity, even if eternity (or at least a few tens of billions of years) sees us back at a singularity.

Be that as it may, the more likely featureless future of the cosmos is its ever-increasing extension, the unbounded growth of its scale. In such a scenario,

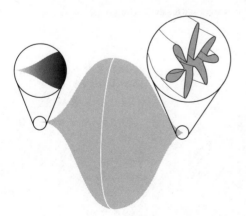

**Fig. 4.13** Even if we are heading for a Big Crunch, we should not expect the entropy to start decreasing again as the universe begins to contract. It is conceivable that there is a gravitational contribution to the entropy, in the sense that the final singularity (right) is far, far more complex than the initial singularity (left), so that the entropy of the universe goes on increasing even though it is contracting.

there is always more room for energy and matter to disperse. Even if all matter were to decay into radiation, the entropy of that radiation would gradually increase as the volume it occupies increases. The real problem, though, is that if all matter were to decay into radiation and all that radiation were to be stretched out into infinite wavelengths, so that in the distant future there was only dead flat spacetime with no energy globally, then at first sight it looks as though the entropy of the universe would be zero. However, physics at cosmological scales of length and time is an as yet uncertain subject, and it could be that even a scattering of fluctuations of energy density in the enormous volume of space then available is sufficient to ensure that the total entropy is extremely large. This is an open question.

Gravitation and entropy are remarkable bedfellows. At first sight there might be thought to be little connection between the theory of general relativity, Einstein's theory of gravitation (which we meet in Chapter 9), and the workings of the Second Law, except in so far as there may be a gravitational contribution to entropy. However, a remarkable fact emerges when we start to think about the structure of spacetime in terms of entropy. In 1995, Ted Jacobson[9] showed that if we combine the Clausius expression for the change of entropy when heat enters a region with an assertion about the relation of entropy to the area of the surface bounding the region (in fact, the two are proportional, as they are known to be for the surface surrounding a black hole), then the local structure of spacetime is distorted in exactly the way predicted by Einstein's equations for general relativity. In other words, in a rather refined mathematical sense, the Second Law implies the existence of Einstein's equations of general relativity!

So, maybe the steam engine is not just inside us, it is everywhere.

[9]  This paper can be found via http://xxx.lanl.gov and the subdirectory gr-qc.

# FIVE

# ATOMS

## THE REDUCTION OF MATTER

### THE GREAT IDEA
### *Matter is atomic*

*I will reveal those atoms from which nature creates all things . . .*
LUCRETIUS

W E  H A V E  seen the outer manifestations of change in the emergence of the
biosphere, and the inner mechanisms of that change in the molecular
basis of genetics. We have seen what doesn't change, energy, and we have seen
why things change, in terms of entropy. Now we examine the material basis of
change in greater detail, so making the transition from elephants to elements.

What does science expose about the nature of matter, the stuff from which
everything tangible is made? We shall explore this enormously important question in two stages. In one, the subject of this chapter, we examine the question
at what will prove to be, but will not seem at the time, a superficial level and
describe the emergence of the concept of atom, the currency of all discourse in
chemistry. We shall see why different atoms have the different personalities we
call their *chemical properties*. Do not be put off by the thought that this is a chapter on chemistry. Chemistry is the bridge between the perceived world of substances and the imagined world of atoms, and despite one's often gruesome
memories of encounters with the subject at school, it is a deeply enthralling
and illuminating subject even when picked at (as here) rather than chewed. I
aim to introduce a little chemistry to open your eyes to the world around you
and simultaneously to deepen your delight. Then, in the following chapter, we
leave the superficialities of atoms and burrow down into the extreme depths of
the well of concepts we call matter. Then we shall move towards understanding what matter truly is in a manner that might even have satisfied the Greeks.

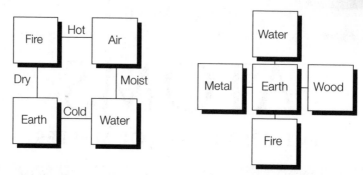

**Fig. 5.1** Two very early periodic tables of the elements. The diagram on the left shows the four elements assumed to be the foundation of all matter by the ancient Greeks and the attributes that those elements bring to the materials they form. The diagram on the right is a Chinese version, which emerged from a Taoist view of being devised by Lao Tzu (*c*.600 BCE) in which the five elements resulted from the struggle between the *yang* (male, positive, hot, light) and the *yin* (female, negative, cool, and dark).

The Greeks thought a great deal about matter and proposed so many different hypotheses about its nature that at least one of them was likely to turn out to be right. Some of their suggestions were completely wrong, but showed a commendable spirit of enquiry. Thus, Thales of Miletus (*c*.500 BCE), regarded in the popular imagination as the father of philosophy, finding fossil sea shells high up mountains, leapt to the conclusion, shortly before falling to his death, that everything was made of water. A thousand years later, this view found itself in the Qur'an:[1]

> We have made of water everything living, will they not then believe?

The grandeur of its source, in the eyes of some, gives this view some authority even today. However, the Greeks moved on in their quest to comprehend, considering that a single substance was insufficient to account for the variety of substances in the world. Thus Heraclitus of Ephesus (*c*.540–475 BCE) elaborated Thales by concluding that an agent of change was necessary, and added fire to the cauldron of being. Soon the Sicilian Empedocles (*c*.492–432 BCE), considering that solidity was difficult to engineer from water, air, and fire alone, elaborated the ur-mixture even further by adding earth and suggesting, and probably believing, that all there is can be built from air, earth, fire, and water (Fig. 5.1). Aristotle (384–322 BCE) was almost entirely at home with Empedocles' reduction of the world to his four elements, arguing that the sublunar terrestrial world, the theatre of change and decay, was quite different from the eternal

---

[1] Qur'an, 21:30.

celestial sphere, and that Empedocles' elements were applicable to the former but not adequate to account for the latter. For the timeless, exquisite celestial sphere, Aristotle considered that a fifth basic component, a *quintessence*, was essential.

All this was quite wrong, of course, as none of these substances is elemental, except perhaps the hypothetical, experimentally inaccessible, and as we know non-existent quintessence of the heavens. But the formulation and elaboration of the concept that complexity is fabricated from simplicity was a profoundly important conceptual step, and this attitude still lies at the core of modern science.

The supposition of the existence of elements, albeit the wrong elements, stimulated the question that lies at the heart of this chapter: is matter continuous or discrete? In other words, can the elements be divided *ad infinitum* into ever small pieces or is it discrete, in which case cutting would bring us finally to the further uncuttable, the atom? With the freedom that comes from the absence of experimental data, Greek speculation was able to flourish and both ideas had their supporters. That one of the ideas, the atomic view, would turn out to be right should not necessarily lead us to admire its proponents, for their support was founded on whimsical speculation and philosophical taste, neither of which is now considered to be a particularly reliable component of the scientific method and the search for truth.

We can trace the line of lucky speculation back to Leucippus of Miletus (*fl.* 450–420 BCE), who envisaged matter as granular, as being composed of atoms that marked the termination of the cutter's cutting. Only if there were an end to cuttability, Leucippus argued, would matter be eternal, for otherwise everything would have dissolved into nothing long ago. His view of atoms, though, was far from what we now regard as orthodox. Thus, he envisaged atoms of a wide variety of shapes and sizes, with different atoms for just about every different substance. This view was elaborated, and the speculated entities named *atomos*, meaning not cuttable, indivisible, by his student Democritus of Abdera (*c.*350–322 BCE), the 'laughing philosopher'. Democritus took the view that here are atoms of milk and atoms of coal, atoms of bone and atoms of water. His imagination flew unfettered by experiment, and he considered that there were also atoms of sight, sound, and the soul. The atoms of the soul he considered to be particularly fine, as befits the soul; those of the colour white were smooth and round, like an interpretation of the colour itself.

These thoughts were part of the system of beliefs of the Epicureans, the followers of Epicurus of Samos (341–270 BCE), who used them to attack superstition by arguing that because all, including the gods, were made of atoms, even the gods—for Epicurus, gods were engagingly disinterested; they were *bon viveurs*, models of contentment and disdain who could not be bothered to intrude on petty human affairs—were subject to natural law. The Epicurean world view, a heady combination of hedonism and atomism, saw sensation as the root of knowledge and sensation as the impression made on the soul by images composed of fine films of atoms emitted by the objects being sensed. This atomic view of structure and sensation was propagated to a receptive tyrant- and god-weary Roman public in an elaborate hexametrical didactic epic by Tito Lucretius Carus (*c*.95–55), whose *De rerum natura* (On the nature of things) could be regarded as the first textbook of physical chemistry. This text was lost until the fifteenth century, but on its rediscovery it encouraged more modern minds to turn afresh to atomism.

Plato and his pupil Aristotle were vehemently opposed to atomism and their powerful yet contaminated view of the world was dominant throughout the Middle Ages, not least on account of the strong whiff of materialism and atheism in the Epicurean view. Aristotle's view was that atomism, which he regarded as pure invention and therefore—in contrast to his own pure inventions—worthy only of scorn, was incapable of accounting for the rich tapestry of sensory experience that characterizes the real world. For him, too, the void—which was necessary if atoms were to move—was anathema, for motion, he thought, could not be sustained in a void, for voids lacked impellers and motion was absent unless impelled (see Chapter 3).

Such was the power of Aristotle's authority that these views shaped what passed for human understanding, with little further encrustation, for two thousand years. They sustained the alchemists in their misguided and largely fruitless endeavours just as his views on motion had stifled physics. Then, as the world woke up to the emptiness of Aristotle's armchair physics in the seventeenth century, so it gradually dawned on people that his armchair chemistry was empty too. However, although we can scoff at Aristotle from our position far downstream in his intellectual heritage, with several revolutions distancing us from his thought, we should not redirect our praise towards the Epicureans, even though they seem at first sight to have been closer to the truth. The Epicureans were armchair-bound too, and their atomism was as wild a speculation as Aristotle's anti-atomism. All early postulations of atoms were pure conjecture: all of it was speculative philosophy, not science.

Science took longer to come to grips with the nature of matter than it did with matter's motion. The nature of the tangible itself was more elusive than the motion of the tangible through space, for although numbers can readily be attached to positions in space and time, and dynamics thereby pinned to the physicist's bench, it was far from clear how numbers should be attached to matter. Indeed, were numbers relevant at all to the properties commonly regarded as chemical? Was the nature of matter for ever to be only a matter of anecdote and speculation?

The balance turned out to be the key (Fig. 5.2). In the hands of Antoine Laurent de Lavoisier (1743–94), who is widely regarded as the father of modern chemistry and 'the spirit of accountancy raised to genius', the chemical balance became a scalpel that could be used to cut deeply into the mystery of matter. Carefully and thoughtfully deployed, the balance could be used to attach numbers to matter and bring chemical reactions into the realm of arithmetic. In particular it could be used to determine the masses of substances that react together. As a result, patterns began to emerge in the data and, as we have already seen, patterns are the lifeblood of science and the seeds of theories.

The pattern of the masses of the elements that combine together was the acorn that grew into the oak tree of Dalton's *atomic hypothesis*. John Dalton (1766–1844), the dour, colour-blind Quaker son of a hand-loom weaver, a school teacher at age twelve, and a meticulous observer of the weather, had no pastimes apart from a game of bowls every Thursday night. Perhaps the

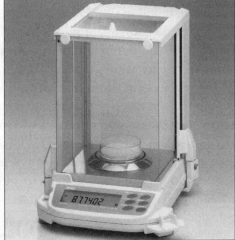

**Fig. 5.2** A classical chemical balance, not unlike that used by Lavoisier in the researches that allowed him to attach numbers to matter and thereby turn chemistry into a physical science, and its modern counterpart.

subconscious memory of those balls suggested to him the theory that he first presented at a lecture at the Royal Institution in December 1803 and published in 1807. His hypothesis was that matter consisted of atoms that cannot be created or destroyed, that all atoms of a given element are identical, and that in a chemical reaction atoms simply change partners. His crucial concept was that each atom had a characteristic mass, and that the chemical balance is therefore an observer of the changes in mass that occur as atoms change their partners. This is the step that philosophers of science call *transduction*, in which a concept at a microscopic level is linked to an observable, macroscopic property. Most of modern physics and chemistry is an elaboration of transduction, with the observed interpreted in terms of the imagined, and specifically measurements made on a human scale interpreted in terms of entities billions of times smaller.

Dalton actually went a little further than it is now considered comfortable to remember. He considered that atoms of different elements are surrounded by different amounts of caloric, the hypothetical (and since discarded) imponderable fluid we experience as heat (Chapter 3). He suggested that the atoms of gaseous elements had the thickest shell of caloric, which enabled them to move about freely. The atoms of solid elements had the thinnest shells, which meant that they stayed in place. This faintly embarrassing distraction from the central idea of the atomic hypothesis has been quietly forgotten.[2]

Through his use of the balance, Dalton was able to compile a list of the masses of his atoms relative to the mass of a hydrogen atom, the lightest element, taken as 1. He termed these relative atomic masses *atomic weights*, and the name is still in use. His experiments were crude and his interpretation of them depended on assumptions about how many atoms of one element combined with another, and here his guesses were often wrong (Fig. 5.3). Thus, using simplicity as his guide, he supposed that water is composed of one atom of oxygen and one atom of hydrogen, and deduced that the atomic weight of oxygen is 7 (in fact, more accurate data would give 8 using his reasoning); we know that in fact water consists of two atoms of hydrogen and one atom of oxygen, so the true atomic weight of oxygen is 16; that is, an atom of oxygen is sixteen times heavier than an atom of hydrogen. Nevertheless, here is the earliest version of transduction in all its glory, with observations in the laboratory revealing properties of the unseeable.

---

[2] Funnily enough, most components of Dalton's atomic hypothesis are wrong, in the eyes of pedants, at least. Caloric doesn't exist. We can make and destroy atoms (but not in chemical reactions). Not all the atoms of an element have exactly the same mass (isotopes differ slightly in mass). But the spirit of his hypothesis is true, and it deserves respect.

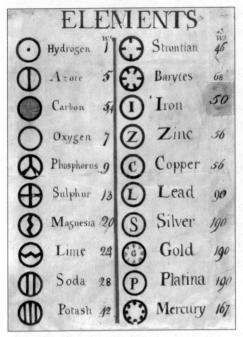

**ELEMENTS**

| | | | | | | |
|---|---|---|---|---|---|---|
| ⊙ | Hydrogen | 1 | ⊕ | Strontian | 46 |
| ◖ | Azote | 5 | ✳ | Barytes | 68 |
| ● | Carbon | 54 | Ⓘ | Iron | 50 |
| ○ | Oxygen | 7 | Ⓩ | Zinc | 56 |
| ⊘ | Phosphorus | 9 | Ⓒ | Copper | 56 |
| ⊕ | Sulphur | 13 | Ⓛ | Lead | 90 |
| ◉ | Magnesia | 20 | Ⓢ | Silver | 190 |
| ⊖ | Lime | 24 | ⓖ | Gold | 190 |
| ◑ | Soda | 28 | Ⓟ | Platina | 190 |
| ⦀ | Potash | 42 | ✺ | Mercury | 167 |

**Fig. 5.3** Dalton's atomic hypothesis emerged in the opening years of the nineteenth century and he lectured on it on many occasions. This is a facsimile of a part of the support he provided for a lecture he gave on 19 October 1835 to the members of the Manchester Mechanics Institution. The typographically awkward symbols for the elements were replaced by a simpler orthography, much to Dalton's annoyance.

Unlike the Greeks' speculation about the atomic nature of matter, Dalton's was a scientific theory. This was no idle, or even energetic, musing; this was experiment-based observation allied with rationalization. Yet, not everyone accepted it as a feature of reality. For many years the view lingered among scientists that atoms were an accounting fiction, useful tokens for doing calculations involving masses, but not in any sense real. Most opposition died away in about 1858, though, when the Italian chemist and revolutionary Stanislao Cannizzaro (1826–1910) published a much more accurate list of the atomic weights of the known elements, but even at the end of the nineteenth century some diehards remained reluctant to nail their colours to the atomists' mast.

**Fig. 5.4** Dalton inferred the existence of atoms from regularities in the masses of the elements that combined with each other. Now we can 'see' them and there is no doubt about their existence. The device used to obtain this image of silicon atoms on the surface of a piece of silicon is called a scanning tunnelling microscope. It almost literally feels its way across a surface, and computers convert the signals sent by the probe into an image with an atomic-scale resolution.

Modern techniques of observation have sliced through the bramble of argument on which Dalton and his immediate successors had to negotiate. Now we can see atoms as individual blobs of matter (Fig. 5.4), and there is no longer any doubt that they exist. Certain sociologists of science, of course, might wave their hyper-pessimistic constructivist flags and say that the apparatus used to obtain these images is more social construct, with the apparatus designed to express the current paradigm; but scientists know better.

So, what are atoms? What do they look like? How are they made? Dalton, like the Greeks, presumed that atoms are the final cut; no atom could be cut asunder, no atom had smaller components. But if that were so, it would be hard to see how the rich properties of the elements could be explained, for variety of properties stems from richness of composition. That atoms do have an internal structure was first demonstrated by J. J. Thomson (1856–1940), who showed in 1897 that electrons could be sparked out of atoms. He announced his discovery at the Royal Institution on 30 April 1897. Electrons were the first of the *sub-atomic particles*, particles smaller than atoms, to be identified and Thomson's work in the Cavendish Laboratory in Cambridge showed that they were a universal constituent of all matter and that therefore atoms did in fact have an internal constitution.

There was at the time (at the very end of the nineteenth century) great perplexity about how the electrons were arranged. Some suspected that a single atom might consist of thousands of electrons. The problem was not helped by the absence of any information about the existence of particles with positive charge to offset the negative electric charge of the electrons. The problem was laid to rest by the work of the New Zealander Ernest Rutherford (Lord Rutherford of Nelson, 1871–1937), then in Manchester, who in 1910 stumbled on the existence of the *nucleus*, a minute speck of positively charged matter lying at the centre of the atom and which, though much smaller than the atom itself, accounted for virtually the whole of its mass.[3]

It will be as well to gather an impression of the sizes and masses of the various entities that have moved on to the stage so far. A typical atom has a diameter of about 3 billionths of a metre ($3 \times 10^{-9}$ m, 3 nanometres, 3 nm). So, a million of these atoms laid out in a row would stretch 3 mm, the length of this dash —.

---

[3] Rutherford first used the term 'nucleus' in 1912.

You might just be able to imagine the size of these atoms. It gets easier to think of the same dash magnified until it is about 3 kilometres long, in which case the diameter of each atom would be about 3 millimetres, like lumpfish roe.

As you can perhaps apprehend, atoms are quite big: they have to be, because so much is stuffed inside them. Most people think of atoms being very small, but that is only because we ourselves are very large: we have to be, because so much is stuffed inside us. If you start thinking of atoms as large, then they become much less daunting. It will be helpful to inflate an atom in our imagination until it is about a metre across.

The nucleus of an atom is also big, because it also has a lot of things stuffed inside it. Most people think of it as being very, very small; but that is not altogether helpful because such thoughts hinder the mind's ability to imagine what it looks like. Some scientists would think that hindrance a very good thing, as the importation of macroscopic ideas into entities as small as atoms, let alone their nuclei, is fraught with danger, as familiar concepts simply do not apply to objects as small as these (as we shall see with a vengeance in the quantum theory described in Chapter 7). Be that as it may, let's at least try to imagine the diameter of a nucleus. Experiments show that the diameter of a nucleus is about one ten-thousandth that of an atom. So if we think of an atom as a ball about a metre in diameter, its nucleus would be a speck only one tenth of a millimetre in diameter. So, to we ponderous creatures, nuclei are indeed very small; even to an atom-sized perceptive entity they would also seem quite small, but discernible. To a nuclear physicist who needs to reflect on the composition of nuclei, the nucleus is rather big.

As we have said, nuclei are big because so much is stuffed inside them. Here is the seat of the positive charge of the atom, the contribution that cancels the negative charge of the electrons that surround it. Here too is the seat of almost all the mass of an atom, for only about 0.1 per cent of the mass of anything is due to the electrons. When you pick up a heavy object, you are really heaving up nuclei. If all the atoms in your body could be stripped of their nuclei, then you would weigh only about 20 grams. Another less well known feature of nuclei is that many also spin on their axis, but some do not. The nuclei of hydrogen and nitrogen do spin; those of carbon and oxygen do not. The spin of a nucleus cannot be changed, it is an intrinsic property, like the electric charge, so every hydrogen nucleus is destined to spin for eternity at the same unchanging rate.

At the beginning of the twentieth century it became clear that the electron was not the first subatomic particle to have been discovered. The very first had been known but not recognized as such for more than a century. The nucleus of a hydrogen atom, the simplest of all atoms, consists of a single subatomic particle, the *proton*. This particle is the entity responsible for the properties of acids, and when you taste the sharpness of lemon juice your tongue is in fact being tickled by protons. With regret, we shall not explore that aspect here, or see why the tongue is a good detector for at least one kind of fundamental particle. A proton is a heavy particle with a positive charge equal and opposite to that of an electron: its mass is about two thousand times that of an electron.

A hydrogen atom consists of a single proton and an accompanying electron: the positive charge of the nucleus is cancelled by the negative charge of the electron. The next simplest element, helium, has a nucleus built from two protons, so it has two accompanying electrons.[4] The number of protons in an atomic nucleus is called the *atomic number* of the element, so the atomic number of hydrogen is 1, that of helium is 2, and so on. For the atom to be electrically neutral, which all atoms are, it follows that the number of electrons present must be equal to the atomic number, for then the total positive charge of the nucleus is cancelled by the total negative charge of the accompanying electrons.

The realization that the nucleus of an element could be ascribed a number, and that that number could be interpreted in terms of the composition of the nucleus, meant that at last a roll-call of the elements could be made. Now a missing element could be identified by noting whether an element had been found with its particular atomic number and speculations about the existence of an element between two others could be ruled out if their atomic numbers were contiguous. Atomic numbers were open to experimental determination by a technique developed by Henry Moseley (1887–1915) shortly before he was conscripted, only to be felled by a sniper's bullet at Gallipoli. As Wilfred Owen wrote, before meeting his own bullet on the evening of the end of the same war,

*Courage was mine, and I had mystery,*
*Wisdom was mine, and I had mastery.*

It was indeed mastery based on wisdom that had dispelled mystery, for now we knew the roll-call of the elements, the presence of the nucleus, and the number of electrons present in each atom.

---

[4]  All nuclei other than that of hydrogen have another subatomic particle present too: this is the neutron, a close relative of the proton but without its positive charge. Protons and neutrons are jointly responsible for the mass of nuclei and hence for most of the mass of matter.

The precise arrangement of the electrons around the nucleus was still a problem. The point to grasp at this stage of the discussion is that an atom is almost completely empty space. All its mass, as we have seen, is due to the minute central nucleus, with the space surrounding it to a distance of about 10 thousand times the diameter of the nucleus occupied by a handful of electrons—six, for instance, in the case of carbon. Your body is this almost-empty space, yet somehow or other you seem substantial. In a real and not sardonic sense, you are emptiness, thinking with an almost-empty brain, clothed in emptiness, eating emptiness, seated on and supported by emptiness. To imagine this emptiness of the atom, think of yourself as standing on an Earth-sized nucleus and looking out into a clear, starry night. The emptiness of space you see around you is not unlike the emptiness of an atom within you.

This extranuclear emptiness, though, is the seat of an element's personality. While the nucleus is a passive bystander merely responsible for marshalling its complement of electrons around itself, a centre of control, the wisp of electrons that occupy the almost-emptiness are the participants in chemical reactions.

Scientists could not resist the temptation to suppose that the electrons were as planets to the nucleus as a star, or a Moon as to an Earth, and this image is still so powerful that it would have been best if I had not had to mention it. The 'Saturnian' planetary model of the atom was suggested by the Japanese physicist Hantaro Nagaoka (1865–1950) in 1904 and became a natural model to presume as a consequence of Rutherford's discovery of the nucleus a few years later. The planetary model, now thought of as planets orbiting a central star rather than Nagaoka's rings around Saturn, was fanned into life in 1912 when Niels Bohr adapted an early version of quantum theory to describe the motion of the single electron in a hydrogen atom and his successful quantitative calculation of the spectrum of the atom. One can only speculate about, and envy, the depths of Bohr's delight when his calculation produced results in almost exact agreement with observation.

Yet it—the planetary model and Bohr's clever, apparently supporting calculation—was wrong. Here lie two lessons for science and life in general. First, we cannot reliably project, without considerable circumspection, from the familiar macroscopic into the hitherto unknown microscopic. Dragons abound in the underworld of reality. Second, even quantitative agreement can in special circumstances be a corrupt umpire of truth. The special circumstances that corrupted the umpire in this instance was the beauty—in a sense we visit in the next chapter, but for now a curious and enigmatic use of the term—of the characteristics of the electric pull of the nucleus on an electron.

You must discard from your conscious mind, and even better your unconscious mind, the image of planets orbiting a central nucleus: that is simply wrong. It is a false model of an atom; it is science fiction, a dead, discarded model. The root of its error is the realization that electrons are not particles in the familiar sense, but have an intrinsic wave-like character. This dual character, which lies at the heart of quantum theory and moves on to our stage in Chapter 7, eliminates the concept of a trajectory, in this case the orbital path of a planetary electron around a central star-like nucleus, and implies that it is wholly inappropriate to picture the electron as a particle in an orbit.

In Chapter 7 we shall see how Erwin Schrödinger (1887–1961) developed the equation that, when solved, tells us the behaviour of electrons. All we need to know at this stage is a few of its implications for atoms. What is currently taken to be a reasonably correct structure of a hydrogen atom—we come to other atoms later—was one of the first outcomes of the application of Schrödinger's equation.[5] In the flood of four famous papers (the first in three parts) published in 1926 in what he called 'a late erotic outburst', and written while on holiday with a mistress, Schrödinger solved his equation for the electron in a hydrogen atom and found, from quite different premises, the same expression for its energy as Bohr had found years before.

To understand the results of Schrödinger's calculation we have to know that the solutions of his equation predict the *probability* that the electron will be found at each point of space, not, as in classical physics, the precise location of the electron at any instant. The solutions are called *atomic orbitals*, the name being intended to convey the allusion to a planetary electron in an orbit but without the strictness of that inapplicable classical concept.

Figure. 5.5 shows the shape of the atomic orbital of lowest energy in a hydrogen atom, the orbital of the electron as the atom would normally be found. The illustration depicts the probability of finding the electron in a region by the density of shading. As you can see, because the cloud is densest close to the nucleus, the electron can be thought of as clustering close to the nucleus, like a wasp round a jam pot, with the location of greatest probability being at the nucleus itself. If you thought of a tiny hollow ball being located at various places in the atom, then you would find the electron inside the ball most often when the ball is located at the nucleus. The cloud of probability is spherically symmetrical (there is no favoured direction), so we can also represent the orbital by the spherical surface that captures most of the cloud. However, you should not think of the orbital as having a sharp edge: as the graph in the illustration

---

[5]   It is an incidental and irrelevant genealogical point that Erwin's grandmother, Emily Bauer, was half English, that side of the family coming from Leamington Spa.

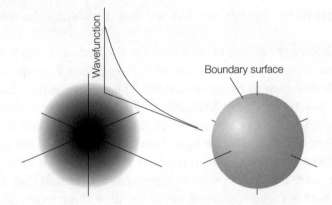

**Fig. 5.5** Here are several representations of the lowest energy *s*-orbital in a hydrogen atom. The diagram on the left shows the probability of finding an electron at each point by the density of shading. The accompanying graph shows how the probability decays exponentially with distance from the nucleus. The diagram on the right shows the 'boundary surface', which is the surface that captures about 90 per cent of the probability of finding the electron.

shows, the probability of finding the electron at a particular point decays smoothly towards zero and—in principle at least—doesn't reach zero until we are an infinite distance from the atom. On this interpretation, all atoms are infinitely large, which is in sharp contrast to the view that atoms are tiny. In practice, of course, the probability of finding an electron far from the nucleus (more than a few hundred trillionths of a metre) is negligibly small. It is best to think of a hydrogen atom as having an electron confined to a region of space very close to the nuc-leus (a region of radius about 100 trillionths of a metre, 100 picometres, 100 pm). This lowest-energy, spherical orbital is called an *s-orbital*. It would be nice to think, and is in any case a useful mnemonic, that the *s* denotes spherical; but in fact it is coined for technical reasons relating to the sharpness of lines in the spectrum of atomic hydrogen.

One feature that will become clear once we know more quantum theory, but which we need to know at this stage, is that the fact that the orbital in the illustration is spherically symmetrical implies that the electron it describes has zero angular momentum around the nucleus. We met angular momentum in Chapter 3, where we saw that it was like linear momentum, but applied to motion in a circle rather than along a straight line. All we need to know at this stage is that the waviness of an atomic orbital, how rapidly the density of shading changes as we travel round the nucleus, tells us the angular momentum. For an *s*-orbital, the shading has constant density on any circular path with the nucleus at the centre, so we conclude that the electron has zero angular momentum round the nucleus. This tiny technical point might seem to be just that: in a short while, though, we shall see that it underlies the splendour of the world.

When Schrödinger solved his equation for the hydrogen atom, he discovered that there are many other atomic orbitals, each one corresponding to an energy higher than that of the ground state. The analogy is the vibration of a sphere, with overtones of its fundamental frequency corresponding to the higher energy states. An electron can be lifted into these orbitals if it is supplied with enough energy, such as by the lightning flash of an electric discharge or the absorption of energy from a pulse of photons we call a flash of light.

There are several features of these higher energy orbitals that we need to know. First, there is a whole series of s-orbitals, all of them spherical but differing in their distance from the nucleus: they form a series of concentric shells like a Russian doll with the nucleus at the centre. An electron in any of these s-orbitals has no angular momentum, so it may be found at the nucleus itself. Once again, don't be fooled into thinking this a pedantic, academic detail: cities and great industries are built on details such as this.

There are also solutions that do not have spherical symmetry, in which the clouds of electron probability are concentrated in pools in different regions around the nucleus rather than being uniformly distributed around it. We need to be aware of the three types of orbital shown in Fig. 5.6. Orbitals in which the probability collects in two pools are called p-orbitals, those with four pools are called d-orbitals, and those with six pools are called f-orbitals.[6] Because the density of shading depicting the probability of finding an electron at a location varies as we walk in a circle around the nucleus, measuring it as we go, p-, d-, and f-orbitals correspond to states of non-zero angular momentum of the electron they describe, with a d-orbital corresponding to a higher angular momentum than a p-orbital, and the even more crinkled f-orbital corresponding to a higher angular momentum still. This angular momentum gives rise to a centrifugal force that flings the electron away from the nucleus. Here is yet another tiny technical point that will effloresce shortly into enormous significance: on account of this centrifugal effect, an electron in any of these orbitals will never be found at the nucleus itself.

We now need to know two further features of the solutions that Schrödinger found for his equation. (I apologize for all this foreplay, but it will soon seem to have been appropriate.) First, the pattern of energies is shown in Fig. 5.7. We see that as the energy increases, more groups of orbitals become

---

[6]   The names are rooted in almost long-forgotten and now irrelevant technical features of spectra, in which lines involving electrons in these orbitals are the *principal* ones, or have an appearance that may be *diffuse*, or are classified for obscure (but known) reasons as *fundamental*.

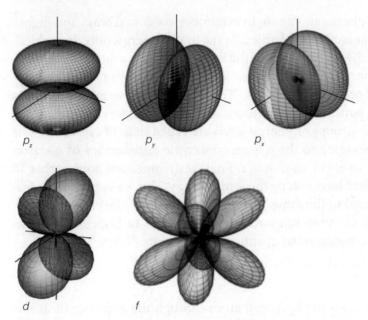

**Fig. 5.6** The two-lobed distribution of electron density (depicted by a boundary surface) is characteristic of a *p*-orbital, the four-lobed distribution is characteristic of a *d*-orbital, and the six-lobed distribution that of an *f*-orbital. Because the orbitals are progressively more crinkled (that is, correspond to waves of shorter wavelength wrapped round a sphere), they correspond to increasing angular momentum of the electron. In none of these orbitals is there any probability of finding the electron at the nucleus itself: it is progressively hurled away from the nucleus as the angular momentum increases.

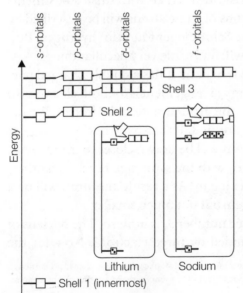

**Fig. 5.7** The energy levels of typical atoms. For hydrogen, with its single electron, all the orbitals of a given shell have exactly the same energy. For atoms other than hydrogen (as depicted here) each shell contains orbitals of progressively higher energy. In all cases, *p*-orbitals first become available in the second shell, *d*-orbitals become available in the third shell, and *f*-orbitals become available in the fourth shell. There are shells higher in energy than the ones we have shown. Each box represents an orbital that can be occupied by up to two electrons. The two insets show the analogous electronic structures of lithium (one electron outside a core) and sodium (one electron outside a core).

available, just as a sphere can vibrate in ever more contorted ways and higher frequencies as it is hit harder and harder. At the lowest energy, only one orbital, the s-orbital of Fig. 5.5, is available. At the next level, one s-orbital and three p-orbitals are available. At the next higher level, there is one s-orbital, three p-orbitals, and five d-orbitals, and so on. There is nothing magical about this pattern, it is just the pattern of the solutions of the Schrödinger equation for the hydrogen atom. The groups of energy levels are called *shells* because, more or less, the orbitals belonging to them form concentric distributions of electron probability, like the layers of an onion. A final (yes!) important point is that all orbitals of a given shell have exactly the same energy. This is a very peculiar feature, and can be traced to the same 'beautiful' character of the electrical interaction between the electron and nucleus that resulted in Bohr's calculation being conceptually erroneous but quantitatively accurate.

I will now lead you from the hydrogen atom through the sequence of atoms corresponding to the other elements. We know the order in which we should discuss these elements because we know their atomic numbers and therefore the number of electrons that must be accommodated in each case. For instance, if the atomic number of the element is 15 (as it is for phosphorus), then its nucleus has fifteen positive charges and, to achieve electrical neutrality, each atom must have fifteen electrons. The basic idea is that, with small adjustments that we must come to shortly, the electrons of these atoms will be described by orbitals and energies that resemble those Schrödinger found for hydrogen. But, in the course of this atom-building, we will find some very peculiar things.

The element with atomic number 2 is helium; it has a doubly charged nucleus and two electrons.[7] The lowest energy arrangement is for both electrons to be described by the same s-orbital as for the ground state of hydrogen. We say that the two electrons *occupy* the same s-orbital. Because the nuclear charge is higher than in hydrogen, the two electrons will be drawn closer to the nucleus; but as electrons repel each other (charges with the same sign repel each other) there will be some resistance to this sucking in. As a result, the atom will be a little more compact than that of hydrogen, but not much smaller.

The element next in line, with atomic number 3, is lithium. The nucleus of lithium is triply charged and is surrounded by three electrons. Now for the

---

[7] A helium nucleus consists of two protons and two neutrons, giving an atomic weight of 4. A small proportion of helium nuclei have only one neutron and therefore an atomic weight of 3. Atoms with the same atomic number but different numbers of neutrons are *isotopes* of the element.

amazing thing. Those three electrons cannot—simply cannot—all occupy the lowest energy *s*-orbital. A feature that has been entirely missing from our discussion so far, but now needs to be introduced, is that an electron has three permanent, intrinsic characteristics: its mass, its charge, and its *spin*. Just as many nuclei have spin, as we saw earlier, every electron in the universe also has this property. For our purposes, we can think of the spin as just like a classical spinning motion, like a planet on its axis. However, we should be aware that in this context spin is a purely quantum mechanical property, and cannot really be thought of classically. For instance, an electron has to turn round twice to get back to its starting state! A second quantum property of spin—one more relevant to our present discussion—is that an electron (to use classical language again) has a fixed rate of spin but may rotate either clockwise or anticlockwise at that rate. No intermediate rates or orientation of rotation are allowed.[8]

The third quantum property of spin—and for this there is no classical interpretation—is the *exclusion principle* proposed in 1924 by the Austrian physicist Wolfgang Pauli (1900–58), which states that:

No more than two electrons can occupy any one orbital, and if two electrons are present in the same orbital, then their spins must be paired.

By 'paired' we mean that if one electron has clockwise spin, then the other must have anticlockwise spin. This principle is the key to understanding chemistry. It is also the key to understanding why objects are solid even though they are mostly emptiness, for the electrons of one atom cannot be found in the region of electrons of another atom. So, even though electrons are spread out very thinly in the region we regard as 'the atom', another atom cannot enter that region. Thus, our bulk and our distinguishability from every other object that surrounds us is due ultimately to electron spin. Turn off electron spin, and all matter—all the inhabitants of the world, all the mountains, oceans, and forests, everything there is—would collapse into a uniform blob of featureless goo. Spin is the source of our individuality.

Now we can complete the lithium story. We imagine adding the three electrons in succession, and accommodating them in orbitals that result in the lowest overall energy, paying due regard to the exclusion principle. The first two electrons occupy the first *s*-orbital. That orbital now contains two electrons, so it is full. The third electron is therefore forced to occupy one of the *s*- or

---

[8]   If they all rotated in the same direction, the total angular momentum of all the electrons in your body would be about the same as that of a ping-pong ball making one rotation a minute. In fact, half the electrons are rotating clockwise and half are rotating anticlockwise, so you have no intrinsic net angular momentum.

*p*-orbitals of the next shell. But which orbital does it occupy, given that all four orbitals have the same energy?

It is not true that they have the same energy. We made that remark about hydrogen, tracing it to some enigmatic 'beautiful' feature of the electrical inter-action between the nucleus and the electron. When more than one electron is present in an atom, this 'beauty' (by which we mean a very special kind of sym-metry) is lost, and *s*- and *p*-orbitals no longer have the same energy. It turns out that *p*-orbitals of a given shell have slightly higher energy than *s*-orbitals of the same shell. The reason for this difference can be traced to the fact that an elec-tron in an *s*-orbital can be found right at the nucleus whereas an electron in a *p*-orbital cannot be found there. In short, an electron in the second *s*-orbital can penetrate through the region occupied by the two electrons in the first *s*-orbital and experience the full attractive power of the triply charged lithium nucleus. On account of the centrifugal effect of its angular momentum, an electron in the *p*-orbital cannot penetrate nearly as close to the nucleus and therefore does not experience its full attractive power. As a result, it lies at higher energy (as shown in Fig. 5.7).

With that energy difference in mind, we can now conclude that a lithium atom consists of two electrons in the *s*-orbital of the first shell surrounded by an electron that occupies the next higher energy *s*-orbital. We can think of the electrons as forming two physical, concentric shells, one clustering close to the nucleus to form a spherical core, the other surrounding it like the shell of a nut (Fig. 5.8).

The next element (with atomic number 4) is beryllium, with four electrons around its nucleus. There is one more electron than in lithium, and that elec-tron can join the latter's outermost electron in the second *s*-orbital. Next comes the fifth element, boron, with atomic number 5 and five electrons. The second *s*-orbital is full, so the fifth electron must enter one of the three *p*-orbitals. The same story holds for the next five elements, for there are three *p*-orbitals and

**Fig. 5.8** A representation of the structure of a lithium atom. There are two electrons in a compact core and one further electron in an encompassing outer shell.

these orbitals can accommodate up to six electrons. Thus carbon (six electrons) has an inner helium-like core of two electrons, two more electrons in a surrounding s-orbital, and then two more electrons in the p-orbitals. As it happens, these two electrons find it energetically favourable to occupy different p-orbitals of the shell, for they are then further apart and repel each other less. Nitrogen (seven electrons) has a further electron in a p-orbital, oxygen (eight electrons), fluorine (nine electrons), and neon (ten electrons) likewise.

At this point, all six p-orbitals of the shell are full, and the next electron (for sodium, atomic number 11) must occupy the next higher atomic orbital, which is another s-orbital. The structure of a sodium atom is like that of a lithium atom, with an inner complete core and a single s-electron forming a concentric outer shell.

This is an extraordinary point on our journey, although I have let a glimpse of the terrain go quietly unremarked. We have seen that the structure of a helium atom consists of a completed shell; we also need to know that helium is an unreactive, monatomic gas (that is, the gas consists of single atoms in free motion). Eight elements further on we come to neon, another inert monatomic gas, each atom having a completed shell of electrons. Immediately after neon, we caught a glimpse of lithium, a highly reactive metal; its atomic structure consists of a single electron outside a completed core. Just now—eight elements on from lithium—sodium, another highly reactive metal, flitted across the scene. The structure of a sodium atom is just like that of a lithium atom, with a single electron outside a completed shell. We have lighted upon the *periodicity* of the elements, the realization that matter is not a random collection of disjoint members, but *families* of members with similar chemical characteristics *and similar electronic structures*.

To understand the impact of this discovery, and to see it in its proper historical and cultural context, we need to return to the nineteenth century, to climb out of the structures of atoms and to see the elements from the outside, through more myopic and empirical nineteenth-century eyes.

By the middle of the nineteenth century, about five dozen elements were known. Some had been known since prehistoric times, but not known to be elements. Thus, iron, carbon, copper, and sulfur were all known to the ancients. These are truly elements in the modern sense rather than the conjectural sense of the Greeks. Elements, in the words of Robert Boyle (1627–91) in *The sceptical chymist* (1661), are certain primitive and simple, or perfectly unmingled bodies; which not being made up of other bodies, or of one another, are the ingredients of which all those called perfectly mixed bodies are immediately compounded, and into which they are ultimately resolved.

In Antoine Lavoisier's less prolix, more operational definition, they are

all substances that we have not yet been able to decompose by any means.

Lavoisier's definition left open the possibility that more strenuous efforts might achieve decomposition and thereby expel that substance from a table of elementhood. Lavoisier made a list of thirty-three elements, according to this definition. Eight were indeed expelled when more vigorous procedures were brought to bear, and two (light and caloric) were completely wrong-headed. The modern definition stands back from this chemical groping, and defines an element in a straightforward way as

a substance composed of atoms of the same atomic number.

The modern age of elements began in earnest when Hennig Brand (*fl.* 1669) of Hamburg discovered phosphorus, the first new element for centuries. His procedure was not one to endear him to his neighbours or encourage other putative searchers of the novel. He collected fifty buckets of human urine, which he allowed to evaporate and putrefy, boiled it to a pasty residue, allowed it to ferment, and heated the black residue with sand, collecting the vapour in a retort.[9] This seemingly magical substance glowed in air and was considered, therefore, to be an instrument to fight disease or at least to make a profit. As in Brand's procedure, the earliest technique for shaking compounds apart into their constituent elements was heat, sometimes heat in combination with other substances, such as carbon to extract iron from its ore, and sometimes heat alone, as for the disputed discovery of oxygen by the action of heat on an oxide of mercury.

Fierce heat was hard to come by before the industrial revolution, one Promethean procedure being to snatch it from the Sun by using a powerful lens. A new tool fell into the hands of the substance-shakers with the invention of the voltaic cell and the availability of electric current. Thus Humphry Davy (1778–1829) turned electrodes on to almost anything lying around at the Royal Institution, and in a week in October 1807 discovered first potassium by

[9] It should not go unremarked that urine and sand are both golden, so Brand's approach was basically an alchemical quest for the principle of their colouration, which he presumed was gold.

electrolysing molten potash (potassium nitrate) and then sodium by electrolysing molten soda (sodium carbonate). John Davy, Humphry's brother, said that Humphry 'danced around and was delirious with joy' at his discovery. In all, Davy discovered six new elements (sodium, potassium, calcium, magnesium, strontium, and barium). The surge in discovery, due largely to the application of electrolysis, brought the number of elements to forty-nine by 1818. The Swedish chemist Jöns Berzelius (1779–1848) himself discovered three elements (cerium, selenium, and thorium) and detached the symbolism of elements from Dalton's faintly alchemical and typographically awkward style introducing the much more practical alphabetical notation we use today, such as Ce for cerium, Se for selenium, and Th for thorium. Dalton was highly vexed by this foreign intrusion into his domain, and had the first of two strokes while arguing about his symbols with a colleague.

Patterns are hard to detect in jigsaws until one has a sufficiency of pieces. The first pattern in the properties of matter began to emerge in the 1820s, when the box of pieces was almost half complete. There were two aspects of this jigsaw, one being the qualitative properties of the elements, their chemical similarities and differences, and the other the emerging quantitative measure of the atoms of the elements, their atomic weights. Johann Döbereiner (1780–1849) of Jena, the untutored but observant son of a coachman who later sank to becoming a university professor, noticed something rather peculiar, which brought these two dimensions of the puzzle into harmony. He noticed that certain triads of chemically similar elements had atomic weights such that that of one of the elements was close to the average of the other two. For example, chlorine, bromine, and iodine are chemically similar and their atomic weights are respectively 35, 80, and 127 (the mean of 35 and 127 being 81). Döbereiner found three such triads, and thus emerged the thought that the elements formed, for whatever reason, a tapestry.

The hunt for organization was under way. I don't intend to give a detailed history of the hunt or give due recognition to all the personalities, for I am more concerned with outcomes than endeavours. But two contributors are worth inviting on to the stage. John Newlands (1837–98) was of Anglo-Italian descent, and like Cannizzaro was sufficiently fired by nationalistic enthusiasms to go off at twenty-three to Sicily to fight with Garibaldi's Red Shirts. His oats well sown, he returned to England and identified another component of the pattern. He saw that whereas Döbereiner had noticed only a scattering of triads, there was a more systematic pattern, for the lighter elements at least. Thus, he found that when the light elements were arranged in order of increasing atomic weight, then similarities of properties repeated each other after every eighth element (the gaseous elements helium, neon, and argon were unknown at the time). In retrospect unwisely, he likened this repetitiveness to the notes of the musical scale and referred to his 'law of octaves'. This fanciful analogy cost him dear, for he was taunted for suggesting anything so outrageous and presumably fortuitous, and others suggested that he tried arranging the elements alphabetically or by some other whimsical criterion.

Yet he was right. The properties of the early elements do repeat like the notes of the musical scale, but not for any musical reason. As we have seen, the structures of the atoms of the elements do repeat periodically as inner shells are completed and the pattern of orbital occupation begins afresh. But such theoretical authority was too far in the future to be of any help in the early nineteenth century when atoms were still in their conceptual infancy and the electron unknown.

The second personality is of course Dimitri Ivanovich Mendeleev (1834–1907), the youngest of eleven, fourteen, or seventeen children according to one's source (but at least a Döbereiner triad of numbers!), the son of a dealer in horses (*mjenu djelatj* = 'make exchange') and a mother of heroic endeavour on behalf of her brilliant and precocious youngest son. By the time that Mendeleev sat down to write his introductory general chemistry text, *Osnovy khimii* (Principles of chemistry), the number of known elements had risen to sixty-one. His problem was how to organize the material to present it in a logical, consistent way to his readers. It is at this point that happy anecdote would seem to diverge from stern truth.

The happy anecdote has it that Mendeleev was imbued with effort for days, perhaps weeks, on end to come to some logical organization. Exhausted by the effort he fell asleep on 17 February 1869[10] and saw 'in a dream a table where all the elements fell into place as required. Awakening, I immediately wrote it down on a piece of paper' (Fig. 5.9). Another agreeable component of the anecdote is that Mendeleev's enthusiasm for playing patience (solitaire) to while away long journeys led him to write the names of the elements down on pieces of card and to play with their organization. Such is the power of these images, and the precision with which the dream is dated, that they are generally thought to be true. However, that appears not to be so: the evidence appears to be contrary to both components of the happy tale. There was no dream, and even the plausible tale of the elements as patience appears to be embroidery rather than fact.

---

[10] Old (Julian) style. The date corresponds to 1 March on the Gregorian calender.

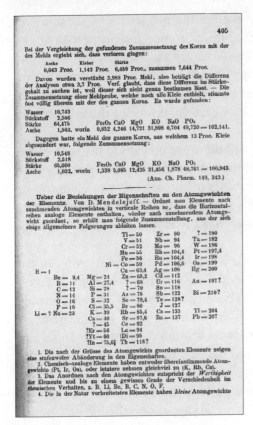

**Fig. 5.9** A facsimile of the page in *Zeitschrift für Chemie* (Journal of Chemistry) in 1869, in which Mendeleev announced an early version of the periodic table.

Whatever the truth may be, it is certainly the case that Mendeleev presented to the world a table, his *periodic table*, that brought the elements together in a kind of genealogical manner. He used atomic weights to order the elements, and found repeated similarities after periods of eight, eight, and eighteen elements. He had to fudge the table here and there (that is normally ascribed to chemical insight, but it sounds more like the method that Procrustes used on the victims too long for his bed). Thus, the order of elements based on atomic weight did not fit the pattern of chemical similarity everywhere, so Mendeleev ignored the order and chose his own. We now know that that procedure is valid, for atomic weight is not the best criterion for ordering the elements: the elements are best ordered by atomic number and, for reasons now fully understood, atomic weight does not quite follow the order of atomic number everywhere. There were also embarrassing gaps. But in this case embarrassment was turned to good account, for Mendeleev was so secure in his formulation of the table that he was able, by interpolation between the properties of the known, neighbouring elements, to predict the properties of these as yet undiscovered elements. Thus he predicted the existence and properties of elements he called eka-aluminium and eka-silicon (*eka* is Sanskrit for one), and these were later discovered by the French as gallium and the Germans as germanium, respectively.[11] He also made mistakes and predicted elements that do not in fact exist; but with the good will of a grateful posterity, those errors are largely forgotten.

We now know about 110 elements, and there are no gaps in the bulk of the table. We know that because the atomic numbers run smoothly from 1 to 110, with no omissions. There are sporadic reports of the discovery of elements up to 114, but some come and go, and 113 is still missing. This is the useless, 'academic' end of the periodic table, so the fact that it is frayed at this edge is of little practical concern.

The modern form of the periodic table is shown in Fig. 5.10. As you can see, it has been rotated through 90° from Mendeleev's arrangement, but the general features of his layout are easy to see. The vertical columns are called groups and the horizontal rows are called periods. Newland's octaves can still be heard in Periods 2 and 3, and Döbereiner's triads are scattered about here and there. The vertical groups contain elements that have considerable similarities, such as in the types of compounds they form, and show systematic variations from top to bottom. Elements in the horizontal periods show a smooth variation on going from left to right. For instance, the metals appear on the left of a period and the non-metals appear on the right. The elements in the long thin central section, such as iron (Fe) and platinum (Pt), are the transition metals, as they represent a transition between the very reactive metals, such as sodium (Na) and calcium (Ca), on the left of the table and the much less reactive metals, such as tin (Sn) and lead (Pb), on the right. The very thin twenty-eight element section placed underneath the table consists of the inner transition metals. This thin strip should really be incorporated into the main table, but that would make the table too long and spindly to be printed conveniently. The inner transition metals are all very similar in chemical properties and were among the most recent elements to be separated and identified. In fact, the very bottom line, following uranium (U), consists only of elements that have been made artificially.

---

[11] Before the intervention of international committees that insisted on sobriety in the nomenclature of elements, jokes—artfully concealed—were to be had. Thus although gallium is a name that its French discoverer, François Lecoq de Boisbaudran, may have chosen to inflate his countrymen's breasts with pride, *gallus gallus* is the Latin for cock, so his own breast was not uninflated.

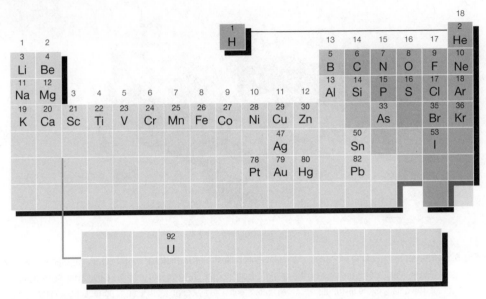

**Fig. 5.10** The modern form of the periodic table. I have shown only a few of the elements, those I judge to be reasonably well known or standing at the heads of their groups (K is potassium, Na sodium, Fe iron, Pb lead, and Sn tin: chemists burst into Latin occasionally). The numbered vertical columns are called *groups* and the horizontal rows are called *periods*. Hydrogen is (somewhat idiosyncratically but in my view sensibly) set off at the head of the table and ascribed to no group. The pale grey marks the metals, the dark grey the non-metals, and the mid-grey the *metalloids*, elements that hover between metals and non-metals in their properties. The two-row block of elements below the table should really by inserted in the position shown, but that makes the table too ungainly. The table is gradually growing as new elements are made.

The periodic table is still growing. Scientists are using particle accelerators to hurl the nuclei of one element against the nuclei of other elements, hoping that the two nuclei will merge and form the nucleus of an as yet unknown element. This is how element 112 (which has not yet been named) was made. However, the nuclei are very unstable, and the few atoms that were made had a very fleeting existence.

I hope you can begin to see why chemists regard the periodic table as their single most important concept. It summarizes the properties of the elements— the variation in their physical properties such as their density, the variation in the properties of the atoms, such as their diameters, and the variation in chemical properties, such as the number and type of bonds they form to other atoms (Fig. 5.11). At a glance we can see whether an element has the properties characteristic of a metal (iron), a non-metal (sulfur), or something in between (silicon). We can anticipate the chemical properties of an element by noting the properties of its neighbours and thinking about the trends expected down

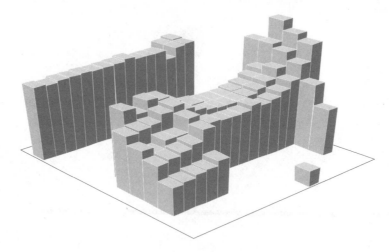

**Fig. 5.11** The periodicity of the properties of the elements is illustrated by this chart, which shows the diameters of the atoms. The smallest atoms are close to the top right-hand corner. The biggest atoms are close to the bottom left-hand corner. The details of the distribution are well understood. The size of an atom is an important criterion for determining the physical properties (such as the density) and the chemical properties (such as the number of bonds that an atom can form) of an element.

groups or across periods. In short, the periodic table is an extraordinarily succinct and useful summary of the properties of the elements and has considerable predictive power. We have come a long way since the original periodic table, with earth, air, fire, and water arranged in a simple square!

Mendeleev compiled his table empirically. He knew nothing of the structures of atoms and could have had no conception of the underlying foundation of the table. We have that understanding. The periodic table, we now know, is a portrayal of the rhythms of the filling of the energy levels of atoms, as depicted in Fig. 5.7.

We had a fleeting image of the roots of periodicity earlier in the chapter when we noticed the similarities between helium and neon on the one hand and between lithium and sodium on the other, and recognized that the electronic structures of their atoms are analogous: helium and neon have atoms with completed shells, lithium and sodium have atoms in which a single electron occupies an *s*-orbital outside a completed shell. This image is the root of the entire table. Thus, as we step from atom to atom, along the path of increasing atomic number, so each step increases the atomic number by one and therefore the number of electrons that must be accommodated. Each additional electron enters the next available atomic orbital subject to the requirements of the Pauli exclusion principle, that no more than two may occupy any one orbital.

This sequence matches the appearance of the periodic table. Thus, the atoms of elements in Groups 1 and 2 (the groups containing sodium and magnesium, for instance) are those for which an *s*-orbital is being occupied. An *s*-orbital can accommodate up to two electrons, which is consistent with there being two groups in this part of the table: in Group 1 there is one electron in the orbital; in Group 2 there are two. On the right of the table there is a block of six groups: in these elements electrons are in the process of filling the three *p*-orbitals of the appropriate shell of the atom: up to six electrons can occupy these orbitals: the elements of Group 13 (such as boron, B) have one such electron, those in Group 14 (such as carbon, C) have two, and so on, until the orbitals are full in Group 18, the almost completely inert and so-called *noble gases*. The narrow block in the middle of the table, the transition metals, consist of elements in which the five *d*-orbitals of the relevant shell are being occupied: five *d*-orbitals can accommodate up to ten electrons, which neatly accounts for the ten elements across each row of this block of groups. The inner transition elements are those for which *f*-orbitals are being occupied. There are seven *f*-orbitals in any shell, which accounts for the fourteen members of each row of this block.

We have come full circle. The chemists of the nineteenth century discerned family kinships among the elements. The full set of relationships was identified—in so far as the elements had been identified—by Mendeleev towards the end of the century. Yet his arrangement was empirical and there could be no understanding of *why* one element should be the cousin of another. How could it be that one sort of matter was related to another? Light flooded over this question once the structures of atoms became understood early in the twentieth century. Once the nucleus had been identified and the rules governing the arrangement of electrons had been established in the 1920s, it immediately became clear that *the periodic table is a portrayal of the solutions of Schrödinger's equation*. The table is mathematics made material. With two simple ideas—that electrons organize themselves so as to achieve the lowest possible energy, and that no more than two electrons can occupy any given orbital—the pattern of matter becomes understandable. Chemistry is at the heart of understanding matter, and at the very heart of chemistry lies its currency of discourse, atoms.

# SIX

# SYMMETRY

## THE QUANTIFICATION OF BEAUTY

### THE GREAT IDEA
*Symmetry limits, guides, and drives*

*Chrysippus holds that beauty does not consist in the elements
but in the symmetry of the parts*[1]

GALEN

C OULD it be that beauty is the key to understanding this beautiful world? The Greek sculptor Polyclitus of Argos, who flourished in about 450–420 BCE, laid down the foundations of our current understanding of the fundamental particles when, in his *Canon*, his guide to aesthetics, he wrote that 'the beautiful comes about little by little, through many numbers'. Polyclitus wrote of *symmetria*, the dynamical counterbalance of the relaxed and tensed parts of the human body and the relative orientations of these parts that results in a harmonious whole. Two and a half thousand years later we resort to the mathematical aspects of symmetry—and the symmetrical aspects of mathematics—to build our understanding of the fundamental entities from which matter is carved and the dynamical counterbalance of the forces that hold these entities together.

Provided by beauty we mean symmetry and the controlled loss of symmetry, Mondrian become Monet, then beauty indeed lies at the heart of the world. Some of this beauty is open to immediate apprehension, as when we look at a pleasing design. Some, though, is deeply hidden and not obvious to the untutored eye. The thousands of years since Polyclitus have been used to disinter this hidden beauty by casting the assessment of beauty into mathemat-

---

[1] Galen of Pergamum (129–199 CE), writing on Polyclitus.

ical form and then using mathematical tools to dig deeply into the landscape of reality. As I have already emphasized, as science has progressed it has increased its depth and reach by increasing the abstraction of its concepts. Nowhere is this transition more elaborated than in the discovery of symmetry and its deployment as an instrument of understanding.

I will now guide you, as carefully as I can, along this path from the immediate to the imagined, and show you the power that symmetry puts into our hands. The path will take us right to the edge of the brink of the unimaginable.

An object is symmetrical if an action—which we call a *symmetry operation*—carried out on it leaves it apparently unchanged. In other words, if you close your eyes for a moment, then when you open them again you are unable to say whether I have carried out an action or not. Think of a plain, undecorated ball; shut your eyes for a second; open them: have I rotated the ball? The actions we consider may be rotation round an axis or reflection in a mirror, but there are many other symmetry operations that we shall have to assess, some being elaborate combinations of more primitive actions, such as movement through space (that is called *translation*) followed by reflection in a mirror. You will find reflection in music. One particularly transparent example is the probably fake two-part composition by 'Mozart', which begins

and concludes:

the second part being the reflection of the first part.[2]

Some objects are more symmetrical than others. A sphere is highly symmetrical—one of the most symmetrical objects that we normally encounter. Think of the number of ways that I can change a sphere while your eyes are shut and

---

[2] It would be appropriate for this composition to be given the Köchel number 609. But the other A. Einstein, Alfred, lists this piece among the *Anhang* of doubtful pieces (Anh. 284dd).

which you cannot detect when you open them again. I can rotate it around any of an infinite number of axes passing through its centre, and the angle of rotation can lie anywhere between 0 and 360°. That's not all. I can think of a mirror passing through the centre of the sphere in any of an infinite number of orientations, and you won't be able to detect the reflection of one hemisphere into the other. There is another act I can carry out in my mind: I can imagine taking every atom of the ball in a straight line to the centre of the sphere and then moving the atom out an equal distance on the other side. In that way I reconstruct the ball by the action known as *inversion*. You can't tell that I have done it, as the ball looks exactly the same after it has been inverted as it did initially.

A cube is much less symmetrical than a sphere. Here are some of the actions I can carry out without you knowing. I can rotate a cube by 90° or 180° clockwise or anticlockwise around an axis passing through the centre of any three of the pairs of opposite faces (Fig. 6.1). I can rotate it by 120° clockwise or anticlockwise around any of the four axes passing through opposite corners. I can reflect it in any of the three planes where I can place the mirror to cut the cube in half. I can reconstitute the cube by inversion through its centre. I could even leave the cube untouched: you wouldn't know. So, doing nothing—which is called the *identity operation*—is also an operation I ought to include when considering the symmetry of an object. That makes many different actions I can do but you can't detect; so a cube is highly symmetrical, but nothing like as symmetrical as a sphere with its infinite number of undetectable actions.

In a certain refined sense, everything is symmetrical. That's because we include the identity in the symmetry operations we have to consider, and even the most unsymmetrical object we can examine—a screwed up sheet of newspaper—looks the same when we open our eyes after doing nothing to it. Now, that might seem like a cheat, which of course it is. However, the inclusion of the identity brings all objects into the scope of the mathematical theory of symmetry, so we can use symmetry arguments to discuss everything, not just

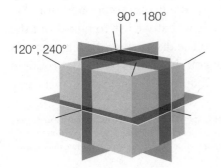

90°, 180°

120°, 240°

**Fig. 6.1** Some of the symmetry operations of a cube. A cube appears unchanged when we rotate it by 90° or 180° around an axis perpendicular to any face and by 120° or 240° about an axis passing through opposite corners. It is also apparently unchanged by reflection in any of the planes shown here. There are two other symmetry operations: inversion through the centre of the cube and the identity (doing nothing).

objects we think of as 'symmetrical'. Mathematics is like that: it generalizes definitions to increase the reach of its theorems as widely as possible. Of course, although everything is symmetrical (in this cheating sense), some things are more symmetrical than others. 'More symmetrical' simply means having more ways of being changed that, when we open our eyes, we cannot tell whether an action has been carried out or not. A sphere is more symmetrical than a cube, a cube is more symmetrical than a palm tree. As you can see, we are now able to order objects according to their degrees of symmetry: the flavour of symmetry is becoming numerical.

The mathematical theory of symmetry, where this flavour gets hardened into precise definitions and takes on a formal structure, is called *group theory*. The theory takes its name from the fact that the symmetry operations we have been talking about form what in mathematics is called a *group*. Broadly speaking, a group consists of a set of things and a rule of combination such that the combination of any pair of those things is also a member of the set. We can see why symmetry operations form a group by thinking about the cube again. Suppose I do two actions in succession, such as rotating the cube by 90° around one perpendicular axis and then rotating the resulting cube by 120° around a diagonal axis. The result is the same as if I had rotated the cube by 120° around one of the other diagonal axes, so the two operations carried out in succession is equivalent to a single symmetry operation. That is true of all the symmetry operations of a cube, so these actions form a group. The groups of symmetry operations of different shapes are given names. The huge symmetry group of a sphere, for instance, is called SO(3). Later we shall meet other groups with names like SU(2) and SU(3).[3]

The concept of group extends far beyond symmetry operations, which is why group theory is such a powerful part of mathematics. For instance, take all the positive and negative integers . . . –3, –2, –1, 0, 1, 2, 3, . . . as the set of 'things' and let the rule of combination be addition. Then, because the addition of any two integers is itself an integer, the integers form a group under addition. So, arithmetic is a part of group theory, and the ideas that we use to talk about the symmetries of actual objects can be applied to the discussion of ideas in arithmetic, and vice versa. I am not going to take you down this particular route in this chapter, but it will have a role to play in Chapter 10. Meanwhile, just carry

---

[3] The names reflect some of the technical properties of the groups which it would be inappropriate to go into, except to say that O denotes 'orthogonal' and U denotes 'unitary'; S denotes a particular 'special' version of these groups. The 3, at least, is easy to understand: it refers to symmetry operations carried out in our normal three dimensions.

away the thought—a thought that pervades this entire book—that a simple idea can have applications of almost unbounded generality.

Let's get back to a consideration of symmetry itself. We need to distinguish groups of symmetry operations that leave one point of an object unchanged from groups that involve motion through space. The former are called *point groups*, the latter *space groups*. All the symmetry operations of a sphere, and those of a cube, leave the point at its centre in the same location as it was initially. If an action moved the centre point of an individual object, such as when a sphere is reflected in a plane that does not pass through its centre, we could tell that something had been done, and the action would not be a symmetry operation. All the symmetry operations of individual objects leave at least one point unchanged, so the symmetries of individual objects are described by point groups.

Patterns that extend through space are described by space groups. Here we have to cheat a bit, and think of the pattern as extending for ever in any direction, or think of ourselves as so short-sighted that we can't see what happens at the ends of the patterns. Patterns that extend effectively for ever in one dimension are called *frieze patterns*, because they show the symmetry properties typical of friezes. The formal definition of a frieze in classical architecture is the horizontal band forming a central part of the entablature, the part of the structure supported by columns, lying between the architrave and the cornice. More informally, a frieze is any horizontal decorated band with a motif that is repeated regularly throughout its length. Here the slumbering giant of group theory opens one eye and provides us with the first of its remarkable insights: *there are only five possible varieties of frieze*. All the friezes that have ever been built and that ever can be built can be classified as one of five different varieties (Fig. 6.2). Of course, the motifs may be different—archers, diamonds, goats, squiggles—but provided the pattern is repeated periodically (which rules out as frieze-like the Elgin Marbles, which are not repetitive) their arrangement in space is limited to these five different varieties.

This is the first glimpse we have of the vertiginous depths that group theory can plumb. By taking a colossal intellectual leap (which I intend to lead you through in smaller steps as this chapter unfolds, but it will be helpful to know were we are heading), we can perhaps begin to accept that just as symmetry limits the number of possible patterns in space, it may be the case that the sym-

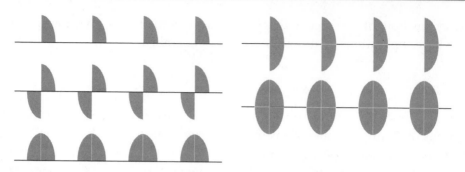

**Fig. 6.2** These shapes symbolize the five frieze patterns that are allowed in one dimension. There are many different designs, because the quadrant shown here in various orientations can be replaced by any feature, but these five patterns are the scaffolding of all possible regular friezes.

metry of spacetime—whatever that means—limits the number of types of elementary particle that can exist. Symmetry limits.

As architecture advanced from Greek temple to bungalow, the demand for classical entablature understandably dwindled and frieze gave way to wallpaper. Patterns on wallpaper extend indefinitely in two dimensions, and varieties of those patterns with different motifs—stripes, roses, peacocks—and colours fill the sample books of interior decorators and wallpaper manufacturers. Group theory, though, reveals an awesome truth: *there are only seventeen varieties of wallpaper pattern.*

We can be a little more precise. By a *net* we mean an array of dots that represents the location of the peacock or whatever taste decrees to be the motif. The wallpaper pattern is a combination of the motif and the net. Thus, alternate dots of the net may have peacocks all upright or alternate dots may have peacocks perched upright and perilously inverted. With this distinction in mind, group theory shows that *there are only five types of net* and *seventeen combinations of net and motif* (Fig. 6.3). It is an interesting exercise to examine the wallpapering of the rooms you visit, the paving of the courtyards you cross, the tiling on roofs, or even the design (if it is periodic) of your tie, to exercise your ability to identify the net (that's normally easy) and the overall pattern (much trickier as some motifs are so elaborate). You will never find a repeating pattern that is not one of the seventeen that group theory identifies as the total universe of wallpaper designs.

Now consider three-dimensional packing patterns filling all space. Everyday examples include one of the simplest of all patterns, in which cubic sugar cubes are packed together in a box or—with slightly lower symmetry because the stacked entities are no longer cubic—matchboxes are stacked together (Fig. 6.4).

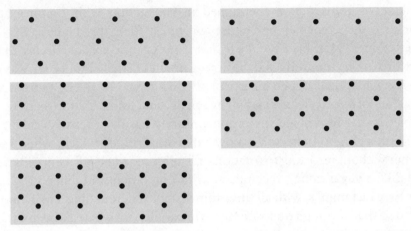

**Fig. 6.3** These patterns show the five nets possible for two-dimensional wallpaper. Images can be attached to each of the points to generate the actual design, but even then it turns out that there are only seventeen possible outcomes.

Here we can see that we can ascribe different symmetries according to the detail that we examine, for featureless matchboxes stack to give one symmetry, but taking account of the design on the box, and perhaps the orientation of the matches within the box, leads us to ascribe a slightly lower symmetry to the packing.

How many three-dimensional patterns are there? Here we can uncover different symmetries by asking different questions. In an early example of the

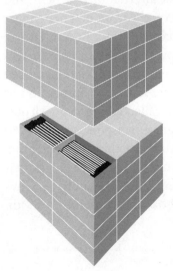

**Fig. 6.4** Two of the stacking arrangements in three-dimensional space. The upper diagram shows cubic unit cells ('sugar cubes') stacked together. The lower diagram shows rectangular unit cells ('matchboxes'). There are seven overall shapes of unit cell that can be stacked in this way to give a periodic structure. The cells themselves may contain entities that affect the overall symmetry: we have shown the interiors of two matchboxes, revealing that alternating boxes have matches pointing in different directions.

technique of transduction mentioned in connection with Dalton's atomic hypothesis, the French mineralogist and priest René-Just Haüy (1743–1822) suggested in 1784 in his *Essai d'une théorie sur la structure des cristaux* that the outer form of crystals reflected the arrangement of smaller units. He was led to this view when he dropped a particularly fine crystal of calcite (a transparent crystalline form of calcium carbonate, chalk) and found that it shattered into tiny pieces that resembled the original. Rarely has destructive accident been turned to such good effect. We now call a small block which, when stacked together without resorting to rotations, fills all space a *unit cell*. Unit cells may be cubic (like a sugar cube), rectangular with one dimension different from the other two, rectangular with all three dimensions different (like a matchbox), or skewed so that although opposite faces are parallel (they have to be, for the unit cell to pack to fill all space), they are not perpendicular to their neighbours. It turns out that there are just *seven* basic shapes of these unit cells.

Just as we identified five nets for wallpaper by noting the location of dots where later the motifs would lie, so we can do the same for unit cells. The resulting arrangements of dots allowed in three dimensions are called *Bravais lattices*, after the French mountaineer, adventurer, and physicist Auguste Bravais (1811–63) who first listed them in 1850. It turns out that there are only *fourteen* of them (Fig. 6.5).[4] Whenever you see objects stacked together to fill all space in a regular manner, such as tin cans in boxes, eggs in layers of trays, and fruit in displays, they all conform to one of these fourteen arrangements.

Just as we can get seventeen basic sorts of wallpaper by ascribing a motif to the net of points in different ways (upright peacocks, alternating peacocks, and so on), so we can attach a motif (such as the design on the front of a matchbox, or the way the matches within are arranged) to each point of a Bravais lattice. Careful consideration of the patterns that arise show that *there are only 230 possible arrangements*. I appreciate that 'only' might seem inappropriate here; but the point is that the number is finite and precisely identified: the number is not 228, or 229; it is exactly 230. These arrangements are called *space groups*. All possible three-dimensional, space-filling periodic designs correspond to these 230 space groups. The packing of uniform, undecorated matchboxes containing matches all pointing in the same direction corresponds to one space group, that of the same matchboxes in the same arrangement but containing matches pointing in alternating directions in neighbouring boxes corresponds to another space group.

---

[4] A site where you can rotate the unit cells to see them from different viewpoints is http://www.minweb.co.uk/bravais/bravais.html.

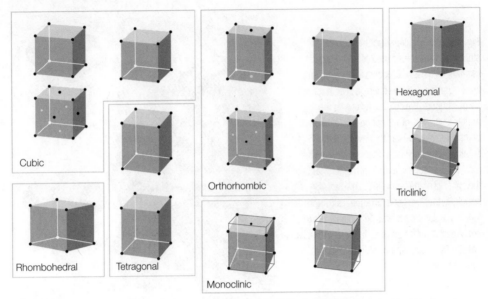

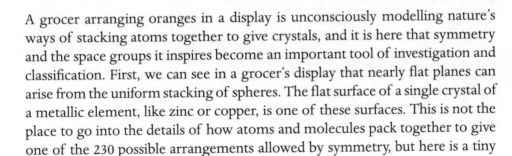

**Fig. 6.5** The three-dimensional analogues of wallpaper nets are the Bravais lattices. There are fourteen Bravais lattices in three dimensions. An entity can be associated with each point in a variety of ways, but no more than 230 arrangements are possible.

A grocer arranging oranges in a display is unconsciously modelling nature's ways of stacking atoms together to give crystals, and it is here that symmetry and the space groups it inspires become an important tool of investigation and classification. First, we can see in a grocer's display that nearly flat planes can arise from the uniform stacking of spheres. The flat surface of a single crystal of a metallic element, like zinc or copper, is one of these surfaces. This is not the place to go into the details of how atoms and molecules pack together to give one of the 230 possible arrangements allowed by symmetry, but here is a tiny taste.

If we think of atoms as hard spheres, like ball bearings, we can imagine a layer of these atoms lying close together, with each sphere surrounded by six neighbours (the maximum number possible for identical spheres). A new layer can be formed by putting an atom into each of the dips in the first layer (Fig. 6.6). A third layer can be formed in either of two ways. In one, we put the atoms in the dips that lie over the positions of the atoms in the first layer; in the second we put them in the dips that lie over the gaps in the first layer. If we denote the

**Fig. 6.6** Two regular structures can be built by stacking hard spheres (representing atoms) together as closely as possible. In the lowest level (light grey), each sphere is in contact with six neighbours. We call this level A. In the middle level (mid-grey), spheres lie in the dips of the first layer. We call this level B. If the spheres of the next layer (dark grey) lie in the dips of the second layer that are directly above the spheres of the first layer, to give an ABA structure, then we get a hexagonal structure (upper part). If the spheres lie in dips that are not directly over the spheres in layer A, then we get an ABC arrangement, which has cubic symmetry.

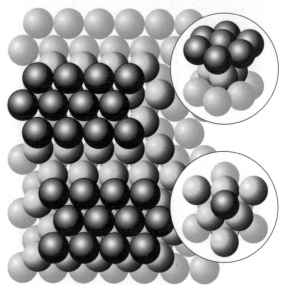

layers by A, B, C, . . . , then the first arrangement is ABABAB . . . and the second is ABCABC . . . . If you look carefully at the first arrangement of spheres, you should be able to pick out a hexagonal arrangement, a hexagonal unit cell. In the second arrangement, you should be able to pick out a cubic arrangement (this is a bit harder to identify as the cube is skewed over the planes). So these two ways of stacking atoms result in crystals with different symmetries. Some of the metals that form hexagonal unit cells are cobalt, magnesium, and zinc. Metals that form cubic unit cells include silver, copper, and iron.

The symmetry of a unit cell affects the mechanical, optical, and electrical properties of solids. For instance, the rigidity of a metal depends on the presence of *slip planes*, which are planes of atoms that can slip over each other when subjected to stress, such as a hammer blow. When the sheets of atoms in Fig. 6.6, or the unit cells, are examined carefully, it turns out that the hexagonal form has only one set of slip planes (they lie parallel to the planes shown in the illustration), whereas the cubic form has eight sets of slip planes in different directions. As a result, metals with the hexagonal structure (zinc, for instance) are brittle, whereas metals with the cubic structure (copper and iron, for instance) are malleable—they can be bent, flattened, drawn, and pounded into different shapes relatively easily. The electrical industry depends on the ductility of copper, and the transport and construction industries depend on the malleability of iron.

As we have seen in other contexts, it is always fun, and often useful, to extend our thinking to higher dimensions. This extension is sometimes essential, such as when we consider the four dimensions of spacetime. The question arises, then, as to the numbers of patterns that can be found in spaces of higher dimension. Mathematicians have explored this question. There are 'only' 4783 space groups in four space dimensions, so five-dimensional beings (who need four-dimensional wallpaper to decorate their hypercubic rooms) will find a much wider variety of patterns in their wallpaper hypermarkets than we three-dimensional beings do.

Not all symmetries are obvious, and it is at this point that I want to edge us into beginning to appreciate the beauty of greater abstraction. It is inevitable that from now on the discussion becomes more abstract and the concepts harder to visualize; but we shall negotiate these tricky shoals slowly and with care, and you will be pleased to discover that you can understand. Here we shall see that symmetry is no longer merely descriptive but becomes powerful, for it is the source of laws. Symmetry guides.

We have already seen an example of the guiding, controlling power of symmetry. We saw in Chapter 3 that the conservation of energy is a consequence of the uniformity of time. That time is smooth, lacking lumps—more formally, that time is translationally invariant—implies that energy is conserved. We also saw that the conservation of linear momentum is a consequence of the smoothness of space—that space is translationally invariant in the absence of forces—and that the conservation of angular momentum is a consequence of the isotropy of space—that space is rotationally invariant in the absence of torques. The lack of lumpiness of space and time is an aspect of their symmetry, so we see that these powerful conservation laws stem from symmetry. Emmy Noether (1882–1935), the most outstanding and influential woman mathematician the world has yet produced, established the hugely important result now known as *Noether's theorem*, that where there is a symmetry, there is always a corresponding conservation law.

Some symmetries are hidden from direct observation, but have consequences nevertheless. At this point, all I am asking you to do is to notice coincidences and wonder if they are the consequence of symmetry. A sign that symmetry lurks hidden beneath the surface of appearances is the exact equality of energies of different arrangements of particles: if two arrangements are related by a symmetry operation, then the energy of those two arrangements is the same. We encountered one particularly satisfying example in Chapter 5 when we saw that in a hydrogen atom, the energy of an electron is exactly the same when it occupies an *s*-orbital as when it occupies any of the three *p*-orbitals of the same shell. An *s*-orbital is spherical and a *p*-orbital has two lobes, so although it is easy to see that one *p*-orbital can be rotated into another *p*-orbital, it is far from obvious that a *p*-orbital can be rotated into an *s*-orbital. I mentioned then that the potential energy—the energy arising from the location of the electron in the electric field of the nucleus, the so-called *Coulomb potential energy*—is especially beautiful, and I can now explain what I meant.

The Coulomb potential energy is spherically symmetrical. That is, wherever we put the electron at a given distance from the nucleus—at the North or South Pole, on the equator, or anywhere in between—its potential energy is exactly the same. The potential energy varies with distance from the nucleus, but for a given distance it is independent of angle. This spherical symmetry tells us that symmetry transformations of the atom include rotations through any angle about any axis, just like the symmetry operations of a sphere. That being so, the three *p*-orbitals can be rotated into each other by a symmetry operation of the sphere, so their energies are the same. However, it still appears that we cannot rotate an *s*-orbital into a *p*-orbital.

Now here's an extraordinary fact: the Coulomb potential energy is gorgeous in the sense that it has rotational symmetry not only in three dimensions (as we have seen) but also in four. This higher symmetry means that there may be a rotation in four dimensions that transforms a three-dimensional *s*-orbital into a three-dimensional *p*-orbital. If that is so, and we can rotate the different varieties of orbitals into each other, then they will have the same energy.

I appreciate that expecting you to think in four dimensions is beyond the call of duty (at least until we get to Chapter 9), and I will not demand that of you now. Instead, I shall use a simple analogy. Think of a sphere resting on a plane. The plane represents our three-dimensional world and the sphere represents a four-dimensional world of which we see only a projection. Suppose we colour the northern half of the sphere black and the southern half white. We can draw a line from the North Pole and project it through the surface of the sphere on to the plane. This projection of the patterned sphere looks like a circle (Fig. 6.7). Now rotate the sphere through 90° into the position shown in the second part of the illustration. The new projection divides the plane into halves, one black, the other white. Another orientation of the sphere is shown in the third part of the illustration, and has a similar projection but rotated by 90°. We Flatlanders find it perfectly plausible that the second and third projections can be interrelated by a rotation, so we are not surprised that these '*p*-orbitals' have the same energy. We find it really puzzling, though, that they can be transformed

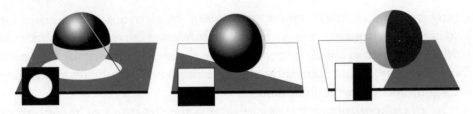

**Fig. 6.7** A representation of how it is possible to rotate *s*- and *p*-orbitals into each other by stepping up a dimension. The orbitals are represented by the patterns in two-dimensional space. If we accept that these patterns are projections of a sphere in three-dimensional space on to the two-dimensional space, then we can see that rotating the sphere interchanges the patterns in two dimensions. The Coulomb potential has four-dimensional symmetry, and permits this type of rotation to be done.

into the first, circular shape, so we cannot see that an 's-orbital' has the same energy as the two *p*-orbitals. The observer in three dimensions has no such trouble: that observer sees all our Flatland patterns as projections of a sphere related by simple rotations. Exactly the same reasoning applies to the orbitals of a hydrogen atom, and we see that the equality of energies of apparently unrelated orbitals is a consequence of there being symmetry hidden away in a fourth dimension.

Here is another very powerful consideration that will bear fruit shortly. The energy of an electron in an *s*-orbital is not *exactly* the same as that of an electron in a *p*-orbital. Scientists know why: there are very weak magnetic interactions between the orbital motion of the electron and its spin, which shift the energies slightly. This is an example of *symmetry breaking*, a process in which although there is a symmetry relation in the background, other weak rogue interactions cause the energies of different states to differ from each other. Another way of looking at the effect of symmetry breaking is to remember that according to Einstein's theory of special relativity, energy and mass are equivalent ($E = mc^2$, Chapter 9), so we could express the difference in energies of electrons in *s*- and *p*-orbitals as a difference in mass. In other words, *mass differences stem from symmetry-breaking interactions*. The energy difference in this case is very tiny, so the mass difference arising from symmetry breaking is exceedingly small, at only $1 \times 10^{-37}$ g; shortly, though, this wholly negligible difference will blossom into a really important point.

The spectacular beauty of the centrosymmetric Coulomb potential energy, which must be the most gorgeous type of potential energy imaginable, is lost as soon as a second electron is present in the atom. As we saw in Chapter 5, the energy levels of a hydrogen atom are a good first approximation to the energy levels of all the other atoms. Then, provided we allow for the changes in energy arising from the electrical repulsions between the electrons (which leads, for instance, to electrons in s-orbitals having slightly lower energy than electrons in p-orbitals), the structure of the periodic table emerges automatically. However, there is another, more sophisticated, symmetry-based way of understanding the significance of the periodic table.

To a first approximation, we can express the structures of the atoms of all the elements in terms of the occupation of strictly hydrogen-like atomic orbitals. Because the energies of the orbitals in any shell are the same, that approach results in a funny kind of periodic table because the s- and p-orbitals (as well as the d- and f-orbitals) of a shell all have the same energy; so we lose the structure of the table and there appears to be no reason for the different chemical personalities of the elements. If you like, you could think of the groups of the table (the vertical columns) as undifferentiated and stacked on top of one another. However, because the electrons do in fact interact with each other and break the four-dimensional symmetry of the Coulomb potential; the s- and p-orbitals of a given shell do not have the same energies. Once we allow for this symmetry breaking, the periodic table crystallizes out into the form that we know (Fig. 6.8). The chemistry portrayed by the periodic table is therefore a portrayal of the underlying four-dimensional symmetry of the Coulomb potential energy broken by the interactions between all the electrons present in each atom. From this point of view, chemistry, at root, is a portrayal of symmetry and its breaking, a loss of perfect symmetry that endows the chemical elements with their personalities. Mendeleev knew little of symmetry, nothing of hidden

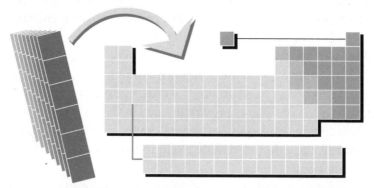

**Fig. 6.8** This is a fanciful depiction of the structure of the periodic table. If we disregard interactions between electrons, each electron experiences the highly symmetrical Coulomb potential of the nucleus, and the periodic table has no structure (the periods are intact): this is represented by the stack of groups on the left of the illustration. Once we allow symmetry breaking (that is, take into account the electron–electron repulsions), the groups fan out into the familiar structure of the periodic table.

symmetry, and even less of symmetry breaking. He would, I hope, have been enthralled by the thought that his table is the portrayal of the consequences of the symmetry-broken Coulomb potential energy.

There is more. We also saw in Chapter 5 that electrons are prevented from crowding into the same orbital by the Pauli exclusion principle, which forbids more than two electrons to enter any one orbital, and if two electrons do occupy one orbital, then their spins must be paired (one spin clockwise, the other anticlockwise). This principle is also rooted in symmetry, so the form of the periodic table, the fact that atoms have bulk, and the observation that we are distinct from our surroundings, are all rooted in symmetry. As we shall now see, the symmetry underlying Pauli's principle is of a rather subtle kind, but one that is not difficult to grasp.

Because, according to quantum theory, we cannot follow the trajectories of electrons, any electron in the universe is completely indistinguishable from any other electron.[5] This indistinguishability suggests that if we were to interchange any two electrons in an atom, then all the properties of the atom ought to remain the same. In other words, electrons should show *permutation symmetry*.

At this point I need to generalize the concept of orbital a little and anticipate one or two aspects of the more extensive discussion in Chapter 7; if the discussion here troubles you, come back to it after reading through the first half of that chapter. We have seen that an orbital tells us the probability of the location of an electron in an atom. An orbital is a special case of a *wavefunction*, which is the solution of the Schrödinger equation for any particle in any sort of environment, not just electrons in atoms. We shall use this more general term from now on. The second point we need to know is that the probability of finding a particle at any point—which so far we have represented by the density of shading—is given by the *square* of the value of its wavefunction at that point.[6] One implication of this interpretation is that a wavefunction and its negative have the same physical significance (because their squares are the same). This leaves open the possibility that the wavefunction might change sign when two electrons are interchanged: we simply wouldn't notice. That is in fact the case. Pauli found that he could account for certain details of the radiation emitted by

---

[5]  Richard Feynman, in a telephone call to John Wheeler, or vice versa, once half jokingly suggested that the reason why all electrons look the same is that there is only one electron in the universe, and what we think of as a lot of electrons is actually a slice through that one electron's path as it whizzes backwards and forwards in time. That would certainly be an economical universe.

[6]  The problem of the interpretation of wavefunctions and of probabilities is discussed at length in Chapter 7.

atoms only if the wavefunction for the atom changed sign when any two electrons were interchanged. We say that the wavefunction must be *antisymmetric* (that is, change sign) under electron interchange. The Pauli exclusion principle, that no more than two electrons can occupy any atomic orbital, follows from this deeper requirement, so the structure of atoms, their bulk, and our own bulk stem from symmetry. Symmetry swells.

Now we are ready to ramp up one more notch of abstraction, and by now I hope your mind is prepared. Almost everything we have talked about so far has concerned symmetry operations that take place in space. But there is more to life than space. At this point we have to direct our attention to the *internal* symmetries of particles, symmetries that refer to actions we can perform to a particle pinned to one point of space, like a butterfly in a display, not able to move through space, to be reflected, rotated, or inverted.

Some of these symmetries—they will turn out to be near-symmetries, broken symmetries—are reasonably easy to imagine. We will begin with the two components of the nucleus we encountered in Chapter 5, the proton and the neutron, which are jointly called *nucleons*. These two particles are suspiciously similar: they have a similar mass (the neutron is slightly heavier; that is, it has a slightly higher energy) and both have the property we know as spin. The principal difference between them is that the proton is charged but the neutron is not. If for the moment we disregard the difference in mass, the two particles are twins. That is, there is a symmetry between them. Particle physicists think of this symmetry in terms of a property called *isospin* (because its properties are analogous to spin itself). Clockwise isospin corresponds to electric charge 'on' (the proton); anticlockwise isospin corresponds to electric charge 'off' (the neutron). The two particles are really the same: one (the proton) is a nucleon with clockwise isospin, the other (the neutron) is a nucleon with anticlockwise isospin. To convert a proton into a neutron, all we have to do is to reverse its isospin.

To a first approximation, the properties of a nucleon are independent of the direction of its isospin. However, the symmetry between clockwise and anticlockwise isospin is not perfect, and is broken weakly by other interactions, such as the interaction of the nucleon with electromagnetic fields. The energy of interaction of an electromagnetic field is different for clockwise and anticlockwise isospin. Consequently, the masses of the two states of the nucleon are slightly different, with the anticlockwise isospin state of a nucleon (a neutron) turning out to be slightly more massive than its clockwise isospin (proton) state.

The identification of isospin (by Heisenberg) is like the discovery of triads of related elements by Döbereiner two centuries earlier (Chapter 5). Döbereiner identified fragments of an overall pattern, a pattern that in due course was identified by Mendeleev and seen to be a portrayal of an underlying symmetry broken by weaker interactions. Could it be that all the elementary particles are related by symmetry, and that their different masses are a consequence of symmetry breaking? Is there a periodic table of the elementary particles, and is that table rooted in symmetry and its partial loss?

We need to step back a little. Mendeleev was able to compile his periodic table because he had access to information about a high proportion of all the elements. Likewise, we need to enter the particle zoo to see what lies within. Döbereiner was unable to make progress beyond his triads with the tiny amount of data he had to hand; we shall make progress beyond isospin only after we have enough data to display a more extensive pattern.

Particle physicists are vandals in spades. In the interests of furthering civilization, they take one piece of matter, hurl it furiously at another, and poke around in the shattered remains that emerge from the collision. As you might suspect, the greater the violence of the impact, the smaller—and presumably more fundamental—the fragments. The particle accelerators used to smash particle into particle are the realization of the dreams of the ancient Greeks, for with them we can hope to cut to the end of matter.

You will be alert to a problem. What splinters are chipped off matter depend on the vigour of the collision. Perhaps we will never be sure that we have reached the end of cutting, for more cutting might be achieved by building a bigger accelerator (in this business, size really does matter, for vigour goes with bigness). Indeed, as we approach the end of this chapter, we shall see that it does appear to be the case that to test our understanding of the ultimate underworld, we should build an accelerator that spans the universe and absorbs in cost and resources more than the sum of the outputs of economies everywhere.

With that thought in mind, we might be at the stage attained by Dalton two centuries ago when he summoned enough energy—puny, chemical energy—to arrive at atoms and was able to build theories based on their personalities regardless of their inner workings. Like climbers of an inverse Everest, science is content to pause at various landing stages on its journey down into the depths, and not to hasten too deeply into the unknown. Atoms were adequately fundamental for the Victorians; our elementary particles will similarly be taken to be sufficiently fundamental for us. In other words, for the time being (but not at the end of the chapter), let's accept that the current zoo of particles is the actual zoo, or at least a sufficiently fundamental zoo, and let's meet the animals our hunter-vandals have captured since atoms were first torn apart in 1897 and nuclei surrendered to attack in 1919.

When we think of particles, we think of their constituent parts and the forces that hold those parts together, the glue. Scientists have found one force to be responsible for all the interactions. Well, that is an exaggeration: to be more precise, they believe there is only one force acting in the universe, which is nicely economical, but that this force manifests itself in five different ways. Three of these manifestations—electrical, magnetic, and gravitational—are familiar to us from everyday life. The two other manifestations—the weak and strong forces—are totally unfamiliar.

One of the great achievements of nineteenth-century science was the demonstration by the Scottish scientist James Clerk Maxwell (1831–79)[7] and published in his *A dynamical theory of the electric field* (1864) that the electric and magnetic forces are best thought of as the two faces of a single *electromagnetic force*. Maxwell based his theoretical work on the results obtained by the experimentally brilliant but mathematically inarticulate Michael Faraday (1791–1867), who had earlier introduced the concept of *field* into physics as the region of influence of a force. Broadly speaking, the electrical force acts between all charged particles and the magnetic force acts between charged particles in motion, such as currents of electrons in neighbouring coils of wire. One of the hugely important fruits of this unification of two previously disparate forces was Maxwell's elucidation of the hitherto puzzling nature of light and his demonstration that it was electromagnetic radiation, his field in flight. This realization was confirmed in 1888 when Heinrich Hertz (1857–94) produced and detected radio waves: the result is the history of modern communication. A second, intellectual fruit was the theory of relativity, which emerged when Maxwell's equations were subjected to Einstein's perceptive gaze (Chapter 9).

A third fruit fell from the same tree in the early nineteenth century when the concept of the photon—a packet of electromagnetic energy—was developed

---

[7] An incidental and irrelevant point of information is that Maxwell's father John was born plain Clerk, but added Maxwell upon inheriting an estate in Kirkcudbrightshire from his Maxwell ancestors.

by Einstein in 1905 (see Chapter 7) and named by the American chemist G. N. Lewis in 1916. The photon was the first of the *messenger particles* to be identified, particles that carry a force between the originating and responding particles, such as two electrons or an electron and a nucleus. The photon is the messenger particle for the electromagnetic field, carrying the force between the interacting particles and travelling at the speed of light.

We need to note two properties of photons at this stage, for they will be relevant later. A photon is massless and, like an electron, it has a spin that can never be extinguished. For technical reasons related to the quantum mechanical description of spin, an electron is ascribed half a unit of spin angular momentum. A photon is likewise ascribed one unit of spin. Particles with half-integral spin (that includes protons and neutrons as well as electrons) are called *fermions* after the Italian physicist Enrico Fermi (1901–54), who discovered how to describe collections of them as well as leading the project to build the first nuclear reactor as part of the wartime Manhattan Project. Particles with integral spin are called *bosons* after the Indian physicist Satyendra Nath Bose (1894–1974) who studied the statistical properties of systems consisting of large numbers of them, such as a box full of light or a sunbeam. It will turn out that all fundamental particles of matter are fermions whereas all messenger particles are bosons. A very deep description of matter, therefore, is that it is a collection of fermions bound together by bosons.

Any reflective star-crossed lover should be able to tell you that the photon is massless, for that we can see the stars is a direct consequence of its masslessness. The chain of argument is broadly as follows. First, we saw at the end of Chapter 3 that particles that live for very short periods have large uncertainties in their energy. Next, for a messenger particle of given mass to come into existence, it must borrow an energy proportional to its mass (from $E = mc^2$): heavy particles correspond to the presence of a lot of energy. A particle can come into existence without being caught by the energy-conservation police only if it lives for a time so short that the theft is concealed by the uncertainty in any audit of the energy. It follows that a heavy particle can come into existence without being caught by the energy-conservation police only if it lives for a very short time (you might get away with stealing a billion dollars for a picosecond). Now for the third link in the argument. During the time it exists, the messenger particle flies along at high speed, and the distance it can travel is proportional to how long it is allowed to survive.[8] A heavy messenger particle, with its very short lifetime, can't travel

---

[8] The range of a force is related to the mass of its messenger particle by *range = (Planck's constant)/(mass × speed of light)*. If a photon were as heavy as an electron, light could travel only $10^{-13}$ m from its source.

far. Conversely, for the messenger to travel infinitely far, it has to live for ever, which it can do without getting caught by the energy-conservation police only if it didn't steal anything in the first place. That is, it must be massless. It follows that, for electromagnetism to have an infinite range, a photon must be massless. If photons had mass, electromagnetic radiation would be unable to travel great distances, and we would not see the stars; our lover would not be star-crossed. If photons were really heavy, atoms would fall apart because the pull of the nucleus would not be able to reach out to grip the electrons.[9]

The third familiar force is gravity. Gravity acts between all particles but is hugely weaker than the electromagnetic interaction. For instance, the gravitational interaction between two electrons is $10^{42}$ times weaker than their electromagnetic interaction. If the gravitational force could move a one-milligram flea, the electromagnetic force could move a million Suns. That we are not overwhelmed by electromagnetism and can experience gravity follows from the fact that the universe consists of equal numbers of positively and negatively charged particles, so the attractions and repulsions cancel on a cosmic scale. Gravitation, though, is entirely additive: there is only gravitational attraction, there is no repulsion, so there is no cancellation. All the particles of the universe conspire weakly together and we experience the strength of their combined pull. Locally, electromagnetic forces are paramount: your shape is determined largely by electromagnetic forces, and the fact that you are not pulled into a formless puddle on the ground is due to the greatly superior strength of electromagnetism compared with gravitation.

There is thought to be a messenger particle for gravitation. At least, it has been named—the *graviton*—but not yet detected for it interacts so weakly with matter. A graviton is a massless boson, like the photon, but spins twice as fast. That gravity pulls over almost infinite space is a sign that the graviton is massless. Any brilliant sailor should be able to tell you that a graviton's spin is 2, for there is a trail of subtle argument that connects this double rate of spin to the fact that in our oceans there are two tides a day.

Now we come to the two unfamiliar forces, the *strong force* and the *weak force*. Unfamiliar they may be, but a thoughtful person should be able to infer the existence of a strong force. The argument is as follows. A nucleus consists of protons and neutrons packed together into a very tiny volume. The electromagnetic force is repulsive between the protons (because they have the same charge, and like charges repel), so there is a strong tendency for a nucleus to explode.

---

[9]  For precision, I need to say that this argument applies only to so-called *virtual particles*, the particles that convey force; actual particles of any mass can travel over huge distances and convey information.

(Some—those of the radioactive elements, like radium—do; for exactly that reason.) What holds the protons in nuclei? Moreover, why don't the uncharged neutrons just fall out? What holds them in? Neutrons are not affected by any electrical force, so they must be attracted by something else. In short, because most nuclei don't explode and because most hold on to their neutrons, there must be a force that is stronger than the electromagnetic force and which acts between protons, between neutrons, and between protons and neutrons. Furthermore, because all matter in the universe has not been wound down into one huge nucleus, this strong attractive force must have a very short range—no more than about the diameter of a nucleus.

Here I have to inject a note of caution. Neutrons and protons are composite particles built of *quarks* (see below).[10] What we should really consider is not the net interaction between nucleons—the overall outcome of the attractions between some components and the repulsions between others—but the more detailed interaction between their individual components. There may be a profound difference. For instance, you and I, even in a close embrace, have virtually zero net electromagnetic force acting between us even though the nuclei of our atoms repel each other strongly and our electrons also repel each other strongly: these strong repulsions are cancelled by the strong attractions between your and my electrons and nuclei (Fig. 6.9).[11] So, if we think of ourselves as two composite particles, the fact that we have zero net electromagnetic interaction conceals the fact that our components have a very strong, long-range interaction. Likewise, the net interaction between nucleons, which are composite particles, may be quite different from the force acting between their component quarks. This is in fact the case. The *residual* force between nucleons has a very short range, the diameter of a nucleus. The force between individual quarks though, the true strong force, has an infinite range and its messenger particles are massless bosons called *gluons*. Unlike the familiar forces, the true strong force gets stronger with increasing separation of the quarks. We shall examine gluons and the 'strong charges' of this topsy-turvy world more closely later.

I would not expect you to infer the existence of the weak force or any of its properties. The weak force was proposed to account for certain kinds of radioactive decay. Although it is best to think in terms of quarks, the net effect

---

[10]  Murray Gell-Man, who assigned the name to his invention (in 1963), having only heard it, pronounces it 'kwork', presumably not having seen the word in context ('Three quarks for Muster Mark', in *Finnegans Wake*). My impression is that most people now pronounce it to rhyme with Mark.

[11]  A rough estimate of the repulsion between our electrons in such an embrace is about $4 \times 10^{27}$ newtons, a force which, if applied to the Earth in orbit around the Sun, would bring it to a standstill in less than ten seconds. Actual embraces are exquisitely balanced affairs.

100 000 000 000 000 000 000 000 000 N

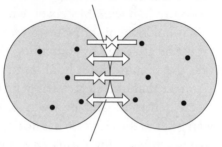

100 000 000 000 000 000 000 000 000 N

**Fig. 6.9** Here is an indication of the extraordinarily fine balance between two electrically neutral bodies made up of electrons (the grey background) and nuclei (the black dots). The repulsive force between the electrons in two such spheres of water (representing human bodies in a close embrace) is trillions and trillions of newtons (N, the unit of force; one newton is about the gravitational force experienced by a 100-gram apple on a tree). The repulsive force between the nuclei is the same. However, the attraction between the electrons in one body for the nuclei in the other body is also trillions and trillions of newtons, and by happy chance the attractions and repulsions cancel exactly. That means that there is no net attraction or repulsion between us.

of the force can be imagined as an influence that slowly wrenches a neutron apart and strips out an electron, leaving behind a proton. The electron is spat out of the nucleus, and gives rise to the form of radioactivity known as β-radiation (beta-radiation). The weak force has a very short range, of less than the diameter of a nucleus. It is mediated by particles called W and Z *vector bosons* with masses about eighty and ninety times that of a proton, respectively.

In general, messenger particles are called *gauge particles*. The origin of this peculiar and cold name will shortly become clear. Suffice it to say that the photon, the graviton, the vector bosons, and the gluons are also gauge particles, which is the first hint we have that the fundamental forces have a common origin. In fact, the unification of the forces begun by Maxwell has been carried as far as combining the electromagnetic and weak interactions into a single force called the *electroweak interaction*. This unification is an aspect of symmetry, and we will return to it once we have examined the particle zoo more closely.

The zoo is divided into two great parks. In one park roam the 'hadrons', in the other the 'leptons'. *Hadrons* are particles that interact by the strong force. In the hadron park, we will consider only the quarks themselves, for all the weird creatures that roam there (protons, neutrons, and many peculiar oddities) are built from these quarks by using laws based on a special kind of symmetry. You may

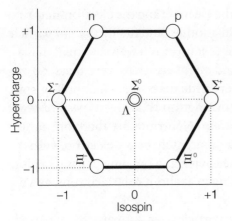

**Fig. 6.10** The eight-fold way is a way of classifying and coordinating elementary particles rather like the periodic table of the chemical elements. Here we see a plot of eight particles (only the proton, p, and the neutron, n, are likely to be familiar, but they have exotic siblings), where one axis is isospin (discussed in the text), and the other axis is another version of an inner symmetry called hypercharge. In this way, eight particles have been shown to be related. More elaborate schemes capture the other particles.

have heard of the *eight-fold way* in this connection (Fig. 6.10). The eight-fold way is a kind of periodic table of the hadrons in which they are classified using this special group of symmetry operations. The proton and the neutron are in one family, so we can think of their isospin kinship as a particle analogue of a Döbereiner triad (in this case a duplet), a fragment of the pattern of the overall classification. The *leptons* are the rest of the particles: they are particles that do not interact by the strong force.

Rather curiously, and it is something that calls out for an explanation, there are three families of hadrons and three related families of leptons (Fig. 6.11). Not unlike typical families in real life, the three families of particles each consist of two groups of particles that are from two *generations*.

|  | First family | | Second family | | Third family | |
|---|---|---|---|---|---|---|
|  | Particle | Mass | Particle | Mass | Particle | Mass |
| Leptons | Electron | 0.000 54 | Muon | 0.11 | Tauon | 1.9 |
|  | Electron-neutrino | $< 10^{-8}$ | Muon-neutrino | <0.0003 | Tau-neutrino | <0.033 |
| Hadrons | Up | 0.0047 | Charm | 1.6 | Top | 189 |
|  | Down | 0.0074 | Strange | 0.16 | Bottom | 5.2 |

**Fig. 6.11** A table of the three families of fundamental particles showing the two generations of leptons and hadrons (quarks) in each case. The masses are multiples of the mass of a proton.

Let's look at the leptons first. There is the electron and the electron-neutrino in one family, the muon and its neutrino in another, and finally the tau particle (or tauon) and its neutrino in a third family. Neutrinos have very small mass—much smaller than that of the electron—and might even have zero mass; no one is quite sure. Neutrinos have no charge, very little mass, but a half unit of spin. They must have some other property to distinguish the three types; for want of a better word, this property is called *flavour*. Neutrinos are therefore almost massless spinning flavour. A muon is like a moderately heavy electron, with the same charge and spin but about 204 times heavier, as a bowling ball is to a ping-pong ball. A tauon is much heavier, weighing in at about 3500 electrons, as a big dog is to a ping-pong ball.

There are also the *antiparticles* of these particles. An antiparticle—a particle of antimatter—is of special interest to writers of science fiction for it sounds so exotic. It's not exotic: it's just rather rare. An antiparticle has all the properties of the corresponding particle, but the opposite sign of charge. For instance, the antiparticle of an electron is the positively charged *positron*, with the same mass and spin as the electron itself. One of the questions we will have to consider is why there is so little antimatter around, why the universe is asymmetric in matter and antimatter.

As we see in the illustration (Fig. 6.11), the six quarks that make up the hadrons also fall into three families of two generations each. As for the leptons, we can distinguish the families by their mass. The quark counterparts of the electron and its neutrino are the *up quark* and the *down quark*, weighing in at 8.7 and 13.7 electrons, respectively. The quark counterpart of the muon and its neutrino are the *charm quark* and *strange quark*, weighing in at 3000 and 300 electron masses, respectively. The counterparts of the tauon and its neutrino are the *top quark* (the last to be discovered, in 1995) and the *bottom quark*, weighing in at an elephantine 350 thousand and 10 thousand electron masses, respectively. These different varieties of quark—up, down, strange, and so on—are said, like the various neutrinos, to have different flavours. Most of our familiar matter (specifically protons and neutrons for nuclei and the surrounding electrons of atoms) are made up of the first family of leptons and quarks (the electron, its neutrino, the up and down quarks), and the other families contribute only to more exotic forms of matter. Quite frankly, the existence of the second and third families seems to be a waste; but no doubt there is a reason, because there is a reason for everything. Does the reason lie in symmetry? We shall see that the answer is probably yes, with the concept of symmetry suitably stretched.

None of the quarks has ever been detected individually. That leads me to make a remark to prepare your mind for appreciating another migration of science's paradigm that will occur towards the end of this chapter. The Greeks for the most part failed as scientists because they eschewed, or had not invented, experimentation: theirs was all theory, with no controlled and shared experience. That the quarks have not been directly identified but are believed to exist because they are required by a currently successful theory, and their existence is supported by overwhelming *secondary* experimental support, is perhaps a dangerous step back towards the Greeks and no doubt leaves positivists upset. Here theory is rather clever, and not a little subversive, because it even predicts that isolated quarks *won't* be found because, as already mentioned, the strong force between quarks increases with distance, so they can never escape from combination with each other. So *not* finding them is a part of the proof of their existence! Should we then believe in quarks, or dismiss them as atoms were once dismissed as accounting tokens? They explain so much, including experimental consequences of their existing, so perhaps we should. If you are content with that kind of belief, that kind of reality, then you may find it possible to accept what is coming later.

That's all there is to matter: three families of fermions with similar properties apart from their spins and their different abilities to participate in the various forces, particularly the strong force. Everything there is, as far as we know, is built from these components bound together, as they are, by the four types of gauge boson. The world, at root, is extraordinarily simple.

But our description is not simple enough. Although minuscule, the number of particles—four fermions (if we concentrate on the first family) and a few gauge bosons—is still a colossal number if we seek true simplicity. We have already indicated that the W and Z bosons of the weak interaction and the photons of the electromagnetic interaction are different faces of the messenger particles of the electroweak interaction. Could it be that all the fermions are different faces of but one entity, and likewise the bosons too? Could it be that, at root, the fermions and the bosons holding them together are actually just different faces of a *single* entity? Now, that would be something, something close to perfect simplicity.

It looks as though that is the case. However, for us to understand what it means, we have to return to the theme of this chapter, symmetry, and see how symmetry provides a framework for the deep comprehension towards which

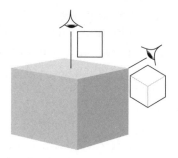

**Fig. 6.12** Keep in mind this analogy for the rest of the chapter: it shows that two apparently unrelated two-dimensional shapes (a square and a hexagon) can be thought of as different views of a single object in a higher dimension, a cube.

we seem to be edging. For a concrete image of how symmetry can relate seemingly unrelated things, you might like to keep in mind a cube. From above it is a square. From above one corner (and with one eye closed) it looks like a hexagon (Fig. 6.12). The rotation of a cube turns a square into a hexagon. Now, that is a very peculiar transformation for a two-dimensional onlooker, but is simplicity itself for we who have access to a third dimension. It will be helpful to bear that image in mind when we are talking about symmetry operations that relate apparently unrelated things.

There is a remarkable feature of nature called *gauge symmetry*. This sombre, unhelpful, and off-putting name was adopted for historical reasons, before particle physicists became high-spirited during the swinging '60s and adopted names like strangeness and charm, and long before they became sober again, when hair shortened, hip became passé, the iridescence of their language faded, · and they fell back on names like 'intermediate vector boson'. Gauge symmetry is one of the abstract, internal symmetries that I warned you to expect. However, it is powerful when interpreted judiciously, for it is a symmetry that reveals the origin of force.

To understand gauge symmetry, we need to go back to the Schrödinger equation for an electron and its solution, the wavefunction. A wavefunction has a property, its *phase*, that can be adjusted without there being any discernible physical outcome. This symmetry arises from the point we made earlier: that only the square of the value of a wavefunction at any point has physical significance, so we can modify the wavefunction itself provided that its square remains the same. It will be convenient to illustrate the change of phase of the wavefunction of a free particle by a rotation of the wave round its direction of travel (Fig. 6.13).[12] Adjusting the phase in this way is an example of a *gauge*

---

[12]  In general, changing the phase means multiplying the wavefunction by a factor $e^{i\phi}$. I should really be drawing complex wavefunctions as helices around the direction of travel, and drawing the phase shift by advancing the helices a bit.

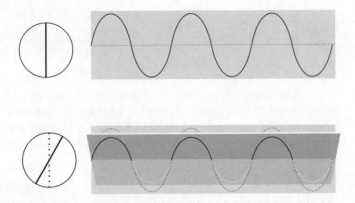

**Fig. 6.13** A representation of gauge transformations. The top diagram shows the wavefunction of a free particle. The lower diagram shows how the wavefunction changes when its phase is modified by the same amount everywhere. We have adopted the convention of twisting the wave around its propagation direction to indicate a change of phase. The amplitude of the wavefunction is unchanged by this modification, so the wavefunction conveys the same information about the location of the particle. The gauge transformation is therefore a symmetry of the system.

*transformation.* This is one of the internal symmetry operations I mentioned, for if you had closed your eyes while I was busily adjusting the phase, you wouldn't know from physical measurements (which depend on the square of the wavefunction, not the wavefunction itself) whether I had done anything or not. If we change the phase of a wavefunction by a constant amount everywhere, then the Schrödinger equation itself remains unchanged, because all waves with shifted phases are also solutions. In other words, gauge transformation by a constant amount is a symmetry of the Schrödinger equation. This group of symmetry operations—changing the phase by anything between 0 and 360°—is called U(1), the 1 signifying that there is only one property being changed.[13] The term 'the symmetry group U(1)' is just a fancy way of referring to our ability to adjust one parameter, the phase of a wave, by any amount without anyone being any the wiser.

In general, a gauge transformation can take on a different value at each point in space; in other words, we can adjust the phase of a wavefunction by a different amount at each point (Fig. 6.14). Suppose we do that and still require the Schrödinger equation to be left unchanged; that is, we require the equation to be *gauge invariant* under all operations of the group U(1), allowing for different shifts in phase at each point in space. Now something remarkable turns up. *To ensure gauge invariance in this more general sense, we need to introduce another term*

---

[13]  As we indicated earlier, the U signifies 'unitary'. This mathematical characteristic of the group springs from the physical requirement that particles are neither created nor destroyed when the symmetry operations are carried out.

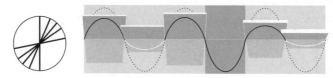

**Fig. 6.14** In this diagram we have attempted to convey a more general gauge transformation in which the phase is changed by a different amount at each point, so the angle of twist away from the vertical is different at each point (as indicated in the inset). We have simplified the representation by supposing that each half-wavelength is twisted by the same amount: in practice, the variation would be continuous. Invariance under this type of gauge transformation implies the existence of a force.

*into the equation. That term is equivalent to the effect of an electromagnetic force acting on the electron.* In other words, the requirement of gauge invariance implies the existence of an electromagnetic force. It is in that sense that the requirements of symmetry demand the existence of a force. Symmetry drives.

We have seen that gauge invariance of the Schrödinger equation under the group of symmetry operations we have called U(1)—fiddling around with the phase—implies the existence of an electromagnetic force. The question that should spring to mind is whether the other forces are also consequences of gauge invariance. That is, is there a sufficiently complicated way of fiddling around with the wavefunctions of particles that, for their equations to remain unchanged, requires the presence of additional terms that we can interpret as the other kinds of forces? Success in this endeavour would show that all the forces had a common origin.

Steven Weinberg (b. 1933), Abdus Salam (1926–96), and Sheldon Glashow (b. 1932) achieved this synthesis of the electromagnetic and weak forces in 1973, and their work has led to the formulation of the currently accepted *standard model* of unified forces. The group of symmetry operations they identified is a combination of the U(1) that results in the electromagnetic force and another more complicated set of transformations called SU(2), which accounts for the weak force. The fact that the overall symmetry group is a combination of U(1) and SU(2), written U(1) × SU(2), tells us that the two types of forces have a common origin. They are the two faces of the electroweak interaction. Recall the cube analogy: the electroweak force is like the cube, the electromagnetic force is like seeing a square in one orientation of the cube, and the weak force is like seeing a hexagon when the cube is rotated into a different orientation.

When the electroweak force is quantized, the U(1) part of the theory gives rise to photons. The SU(2) part gives rise to three particles, the 'intermediate vector bosons' of the theory, consisting of two *W particles* (one of each charge) and one electrically neutral *Z particle*. All four particles have spin 1 and are examples of gauge bosons. The photon was effectively discovered in 1905 when Einstein elucidated the photoelectric effect (Chapter 7); the W and Z particles were identified in 1983 in accelerator experiments at CERN.

The gauge symmetries we have been discussing cannot be complete, because the W and Z particles have mass—a lot of it, as they weigh in as respectively eighty and ninety times as heavy as a proton—whereas the photon is massless. As in our discussion of the isospin symmetry of the nucleon and the hidden symmetry of the periodic table, the difference in mass must come about from an interaction that breaks the underlying symmetry of the particles. This symmetry breaking is ascribed to the interaction of the W and Z particles with another field, called the *Higgs field*, just as the difference in mass of the proton and the neutron is ascribed to their different interaction with the electromagnetic field. The Higgs mechanism for the acquisition of mass takes its name from Peter Higgs (b. 1929) who suggested it; a similar mechanism had been proposed independently by Robert Brout and François Englert at the University of Brussels in 1964. Fields, of course, are quantized, so interaction with the electromagnetic field actually means interaction with the particles of the quantized field, photons. We can think of photons as condensed on the proton more strongly than on the neutron, so lowering its energy and therefore its mass. Much the same occurs with particles immersed in the Higgs field, for we can think of the quanta of the Higgs field, which are called *Higgs particles*, as condensed to different degrees on the mediators of the electroweak interaction. As a result, the W and Z particles acquire mass but the photon doesn't.

The validity of this explanation of symmetry breaking and the acquisition of mass depends on the existence of the Higgs particle. So far, no one has seen one. There are two possible explanations. One is that Higgs particles don't exist. That explanation would be very hard for particle physicists to bear, as the symmetry arguments that imply the existence and unification of the electromagnetic and weak interactions is so compelling. If that argument is correct, then there must be a symmetry-breaking mechanism to endow some of the gauge bosons with mass. So, something like the Higgs mechanism really ought to be active, for otherwise the whole approach collapses. Perhaps it should. Alternatively, Higgs particles could be of such great mass that no accelerator has yet been able to attain the necessary energy to bring them into being. The world of particle physics is currently awaiting the upgrading of two accelerators, one at CERN and the other at Fermilab, just to the west of Chicago, to sufficient energies to search more vigorously for the Higgs particle. At some stage it will be found, or particle physicists will have to revise one of their most cherished models. I hope you can appreciate the importance of the search, for our confidence in our current understanding of matter hangs from it.

○

The strong force also turns out to be a manifestation of gauge symmetry. In this case, we note that quarks possess, as well as flavour, a special kind of charge that enables them to interact with each other by exchanging gluons. Each quark can have any one of three of these 'strong charges', and physicists have fallen into the pleasing convention of referring to these charges as *colour*. This colour has absolutely nothing to do with actual colour: it is just a neat way of referring to the strong charge. Thus, the colour charge of a quark may be red, green, or blue. All known combinations of quarks (the triplets that make up the proton and the neutron, and the quark–antiquark combinations that make up gluons) are 'white': they are mixtures of colour charge that result in whiteness, with no residual colour charge showing, just as the actual colour white is a mixture of actual red, green, and blue.[14]

Now we get to a new version of gauge symmetry. If we change the colours of the quarks systematically, varying the tint from place to place, we have the equivalent of changing the phase of a wavefunction. In this case, three values, the colours, are involved rather than a single phase. Instead of the simple group U(1) of the electromagnetic interaction and the slightly more complicated group SU(2) of the slightly more complicated weak interaction, we have to consider the much more elaborate group of symmetry operations called SU(3). However, just as we saw for the other forces, it turns out that for the equations to remain the same under this more complicated gauge transformation, we need to include a term representing a force into the equation. That additional term has exactly the properties of the strong force. Moreover, when we quantize this interaction, the gauge bosons that drop out of the equations, the spin-1 massless particles responsible for carrying the force between coloured quarks, are the gluons! Here again, we see how respecting a symmetry of nature—this time a rather complex, hidden symmetry—entails the existence of a term that we recognize as a force.

We must now tread into the foggy intellectual swamp where, lurking in some kind of abstract mud, we expect to come upon the unification of the strong and electroweak forces and, correspondingly, the unification of the lep-

---

[14] The phosphors on a TV screen are red, green, and blue: when all three are illuminated by the impact of the electron beam, we perceive white.

tons and hadrons into a single park of the zoo. Once again, symmetry is likely to be our guide. We can suspect that a really elaborate group of symmetry operations will succeed in showing that the strong force and the electroweak forces are just different faces of a single force. If you want a concrete analogy, instead of a rotating cube showing square and hexagonal shapes, think of a more elaborate polyhedron that shows squares and hexagons in some views, but octagons or other shapes in others: all the shapes are aspects of a single object.[15]

The unified theory is called a GUT, standing for *grand unified theory*. At the moment, people are unsure about the identity of the richer symmetry group, and several different proposals have been made. Experiments help to guide and assess the choice. For instance, because the quarks and the leptons are being herded into a single region of the zoo from their separate parks, there is the possibility that a quark may turn into an electron, and therefore that a proton may decay. The simplest choice of the bigger group, which is called SU(5) and is a combination of SU(3), SU(2), and U(1) of the strong force, the weak force, and the electromagnetic force, respectively, suggests that the lifetime of the proton is in the range $10^{27}$–$10^{31}$ years. Experiments, however, show that the lifetime is at least $10^{32}$ years. That discrepancy indicates that the simplest choice of the richer symmetry group is inappropriate, and more elaborate symmetries are currently being studied. Should the programme succeed (and, scientists being the optimists that they are, there is little doubt that it will), the finite lifetime of the proton will have profound implications for the long-term future of the universe, as we explore in Chapter 8.

Our fermion zoo consists of the leptons and hadrons, now released to graze in a single park. There is also a boson zoo, inhabited by the messenger particles of the forces that bind the fermions together into protons and people, and ultimately allow collections of fermions to express opinions. These forces are aspects of a single force. Could it be that there is an even bigger, more elaborate group of symmetry operations in some kind of abstract inner space—an even bigger, more elaborate polyhedron—that rotates an entity so that with one face it appears as a fermion and with another it appears as a boson? There are certain experimental suggestions that such a *supersymmetry group* does really exist, with each particle—electron, meson, neutrino, quark, gauge boson, photon— the different face of a single entity. Of course, there would have to be a lot of symmetry breaking, because the particles have widely different masses, but the periodic table had the same problem, and we know how to handle the

---

[15]  A wonderful site for viewing and rotating all kinds of polyhedra is http://www.georgehart.com/virtual-poly-hedra/vp.html.

acquisition of different masses, such as by letting Higgs particles stick to the massless particles to different extents. If supersymmetry succeeds in showing the equivalence of fermions and bosons, then forces and particles will be intrinsically indistinguishable and everything will be just one thing. Symmetry economizes, and supersymmetry economizes superbly.

When this idea is explored, it shows strong signs of working. However, the theory also predicts the existence of partners of the known particles. These *supersymmetric partners*, which include selectrons, squarks, sneutrinos, photinos, Winos, Zinos, and gluinos, all differ from their conventional partner by half a unit of spin. So, a selectron, for instance, has zero spin and a photino has spin one half. The problem is, where are they? The usual answers can be rolled out: either they don't exist (because the universe is not supersymmetric), or they are so heavy that no accelerator is capable of producing them. No one yet knows the answer, but if you relish beauty, you are probably inclined to believe that the universe is as beautiful as possible, and therefore supersymmetric. Faith, though, is a guide, not a criterion, in science.

There are several outstanding questions that we must now confront. You might have noted them as we went through. One was why matter dominates antimatter. Another is why there are three families of fermions. A third is why there are so many fundamental particles anyway. A fourth question is why gravity seems so elusive a force in our journey to the unification of all forces. Do the answers to all these questions lie in the symmetry of the universe? Is the universe far more beautiful than we currently suspect? Is it infinitely beautiful, perfectly symmetrical?

Well, it may be supersymmetrical but it is certainly not perfectly symmetrical, or there would be equal amounts of matter and antimatter. There are other indications that it is lop-sided, too. There are some obvious, macroscopic lop-sidednesses. For instance, most of the human population is right-handed. No one really knows why that is so: it may be connected with the heart being slightly on the left of the body,[16] but resolving the problem is unlikely to offer

---

[16] Preferential right-handedness in humans (it is much less developed in other animals) may have its evolutionary origin in the tendency of human mothers to hold infants on their left side in order to impart the soothing effect of the sound of the mother's heartbeat. Such mothers would be more skilful at manipulation of objects with the free hand and would be selectively favoured. Studies have been made on medieval skeletons in an attempt to distinguish modern cultural pressures from innate tendencies. A rival theory is that right-handedness arose from the need to give space in the brain for evolutionarily advantageous speech to develop. http://www3.ncbi.nlm.nih.gov/htbin-post/Omim/dispmim?139900.

profound insight into the nature of the universe. Slightly deeper in our structure lie the amino acids that, when strung together into coils and sheets, form the all-important proteins that govern the processes of life (Chapter 2). Amino acid molecules come in two forms, one being the mirror image of the other. It is a fact of life that, on Earth at least, the only amino acids that occur in our proteins have the same handedness (they are left-handed, according to certain technical criteria). No one knows why. It could be pure chance: a distant common ancestor of all life forms used left-handed amino acids, and all living things have descended with that handedness. Some, though, have speculated that the dominant left-handedness of amino acids is related to the cosmic lop-sidedness of the universe, with the left-handed acids being just a little more stable, and thus having the edge, than their right-handed mirror images. No one really knows, but it would certainly be engaging if this chain of handedness could be traced back to something fundamental. It would greatly help to resolve the argument to know whether the proteins of organisms that presumably exist elsewhere in the universe have the same handedness as those of the organisms on Earth.[17]

What do we mean by the lop-sidedness of the universe? In a perfectly symmetrical universe, events observed in a mirror would be indistinguishable from the events themselves. In fact, we could never tell whether we were looking directly at the universe or at its reflection in a mirror. The technical term for this ideal case is the *conservation of parity*. It turns out, though, that the results of some experiments carried out in 1957 can be distinguished from their mirror images, so parity is not conserved. The universe is not the same as its reflection, it is spatially askew.

That the universe is spatially lop-sided raises the possibility that it is lopsided in time, too. In a time-symmetric universe, the laws of nature will be the same for the universe running backwards as for the universe running forwards. Thus, we could not tell whether the universe started at time zero and is running forwards or started at time zero and is running backwards. More specifically, and on a smaller scale, a collision between two particles to form new particles is equivalent to the reverse process in which these product particles collide and give rise to the original particles. The technical term for this symmetry is *time-reversal invariance*. It turns out, though, from experiments conducted in 1964, that in one quiet little corner of the particle zoo[18] the direction of time does matter. The lop-sidedness is closely related to the asymmetry in matter and

---

[17]  Here is one more hazard of interplanetary travel. If, on an alien planet, life is built from right-handed amino acids, hungry Earthling travellers would have to bring their own beefburgers.

[18]  The decay of neutral *K* mesons, kaons.

antimatter, which emerged during the initial moments of the universe's history, and we continue this story in Chapter 8.

Experiments show, therefore, that the universe is lop-sided in space and time. But the lop-sidedness is not just a random asymmetry, for the lop-sidedness in space is linked to the lop-sidedness in time. To understand the link, we need to know that there is a third type of lop-sidedness, called *charge conjugation*, in which every particle is replaced by its antiparticle. We might expect a universe in which particles and antiparticles are interchanged to be indistinguishable from the original version. That is not so. The weak interaction does not respect charge conjugation invariance, so a universe with antimatter replacing matter would behave differently from this universe. (That difference gives us the chance to identify an antimatter region of the universe before flying catastrophically into it.)

With this breakdown in symmetry in mind, it turns out that the universe is symmetrical (as far as we know) if *simultaneously* we change particles for antiparticles (we'll denote that C), reflect the universe in a mirror (call that P), and reverse the direction of time (written T). That is, according to a theorem proposed by Wolfgang Pauli, the universe is *CPT invariant*. So the universe is lop-sided under the individual changes, but perfectly formed if we think in terms of this composite action.

The biggest question left to tackle is the ultimate nature of all the particles we have rolled out on to the stage. Currently in particle physics, a huge amount of effort is being put into a theoretical project that may hold the answer but which might never be directly testable experimentally. If we go back to the beginning of this pair of chapters, where the Greeks were imagining cutting up matter and speculating about how far they could go, the tacit assumption was that they would end up at tiny little point-like entities. For them it was atoms; for us we think of the apparently structureless leptons and quarks as points. But suppose they are not. Suppose the ultimate end of cutting is not a point but a line. This is the starting point of *string theory*,[19] which promises to illuminate many of the questions we have raised. String theory is an extension of the symmetry arguments we have been meeting in this chapter because it involves the topology of spacetime, its stretchiness and the possibility that it is riddled with holes, in addition to the rigid geometrical transformations we have considered so far.

[19] String theory is also called *superstring theory*, because it includes aspects of the fermion–boson linking supersymmetry we have mentioned.

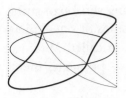 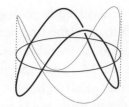

Fig. 6.15 Two modes of oscillation of a string: each zero-point oscillation of the mode corresponds to a different elementary particle.

In string theory, we think of a small string-like circle as the ultimate building block of nature. The string is very small: the radius of the circle is only about the Planck length (about $10^{-35}$ m, Chapter 7). Very, in this instance, does mean very. If a nucleus were magnified to the size of the Earth, the string would be a circle of radius not much bigger than the original nucleus. The string is very taught: its tension is equivalent to hanging from it a weight equivalent to $10^{39}$ tonnes, which is equivalent to a trillion Suns. We are talking seriously small, stiff strings.

Stiff springs vibrate. Each different mode of vibration, according to string theory, corresponds to a fundamental particle. So, there is only one kind of string, but its different modes of vibration correspond to all the various particles we have met so far (Fig. 6.15). I do not mean that as we increase the frequency of vibration, like plucking different notes on a guitar: that takes much too much energy. Even to excite to the first overtone would take so much energy that it would correspond to a particle of hugely greater mass than any known fundamental particle—about the mass of a small bacterium.[20] Here the vibrations are the so-called *zero-point vibrations* of the string. According to quantum mechanics, an oscillator can never be completely still, but always has at least a tiny residual energy, its zero-point energy. Just think of the string as pulsing away quietly, like a human heart, each mode of its pulsing corresponding to a different particle.

When string theory was first mooted it could account only for bosons, and suffered from the following embarrassment: strings had to exist in twenty-six-dimensional spacetime. This *embarrasse de dimensions* was partially relieved when supersymmetry was brought to bear and fermions were welded into the theory. With the constraints implied by supersymmetry, it was found that strings could thrive in only ten dimensions of spacetime, nine of space and one of time. Several ways of organizing those dimensions were devised, and currently it looks as though the different theories can be united into one super-theory if the dimensionality is allowed to rise to eleven. We shall adopt this number, and suppose that string theory is all about strings vibrating in ten

20   It would be about the Planck mass (Chapter 8), about $10^{19}$ proton masses.

dimensions of space and one of time. The currently emerging version of string theory, in eleven dimensions and with more elaborate versions of one-dimensional strings that include two-dimensional membranes, is called *M-theory*. People seem to have forgotten what the M stands for: it is probably 'membrane', but it could also be for the 'mother of all theories'.

The immediate question that springs to mind is where are all these dimensions? We are brought up to believe that we inhabit a four-dimensional world (three of space plus one of time), so where are the missing seven? It is supposed that they are furled up. Or, rather, that they failed to unfurl when the universe was formed: the initial expansion of the universe being so rapid (as we shall examine in Chapter 8) that seven spatial dimensions simply didn't wake up until it was too late. The analogy that is widely used to ease people over the conceptual gulf of confronting 'compactified', furled dimensions is to imagine a hosepipe lying on a lawn. From afar, the pipe looks like a one-dimensional line, but close up we see it has width.

To imagine *one* compactified dimension, we could think of a little circle—with positions on it denoting locations along that dimension—attached to each point of space (Fig. 6.16). To imagine a collision in this space we no longer think of points colliding: we think of rubber bands wriggling along the pipe, and bouncing back and away from each other. In actuality, there are seven dimensions compactified in this way at each point, with the strings somehow wrapped round them, like a rubber band wrapped round a pipe. The compactified dimensions are thought to adopt a special shape at each point. The shapes in question are called *Calabi–Yau spaces* after the two mathematicians, Eugenio Calabi and Shing-Tung Yau, who studied them. Physicists are always grateful to mathematicians, who in their admirably noetic manner have so often studied seemingly useless concepts in the abstract, only to discover later that they have unwittingly been preparing exactly the tools needed to cope with advances in physics. From a Platonic point of view (see Chapter 10), mathematics is out

**Fig. 6.16** What appears to be a one-dimensional line with two point-like particles on it is actually a tube with two circular strings on it. The additional dimension is compactified, and we do not realize it is there until we come to enquire about the deep structure of reality. A collision between two particles is actually the collision between two strings.

**Fig. 6.17** A Calabi–Yau space. Instead of a line in space being a simple tube, like that shown in Fig. 6.16, it is possible that each point on a line is in fact a multi-dimensional space, of which a slice through one candidate is shown here. Think of a structure like this (but in more dimensions) as attached to each point of space.

there waiting to be discovered, so Calabi and Yau were perhaps excavating the pre-existent rather than, as they must have thought, just inventing. Figure. 6.17 shows one such space. Shapes like these—in seven dimensions—are the hose-pipes of string theory, for the strings wind round them and through their holes.

It seems that M-theory is edging towards answering one of the big questions: why are there three families of particles? The answer appears to lie in symmetry. This time, though, the symmetry is that of the Calabi–Yau hosepipes and related to the dimensionality of the holes in these spaces, the holes through which the strings are threaded. The symmetry is the most subtle we have found so far. If a Calabi–Yau space is manipulated in a certain way, it turns out that the number of even-dimensional holes in the new space is equal to the number of odd-dimensional holes in the original space. The number of families is determined by the number of threading patterns, and hence to the number of holes. There is a hint here—and currently it is no more than that—that the number of particle families is intimately related to the manner in which spacetime is compactified and that the number three is emerging as possibly significant.

The other big question is why only three spatial dimensions unfurl to give us our three dimensions of space. String theory can even suggest an answer to that . . . but that is the material of cosmology and Chapter 8.

String theory, and its more elaborate M-theory, has an astonishing power. But it might not be science. Some time back, I warned that I was preparing your mind for the possibility that science will have to modify its criteria of acceptability. That was in connection with quarks: quarks have not been seen, and perhaps cannot be seen, yet we are increasingly confident of their existence as so much that can be verified flows from it. That is verification by implication, rather than verification by experimentation: verification by hearsay, rather than verification by direct experience. Perhaps there comes a point where the line can be crossed, but it is a Rubicon of science to be crossed with considerable caution.

With M-theory, the apotheosis of the symmetry arguments at the heart of this chapter, we take a further step along this perilous path. There is no direct experimental motivation for M-theory: it is a gorgeously beautiful idea, with suggestions of how it can resolve deep questions, but it has not made a single numerical prediction. It suggests ways of accounting for broad issues, such as the number of families of particles, but as there are tens of thousands of Calabi–Yau spaces, there is a hint of gerrymandering post-diction rather than prescient prediction. What *direct* experimental tests it suggests require apparatus of a galactic and even cosmic scale, likely to be forever outside our technological capabilities. The indirect rationalizations it suggests are highly interesting, and awesome in their reach. For instance, M-theory predicts the existence of a massless boson with spin 2, the graviton. So gravity falls within its reach, and we can cautiously believe that the last and most elusive of the forces can be unified with the other forces by the theory. Scientists working on M-theory rightly yearn for it to be true, as it is so beautiful; but I have said before, and must emphasize again, that the satisfying warmth of faith alone is insufficient in science.

# QUANTA

## THE SIMPLIFICATION OF UNDERSTANDING

### THE GREAT IDEA
*Waves behave like particles and particles behave like waves*

*If anyone claims to know what the quantum theory is all about,
they haven't understood it*

RICHARD FEYNMAN

W E   H A V E been hovering on the edge of quantum theory, with one toe dipping into its dangerously infested pool. It is time to take the plunge. To appreciate the significance of the impact of this extraordinary theory we need to note that until the end of the nineteenth century waves unambiguously waved and particles were unambiguously particulate. Unfortunately for naive understanding, though, this distinction did not survive the turn of the century. From a scatter of observations towards the end of that century a virus entered classical physics. A few decades into the twentieth century the disease it brought had destroyed classical physics completely. Not only did the virus eliminate some of the most cherished concepts of classical physics, such as particle, wave, and trajectory, but it also tore to shreds our established understanding of the fabric of reality.

In place of classical physics—the physics of Newton and his immediate successors (Chapter 3)—there arose *quantum mechanics*. Never before have we had a theory of matter that has caused so much consternation among philosophers. And never before have we had a theory of matter that, in the hands of physicists, has proved so reliable. No exception to the predictions of quantum mechanics has ever been observed and no theory has been tested so intensively and to such high precision. The problem is that although we can use the theory with great skill and authority, despite a hundred years of argument no one quite knows what it all means. Nevertheless, it has been estimated that 30 per cent of the GNP of the USA depends on applications of quantum mechanics in one form or another. That's not bad for a theory that no one understands. Think of the potential for growth and life enhancement (and inevitably death enhancement through the development of quantum weaponry) that might accrue if we did understand it!

The virus that was to destroy classical physics was first identified in the late nineteenth century by physicists studying a somewhat recondite problem to do with the light emitted from a hot body. To understand what happened, we need to know that light is a form of *electromagnetic radiation*, which means that it consists of waves of electric and magnetic fields propagating at the speed of light, $c$.[1] The *wavelength* of the radiation is the distance between peaks of the waves, and for visible light is about 5 ten-thousandths of a millimetre. Everyone says that that is very small: it is, but it is *almost* imaginable—just think of a millimetre divided into a thousand slices, and then slice one of those slices in half. Different colours of light correspond to different wavelengths of radiation, with red light having a relatively long wavelength and blue light a relatively short wavelength (Fig. 7.1). White light is a mixture of all colours of light. Small changes in wavelength have considerable consequences: as traffic lights change from red, through yellow, to green, the wavelength decreases from 7.0 to 5.8 and then to 5.3 ten-thousandths of a millimetre, and car drivers respond accordingly to these minuscule differences. The microwave radiation used in microwave ovens is also electromagnetic radiation, but its wavelength is several centimetres, so that's easy to imagine.

We also need to be aware of the term *frequency*: if you imagine standing at a point with a wave rushing past, then the frequency is the number of peaks that pass you per second. Long wavelength light has a low frequency, because only a few peaks pass you per second; short wavelength light has a high frequency, because a lot of peaks pass you per second. For visible light, about 600 trillion ($6 \times 10^{14}$) peaks rush by per second, so the frequency is reported as $6 \times 10^{14}$ cycles per second ($6 \times 10^{14}$ hertz, Hz). Red light is relatively low frequency, only about 440 trillion cycles per second; blue light is relatively high frequency radia-

[1] The value of $c$ is $2.998 \times 10^8$ m s$^{-1}$, about 186 000 miles per second, or 687 million mph.

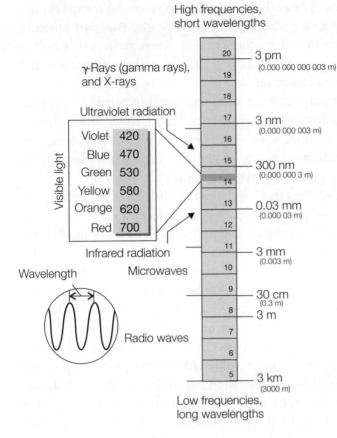

High frequencies,
short wavelengths

**Fig. 7.1** The electromagnetic spectrum showing the classification of different regions. The visible region of the spectrum spans a very narrow range of wavelengths, and the wavelengths (the distance between neighbouring peaks in the wave, as shown in the inset) of the corresponding colours we perceive are given in nanometres (billionths of a metre) in the 'Visible light' box. The numbers on the tall vertical grey bar are the powers of ten of the frequency in cycles per second (hertz, Hz), so 8, for instance, indicates a frequency of $10^8$ Hz (a hundred million cycles per second). The classification of the regions is not rigid, and there is neither an upper nor a lower bound to the spectrum.

tion, at about 640 trillion cycles per second. We perceive this radiation as different colours because different receptors in our eyes respond to different frequencies. The actual numbers in the illustration will not play a role in what is to follow, but it is a part of general culture to be aware of their typical magnitudes and the various regions of the electromagnetic spectrum.

Two characteristics of light emitted from an incandescent object, so-called 'black-body radiation', had been identified in the late nineteenth century and expressed as laws. In 1896, the German physicist Wilhelm Wien (1864–1928) noticed that the intensity of black-body radiation—the brightness of the incandescent body—was greatest at a wavelength that depended in a simple way on the temperature. This feature is qualitatively familiar to us in everyday life, because we know how an object glows first red hot as it is heated and then white hot as its temperature is raised still further. This shift in colour indicates that more and more blue (short wavelength) light contributes to the initially red (long wavelength) incandescence as the temperature is raised, so the maximum in the intensity shifts to shorter wavelengths. In 1879, the Austrian physicist Josef Stefan (1835–93) investigated another familiar everyday feature, how the total intensity of the emitted light increases sharply with temperature, and expressed the dependence quantitatively as a law.[2]

Neither Wien's nor Stefan's law could be explained in terms of classical physics despite strenuous efforts by highly talented theoreticians. In a lecture at the Royal Institution on 27 April 1900 Lord Kelvin identified the failure to account for black-body radiation as one of the two little black clouds then apparent on the horizon of classical physics (the other black cloud was the failure to detect motion through the ether). Kelvin's two little clouds were to grow into a tumultuous storm that would sweep away our conceptions of the world, the manner in which we carry out our calculations and interpret our observations, and our understanding of the deep structure of reality.

In exasperation, Max Planck (1858–1947) unwittingly and unwillingly gave birth to quantum theory. On 19 October 1900 he proposed an equation that seemed to account for Wien's and Stefan's laws, and struggled for the following weeks to provide a theoretical foundation for his expression. In a lecture given before the German Physical Society on 14 December 1900, now regarded as the official birthday of quantum theory, he presented his solution. First, he pictured the radiation as being driven by the vibration of oscillators atoms and electrons—in the hot body, with each frequency of vibration corresponding to the presence of a particular colour of light in the radiation. That was the standard view, and his contemporaries had all done the same. His contemporaries had also tacitly assumed that the energy of each of these oscillators was continuously variable, just as the swing of a pendulum can (they supposed) have any amplitude. Planck, however, took a radically different view. He proposed that the energy of each oscillator could be changed only in *discrete* steps, a staircase of energy rather than a ramp. Specifically, he proposed that the energy of an oscillator of a given frequency is an integral multiple of $h \times frequency$, where $h$ is a new universal constant, now called *Planck's constant*.[3] That is, he proposed

---

[2] Wien's law is that the product of absolute temperature and the wavelength of maximum emission is a constant ($\lambda_{max}T = constant$); Stefan's law, which is also called the Stefan–Boltzmann law, is that the total intensity emitted is proportional to the fourth power of the absolute temperature (*intensity = constant $\times T^4$*). Switching on a light bulb and increasing the filament's temperature from 300 K (room temperature) to 3000 K (its working temperature) increases the emitted intensity by a factor of 10 000, which is why it suddenly glows so brightly.

[3] Planck, by all accounts a courteous and amiable man, did not have the hubris to name it after himself: he left that to others. For him, it was the *quantum of action*. He was able to estimate its value by fitting his equations to observations on black-body radiation. Its modern value is $6.626 \times 10^{-34}$ J s.

that the staircase of allowed energies of any given oscillator is 0, 1, 2, . . . times the quantity $h \times$ *frequency*.

The value of $h$ is so small that the steps in energy for most forms of electromagnetic radiation (especially the radiation we call visible light) are so small that they are undetectable except by very sophisticated methods, so it is easy to understand how physicists were led to think that energies could be varied continuously. Look at a pendulum, can you see that its amplitude of swing can be changed only stepwise?[4] The stepwise variation of energy, however, is the only way in which the properties of black-body radiation can be explained, and the stepwise variation of energy—its *quantization*—is now an established fact.

In private, Planck confided to his son that he thought he had made a discovery comparable to those of Newton. Nevertheless, for much of the rest of his life he tried desperately but fruitlessly to explain quantization in the context of classical physics. There are two lessons here for our comprehension of the scientific method. One is that revolutionary ideas gather strength from resistance to continuous attack. Unlike in some other fields of human endeavour, where crazy ideas are embraced unquestioningly as engaging and welcome friends, in science a crazy idea is subject to constant attack, especially—really especially— if it overthrows an established paradigm. The second lesson is that old men (and old women, although for them there is perforce and regrettably currently less empirical evidence) are not the best evangelists of radical science, for deeply imbued as they are in their conventional upbringing they commonly resent the passing of their learning. Like new mores, new paradigms become accepted only as old generations die.

Be that as it may, Planck's revolutionary, crazy idea that energy came in lumps, that it is granular rather than smooth, that it is as sand rather than as water, an idea that was to transform our perception of reality, was met by silence. At first it was regarded as a mathematical ruse. The physical reality of his proposal emerged only in about 1905 when the gladiator Einstein stepped into the arena, unsheathed his mathematical sword, and slew another classical dragon.

To identify the dragon, we need to put ourselves back into the milieu of late nineteenth-century physics, the dragon's lair. Everyone had become convinced in the course of the century that light—more generally, electromagnetic radiation—is undulatory: it is propagated as a wave. That conviction had not always

---

[4] If you answer yes, you are fibbing. The steps in amplitude of a 1-metre pendulum with a 100-g bob swinging with an amplitude of about 5 cm, like a pendulum in a long-case clock, differ by only $10^{-30}$ cm from the vertical, which is fifteen orders of magnitude smaller that the diameter of an atomic nucleus.

been the case. Newton, supported later by Laplace, had insisted that light is a stream of particles, but the experimental evidence that grew through the nineteenth century convinced everyone that light is a wave. The most compelling evidence was the phenomenon of *diffraction*, first reported by that meticulous observer Leonardo da Vinci (1452–1519) and investigated exhaustively and quantitatively by such magisterial physicists as Huygens, Young, and Fresnel. One of the most dramatic confirmations of the wave theory of light was the prediction that there should be a spot of light at the centre of the shadow of a sphere or circular screen that was illuminated from the other side (Fig. 7.2). In 1818, Augustin Fresnel (1788–1827) had submitted a paper on the theory of diffraction for a competition sponsored by the French Academy. The mathematician Poisson, a member of the judging committee, was highly critical of the wave theory of light and deduced from Fresnel's theory the apparently absurd prediction that a bright spot should appear behind a circular obstruction. However, another member of the committee, François Arago, decided to look for Poisson's bright spot, and found it experimentally. As a result, Fresnel won the competition, and in due course the wave theory of light became the accepted and seemingly unassailable paradigm. The dragon, then, is the wave character of light.

Einstein slew the dragon in 1905, when he showed that light should, after all, be regarded as composed of particles. Einstein's demolition of the paradigm was in two parts. First, he analysed the thermodynamic properties of the electromagnetic radiation inside a heated cavity, and showed that to be consistent with Planck's observations, the radiation must consist of particles rather than

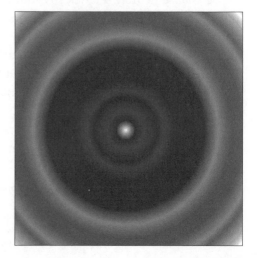

**Fig. 7.2** The Poisson spot. According to the wave theory of light, when an opaque disc is placed in front of a lamp, a white spot is predicted to occur at the centre of its shadow. Diffraction phenomena like this are compelling evidence for the wave nature of light.

waves. These light particles were called *photons* a decade later, and we shall use that name.

Einstein's proposal seemed to meet with immediate experimental support in the form of the *photoelectric effect*, in which electrons are expelled from the surface of a metal exposed to ultraviolet radiation. The photoelectric effect has a number of peculiar features which seemed outside the competence of the wave theory of light to explain. They were immediately explained, however, once the effect was pictured as the result of a collision between an electron and an incoming photon. This model led to a quantitatively accurate account of the photoelectric effect, and was one of the achievements cited when Einstein was awarded the Nobel Prize for physics in 1921. It is a minor quirk of both fate and physics that we now know how to account for the photoelectric effect in terms of electromagnetic waves, so this particular support for the existence of photons, though still represented in textbooks (including my own) as unassailable evidence, is flawed. However, the existence of photons is now not in question, for there is abundant evidence of a different kind.[5]

The reconciliation of the new and experimentally incontrovertible view that light consists of particles with the old and experimentally incontrovertible view that light consists of waves was, as may be imagined, a difficulty when first proposed. It has remained a difficulty ever since, and we shall come back to it later.

By now the quantum virus had entered the body of classical physics and the disease was beginning to spread. Einstein's second contribution to the establishment of quantum theory was also made in his *anni mirabili* of 1905–7. This contribution resolved a puzzle of a more homely kind, relating to the rise in temperature of materials when they are heated. The property investigated was the *heat capacity* of a substance, which is a measure of the heat required to raise the temperature by a given amount.[6] Way back in 1819, with the carefree confidence that comes from sparse experimental results and a peer-review system still in its cradle, the French scientists Pierre-Louis Dulong (1785–1838) and Alexis-Thérèse Petit (1791–1820) had announced that, when corrected for the number of atoms in a sample, all substances have the same heat capacity. Everyone believed them, although it is manifestly not true. Fifty years later, as more data became available and physicists started to measure heat capacities at

[5] A related quirk is that Einstein was awarded the Nobel Prize for a false analysis but correct conclusion and was not awarded it for his most substantial work, the theory of relativity, which at the time was controversial but has proved to be correct (as far as we know).

[6] For instance, the heat capacity of water, $4 \, \text{J K}^{-1} \, \text{g}^{-1}$, its 'specific heat', indicates that $4 \, \text{J}$ of heat is required to raise the temperature of $1 \, \text{g}$ of water by $1\,^\circ\text{C}$.

low temperatures, it became unavoidably obvious that Dulong and Petit's law was a poor summary of the world, and in particular that all heat capacities fall towards zero as the temperature is lowered.

Classical physics could explain Dulong and Petit's law with triumphant ease by supposing that heat is taken up by atoms as they oscillate more and more vigorously. It was therefore somewhat disheartening for classical physicists to be forced to admit that the law was invalid at low temperatures and in many instances at room temperature too. The problem remained unsolved until Einstein turned his extraordinary mind to it in 1906. He accepted the role of oscillating atoms, but crucially he proposed, echoing Planck, that atoms vibrate with energies that increase stepwise, rather like jerkily ascending a staircase of energy levels. At low temperatures, there is not enough energy around to get the atoms oscillating, so the heat capacity is very low. At high temperatures there is enough energy to get all the atoms oscillating and the heat capacity rises to its classical, Dulong and Petit, value. Einstein was able to calculate the temperature dependence of the heat capacity and got reasonably good agreement with observation. His model was refined a few years later by the Dutch physicist Peter Debye (1884–1966), and the refinements, which did not affect the essential idea, resulted in excellent agreement with experiment.

Einstein's contribution was of crucial importance because it extended concepts that had emerged from the study of electromagnetic radiation to a purely mechanical system of vibrating atoms. The virus had made its inter-species transition from radiation to matter.

Once the virus was established in matter as well as in radiation, the disease gnawed at the whole structure of classical physics. There are dates and achievements along the line stretching forward from 1906, especially the imaginative but untenable model of the hydrogen atom proposed in 1916 by the celebrated Danish physicist Niels Bohr (1885–1962), which initially appeared to verify the applicability of quantum concepts to systems of particles. However, the crucial date for our present purposes is 1923, when the virus reached the very heart of matter and dissolved the concept of particle.

Although scientists as serious as Newton had held the view that light consists of particles, so the introduction of photons was not a complete surprise, no serious scientist—other than a few engagingly enterprising and wildly speculating ancient Greeks—had held the view that matter is wavelike. Yet, while society flapped during the 1920s, that is exactly the concept that emerged and took root. The originator of this view was Prince Louis de Broglie (1892–1987), the descendant of a family ennobled by Louis XIV.

De Broglie's contribution to this revolutionary view was based on his identification of the analogy between the propagation of light and the propagation of particles. He argued relativistically, but we can get to the core of his argument without that complication. The central feature of *geometrical optics*, the version of optics that traces out the paths of rays of light as straight lines as they reflect off mirrors and refract through lenses, is that they travel along paths corresponding to the shortest journey time between the source and the final destination. This statement is essentially the *principle of least time*, proposed in 1657 by the French magistrate and amateur but outstanding mathematician Pierre de Fermat (1601–65) as a generalization of an observation made by Hero of Alexandria in about 125 BCE in the latter's *Catoptrics*. A more accurate name is the *principle of stationary time*: the odd phrase 'stationary time' simply means that the time to traverse the path may be either a minimum or, as in certain cases, a maximum. We shall confine our discussion to paths of least time, but the remarks can easily be extended to paths of greatest time too. The puzzle that should immediately occur to us is the following: how does light know, apparently in advance, the path that will result in a journey of least time? If it starts out along the wrong path, might it not be more economical of time to continue rather than to come back to the source and start again?

The wave theory of light comes to the rescue in a particularly elegant way. Suppose we think of an arbitrary path between two fixed points and imagine a wave snaking its way along the path (Fig. 7.3). Then we think of paths lying very close to the first path, with waves snaking along them too. At the destination, the peaks and troughs of the waves arriving by these different paths will annihilate each other: this mutual annihilation is called *destructive interference*. Interference is a characteristic of wave motion: it is seen on the surface of water when the trough of one ripple coincides with the peak of another and the displacements of the water cancel. There is one path, though, that has neighbours for which the positions of the peaks differ so little that they reinforce rather than destroy each other: this mutual reinforcement is called *constructive interference*. This effect is also seen in the ripples on water where peaks coincide and the displacement of the water is enhanced. The paths that interfere constructively are those very close to being a straight line—in general the path of least time—between the source and the destination.

Now we come to the core of this argument. Light does not know in advance

**Fig. 7.3** In the upper illustration, we see a curved path between two fixed points and a nearby curved path. On each path we have drawn a wave of the same wavelength. Although they start out with the same amplitude, when they arrive at the end point the amplitudes are very different. If we imagine a whole bundle of waves travelling by nearby paths, we should be able to appreciate that the amplitudes at the end points are all very different and interfere destructively, to give zero total amplitude. In the lower illustration we see the same thing for a straight-line path and one of its nearby paths. In this case, all the waves arriving at the end point have a very similar amplitude and they do not interfere destructively. We conclude that, given perfect freedom to travel by any route, the only surviving routes are those close to a straight line.

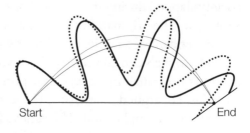

Start    End

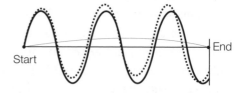

Start    End

nor does it need to know which path will turn out to be the one of least time: it tries *all* paths, but only paths very close to the path of least time do not eliminate each other. Destructive and constructive interference become much more stringent the shorter the wavelength of the light, and only geometrically straight lines survive in the limit of infinitely short wavelengths, which is the limit at which physical (wave) optics becomes geometrical optics. *Total freedom of action has resulted in an apparent rule*. That is the finest kind of scientific explanation, where the wolf of total lack of constraint emerges as the sheep of systematic behaviour, anarchy emerging as rules, disorder underlying order, and freedom the foundation of control.

With that explanation in mind, we turn to consider particles. The path of a particle, according to classical mechanics, is determined by the forces acting on it at each instant (as we saw in Chapter 3). However, as for the propagation of waves, we can cast this description into a statement concerning the entire path. In 1744, the French mathematician and astronomer Pierre-Louis Moreau de Maupertuis (1698–1759) announced that the path taken by a particle was such that a quantity he termed the *action* associated with a path is a minimum. Maupertuis was driven to his *principle of least action* more by theological considerations than by physics, for in his *Essai de cosmologie* (1759) he argued that the perfection of a Divine Being is incompatible with anything other than extreme simplicity and least expenditure of effort. Unfortunately for this view, the modern version of the principle recognizes that in some cases a particle takes the path of greatest action, so a better name is the *principle of stationary action*. For simplicity, we shall stick to paths of least action.

Maupertuis' definition of 'action' was obscure and varied with the problem he was tackling; nevertheless, it was the germ of a correct idea and was expressed almost simultaneously in a mathematically rigorous but restricted form by the Swiss mathematician Leonard Euler (1707–83) and then in its final version in about 1760 by Joseph Louis Lagrange (1736–1813). These historical intricacies, however, need not delay us: the essential point is that there is a well-defined quantity called the *action*—think of it as akin to 'effort'—and a particle adopts a path that corresponds to the least action, the least effort. The puzzle that should immediately occur to us—and now I paraphrase my words above—is how does a particle know, apparently in advance, the path that will result in a journey of least action? If it starts out along the wrong path, might it not be more economical of action—of effort—to continue rather than to come back to the origin and start again?

De Broglie was struck by the analogy between the basic laws of optics and of particle dynamics when expressed as principles of least time and least action, respectively. He saw that the problem of a particle apparently knowing before it set out what path would result in the least action could be resolved in exactly the same way as for light, *provided a wave can be associated with a particle*. Then anarchy would result in law: the waves associated with the particle would explore all paths between source and destination, and only those corresponding to a straight line (if there were no forces operating, and to more general paths if forces—the analogues of mirrors and lenses—were present) would undergo constructive interference and survive annihilation by their neighbours. This annihilation would become more stringent as the wavelength of these 'matter waves' decreased, and in the limit of infinitely short wavelengths, we would regain geometrically well-defined paths through space. In other words, Newtonian dynamics should emerge, with particles following precise trajectories.

By examining this analogy, de Broglie was able to deduce an expression for the wavelengths of his matter waves:

$$Wavelength = \frac{h}{momentum}$$

where $h$ is Planck's constant and the momentum of a particle is the product of its mass and velocity (as we saw in Chapter 3). Thus, Planck's constant (remember that Planck called this constant the 'quantum of action') enters the description of the dynamics of matter at a very deep level, touching the very heart of motion. Notice that the mass appears in the denominator of this expression through its contribution to the momentum, so large masses (balls, people, planets) can be expected to have exceedingly short wavelengths. Your wavelength when you are moving at a brisk 1 metre per second, for instance, is only about $1 \times 10^{-35}$ m, so your motion can be treated according to Newton's dynamics and you can travel with little fear of being diffracted and ending up in Padua instead of Pisa.[7] It is hardly surprising that the waves with wavelengths so short went unnoticed and that Newtonian dynamics was so successful when applied to visible, 'macroscopic' bodies. When electrons are considered, however, we enter a different world, for they are so light that their momenta are low and their wavelengths correspondingly long. The wavelength of an electron in an atom is comparable to the diameter of the atom, and for them Newtonian dynamics is no longer an acceptable approximation.

De Broglie truly deserved his Nobel Prize, awarded in 1929 for his 'discovery of the wave nature of the electron'. The Nobel committee, though, was not quite right in its assessment: de Broglie's identification of the wave nature of particles applies to *all* particles, not just electrons. Electrons are the lightest of the common particles, so his suggestion is most evident for them; but there is no particle or collection of particles (including balls, people, and planets) that in principle does not have a wave character associated with it. The existence of this wave character was confirmed experimentally by showing that electrons underwent that most characteristic property of waves, diffraction. In 1927 the American Clinton Davisson (1881–1958) earned his portion of the 1937 Nobel Prize by showing that electrons are diffracted by a single crystal of nickel, and George Thomson (1892–1975), working in Aberdeen, earned his share by showing that they are diffracted when passed through a thin film. Since then, whole molecules have been diffracted. It is an engaging aspect of family science, that G. P. Thomson won his prize for showing that the electron is a wave whereas his father, J. J. Thomson, won his for showing that the electron is a particle. Breakfast at the Thomsons' was perhaps an icy meal.

We are at the point when revolution was in the air, though it was neither fully formed nor understood. Even de Broglie didn't really know what he meant by

---

[7] When you stop, your first reaction might be that your wavelength suddenly becomes infinite and that you are spread throughout the universe, contrary to common sense. But you only seem to stop; in fact, your body continues to pound away in different directions as you quiver on the spot.

his 'matter waves'. What had been established, though, was the *duality* of matter and radiation, their possession of characteristics of both waves and particles. Light, long known to be wavelike, had been shown to have another face and to behave as particles. Matter, long known to be particulate had been shown to have its second face and to behave as though it is a wave. Once again, the image of a cube comes to mind (Fig. 6.12) in which one aspect looks to us like a square and another a hexagon.

The virus that had now destroyed the most cherished concepts of physics reached its full power in 1926 when the nature of de Broglie's matter waves began to be resolved. As we shall see, it gradually became clear that our pejorative term 'virus' is inappropriate, for the gradual brushing away of the obscuring dust of classical physics revealed a much simpler, cleaner, and understandable world within. The old, steeped as they were in classical traditions, could not come to terms with the new simplicity, and as a result they misled the young. In what follows, I hope to reveal to young and receptive minds the simplicity that quantum mechanics has brought to our understanding of the world.

The searchlight of achievement now turns to illuminate two giants of quantum theory, the enigmatic German Werner Heisenberg (1901–76) and the romantically vigorous Austrian Erwin Schrödinger (1887–1961). They jointly formulated equations that enable us to calculate the dynamical properties of particles (as we shall continue to call them), the replacements for Newton's laws of motion. Their formulations, called respectively *matrix mechanics* and *wave mechanics*, looked completely unlike each other and their philosophies were correspondingly different. But it was soon shown that the two formulations are mathematically identical, so the competing philosophies become a matter of personal choice. Mathematics has this chameleon character, mapping itself on to the physical world in different but equivalent ways, so we should always be circumspect when scorning another's formulation, for it might turn out to be equivalent to our own. The blend of matrix and wave mechanics is now commonly called *quantum mechanics*, and we shall use that term from now on.

This is not the place to go into the details of quantum mechanics or to step through its formulation chronologically. Instead, I will mix and match from the two approaches and thereby show you the essence of quantum mechanics without overwhelming you with the details. I will disregard its history, and concentrate on the high points of its content. You must be prepared to meet a number of unsettling, bizarre ideas, but I will lead you through them carefully.

One of the most famous and contentious aspects of quantum mechanics is the *uncertainty principle*, formulated by Heisenberg in 1927. Heisenberg set out to show that, bearing in mind de Broglie's relation between wavelength and momentum, there are limitations on the knowledge that we can have about a particle. For instance, if we want to measure the location of a particle with a microscope, then we have to use at least one photon to observe the particle, and the more precise the location we require, the shorter the wavelength of the photon that we must use. Broadly speaking, we can't pinpoint anything more precisely than the wavelength of the radiation we are using to locate it: so, with visible light, we can't pinpoint anything to less than about 5 ten-thousandths of a millimetre. Sound—with wavelengths close to 1 m, doesn't let us locate the source with greater precision than about 1 m, which is why bats have to use very high frequency, short wavelength sounds for their echo location.[8] There is a price to pay, though, for using short wavelength electromagnetic radiation to locate a particle. When a photon strikes a particle, it imparts some of its momentum to it, and from de Broglie's relation we can infer that the magnitude of the momentum transferred increases with the shortening wavelength of the photon. Thus, as we sharpen our knowledge about the position of the particle, we blur our knowledge of its momentum. By analysing this problem in detail, Heisenberg was able to deduce the celebrated result that

*Uncertainty in position* × *uncertainty in momentum* is not less than *h*

We should regard Heisenberg's uncertainty principle as an experimental result even though the microscope experiment we have described has not been done explicitly: as formulated by Heisenberg, the uncertainty principle is a summary of a careful analysis of an experimental arrangement in the light of current knowledge. Of course, the actual experiment might give a result quite different from what we predict for one of these *gedanken* (thought) experiments; that, after all, is the very essence of the role of experiment in the scientific method. But, provided our understanding is correct, if current science is applicable, then Heisenberg's conclusion is correct.

Classical physics, which was essentially ignorant of the momentum of a photon, for it knew nothing of photons and was unaware of Planck's constant, is based on the view that position and momentum are simultaneously knowable to arbitrary precision. The question that now arises is how can the uncertainty principle—which we should regard as a fundamental description of nature and a profound departure from classical physics—be incorporated into the mathematical description of motion? In classical physics we think of the

---

[8] In search mode, a bat uses a signal at 35 kHz, corresponding to a wavelength of 1 cm. When it locates a meal, it switches to kill-mode high-frequency chirps in the range 40–90 kHz, corresponding to wavelengths spanning 8–4 mm.

position and momentum of a particle as varying with time, with the unfolding of both with time as the well-defined *trajectory* of the particle.

We can work towards the answer as follows. It should be obvious that we can write, at any given instant,

$$Position \times momentum - momentum \times position = 0$$

For instance, if the position is measured as 2 units from a particular point and the momentum is measured as 3 units, then the first term on the left gives $2 \times 3 = 6$ units and the second term gives $3 \times 2 = 6$ units, and the difference is obviously zero. However, obvious this cancellation may be, but in quantum mechanics wrong it most certainly is. Broadly speaking, because we cannot know the position and momentum simultaneously, we cannot be confident that each term is *exactly* 6 units (or whatever our measurements give), so it is possible that the first term in this expression differs from the second by something of the order of Planck's constant. Heisenberg's great achievement was to show that the uncertainty relation between position and momentum, an experimentally verified statement about the world, is obtained only if the right-hand side of this expression is not zero but is in fact Planck's constant, $h$:[9]

$$Position \times momentum - momentum \times position = h$$

Classical physicists assumed tacitly that the right-hand side is zero, and so constructed the wonderful edifice of classical physics. We now know that the right-hand side is not zero, but it is so small that it is no wonder that classical physicists thought it zero. The fact that the right-hand side is not zero has profound implications, and is the feature that brought classical physics tumbling down.

Heisenberg, with the assistance of his colleagues Max Born (1882–1970) and Pascual Jordan (1902–80), found how to incorporate the non-vanishing right-hand side of the position–momentum expression into quantum mechanics. Schrödinger meanwhile had found another way. You will remember that de Broglie had proposed the existence of a matter wave 'associated' somehow with the particle and that, after taking interference into account, the surviving wave unfolded along the path of least action. It is quite easy to find the rules for telling the wave how to grope through space to find this surviving path. Those rules are the content of the *Schrödinger equation*.[10] This celebrated equation

---

[9] We are simplifying things a bit: the precise value on the right-hand side is not $h$ itself but $ih/2\pi$, with $i$ the square root of $-1$.

[10] Just for the record, here is what the Schrödinger equation looks like (give or take a few factors of $\pi$) for a particle of mass $m$ moving in a region where its potential energy is $V$: $h^2\psi''/m = (V - E)\psi$, where $E$ is the energy of the particle, $\psi$ is the wavefunction we are trying to find, and $\psi''$ is its curvature.

shows how the matter wave varies from point to point, and it turns out that to formulate it we have to make use of exactly the same position–momentum expression as Heisenberg had to use to make his break with classical physics. The central role of that relation in both formulations is the essential reason why Heisenberg's and Schrödinger's approaches are mathematically equivalent.

When we solve the Schrödinger equation, we get the mathematical expressions for the shapes of matter waves. The term 'matter wave' is no longer used, nor is de Broglie's interpretation of them. The modern term for 'matter wave' is *wavefunction* (a term we first encountered in Chapter 5), and we shall use that name from now on.

Wavefunctions are not just meaningless mathematical formulae: we can trace the current interpretation of their physical significance back to a suggestion made by Max Born. Born had noted that, in classical (wave) terms, the intensity of light is proportional to the square of the amplitude (the height above zero) of the electromagnetic wave, whereas in quantum (photon) terms the intensity is proportional to the probability of finding a photon present in a region of space. If the amplitude of a light wave is doubled, its intensity is quadrupled (the beam is four times brighter), and we are four times more likely to find a photon present in a particular region. He then suggested that it was natural to extend this relation to wavefunctions and *to interpret the square of the wavefunction of a particle at some point as giving the probability of finding the particle there*. So, if a wavefunction has twice the amplitude at one location than another, then the chance of finding the particle is four times as great at the former location than at the latter. We can conclude that where the square of a wave-

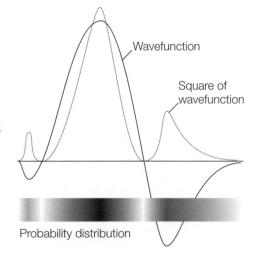

**Fig. 7.4** The Born interpretation of the wavefunction. The full line is an arbitrary wavefunction: note that it passes through zero at several points (these are called nodes) and so has regions of positive and negative amplitude. When we square the wavefunction, we get the dotted line, which is non-negative everywhere but is zero where the wavefunction is zero. According to the Born interpretation, this curve tells us the probability of finding the particle at each point of space. We have indicated this interpretation by the density of shading in the superimposed bar.

Wavefunction

Square of wavefunction

Probability distribution

function is large, there is a high probability of finding a particle, and where it is small, there is a low probability of finding the particle (Fig. 7.4). Note that this interpretation means that the regions where a wavefunction has a negative value—corresponding to a trough in a water wave—has the same significance as a region where it is positive, because when we take the square of the wavefunction, any negative regions become positive too.

The wavefunction may appear to be a somewhat elusive concept despite Born's interpretation. In the next few paragraphs, I shall try to give you an impression of what some look like. I will also show how you can solve the Schrödinger equation in your mind without ever having seen it or having the faintest notion of what it means to solve a second-order partial differential equation.

Broadly speaking, the Schrödinger equation is an equation for the *curvature* of a wavefunction: it tells us where a wavefunction is sharply curved or only slightly curved. The curvature is sharpest where the kinetic energy of the particle is large and is least where the kinetic energy is low. For example, the wavefunction for the bob on the end of a pendulum should look a bit like the one shown in Fig. 7.5: the bob moves fastest at the mid-point of its swing and slowest at the ends, the turning points, where it changes direction, and we see how the wavefunction is curved more sharply near the mid-point of its range. Notice also how the wavefunction has its highest amplitude near the turning points: that is consistent with the familiar behaviour of a pendulum, because it is most likely to be found where it is moving most slowly, which is at the end points of its swing where it is about to change direction.

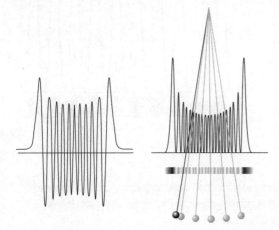

**Fig. 7.5** A typical wavefunction (left). This wavefunction is for a pendulum that has been set swinging with a small amount of energy. The square of the wavefunction (shown on the right) tells us the probability that the swinging pendulum will be found at any displacement. We have depicted this interpretation by the density of shading in the superimposed bar.

Now let's see what some other typical wavefunctions look like. The wave-function of a free particle is very simple. Suppose the particle we are thinking about is a bead able to slide on a long horizontal wire. The potential energy of the bead is the same regardless of its position, so we can suspect that the wave-function will not favour any particular region. A slow particle has a low kinetic energy, so its wavefunction has only a little curvature (Fig. 7.6); in other words, the wavefunction of a slowly moving particle is a uniform wave with a long wavelength, just as the de Broglie relation tells us. A fast particle—with a lot of kinetic energy—has a wavefunction with a lot of curvature, so it snakes up and down many times in a short distance, and is therefore a uniform wave with a very short wavelength. That too is just what the de Broglie relation predicts.

Where are we likely to find the particle? Let's think of the bead as rattling back and forth along the wire between the stops at either end, and we inspect it at random. Because the bead is moving at constant speed, according to classical physics there is an equal chance of finding it at any point on the wire. Quantum mechanics makes a different prediction. To predict where the bead will be found we make use of Born's suggestion: we calculate the square of the wave-function at each position and interpret the result as the probability that the par-ticle will be found at that position. As you can see from the illustration, the particle is most likely to be found at a series of equally spaced regions along the wire, and is not distributed completely uniformly.

Now let's see how the wavefunction of a free particle accommodates the uncertainty principle, that if we know the momentum we cannot know the

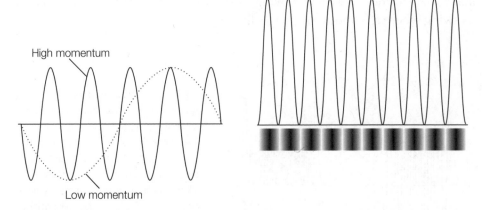

**Fig. 7.6** The diagram on the left shows two wavefunctions for a bead on a long horizontal wire with stops at each end, one corresponding to low momentum and the other to high momentum. The diagram on the right shows the probability of finding the faster moving particle at points along the wire.

position, and vice versa. Wavefunctions like those shown in the illustration are spread out over the length of the wire, so we cannot predict where the particle will be found: it can be anywhere along the wire. On the other hand, we do know the momentum exactly, because the wave has a precise wavelength. Therefore, we known the momentum exactly but can say nothing about position, just as the uncertainty principle requires. Actually, from the wavelength we know only the *magnitude* of the linear momentum: we don't know if the particle is moving to the left or to the right. But because the particle isn't spread completely uniformly along the wire, we are not completely uncertain about where it is, so a little bit of ignorance about momentum (its direction) has opened up the possibility of knowing a little bit about where it is (specifically, where it isn't). You should be beginning to see the subtlety of the relationship between knowing where things are and how fast they are moving.

Suppose, though, that we happened to know that the particle is in fact in a definite region of the wire. Its wavefunction would look a bit like that in Fig. 7.7, with a strong peak where the particle is most likely to be. If we wanted to specify the momentum of this particle, we would have to identify the wavelength of this wavefunction. But a strongly peaked wavefunction doesn't have a definite wavelength because it is not an extended wave, just as a pulse of sound—a bang—doesn't have a definite wavelength. What does it mean to speak of the momentum of the particle?

We can think of the peaked wavefunction in the illustration as being built up by adding together—the technical term is *superposing*—a lot of waves with definite but different wavelengths, each one corresponding to a definite linear momentum. As shown in the illustration, these waves add together where all their peaks coincide to give the peak of the actual wavefunction and cancel everywhere else where their peaks and troughs coincide. Such a superposition

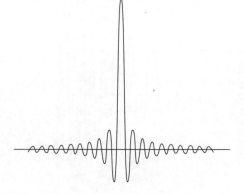

**Fig. 7.7** A wavepacket formed by superimposing thirty wavefunctions like those in the preceding illustration, but with different wavelengths. Although the particle is likely to be found in a reasonably well-defined region of space, we can say nothing about which of the thirty values of linear momentum it possesses. We see in the discussion later that this wavepacket moves in a manner resembling a classical particle.

of wavefunctions is called a *wavepacket*. When we ask for the value of the momentum of the particle with a wavefunction like that in the illustration, we have to say that it is *any* of the values represented by the wavelengths that have been used to form the wavepacket. That is, our partially localized particle has an indefinite momentum, just as the uncertainty principle requires.

If we knew exactly where the particle was at any instant, its wavefunction would be a very sharply peaked spike, with zero amplitude everywhere except at the location of the particle. Such a spike is also a wavepacket, but to get infinite sharpness of location we have to superpose an infinite number of waves of different wavelengths and therefore momenta. We conclude, just as the uncertainty principle tells us, that knowing exactly where a particle is eliminates all possibility of stating its momentum. The uncertainty principle is the quantum version of being lost: you either know where you are, but not where you are going, or you know where you are going, but nothing about where you are.

The concept of a wavepacket helps us to build a bridge between quantum mechanics and the comfortable familiarity of classical mechanics, for it conveys some of the features of classical particles. To see the connection, let's think about a bead on a wire that is not horizontal but slopes down from left to right. Classically, we expect the bead to slide down the wire, going faster and faster. What does quantum mechanics say?

First, we need to formulate the bead's wavefunction, and to do so we can use our knowledge of what the Schrödinger equation tells us about curvature. Because the bead's energy is constant (energy is conserved, Chapter 3), but its potential energy falls from left to right, its kinetic energy rises from left to right along the wire. Increasing kinetic energy corresponds to increasing curvature. We can suspect that the wave will have a wavelength that gets shorter from left to right. Such a wavefunction, for a particle with a perfectly defined total energy, will look something like that in Fig. 7.8.

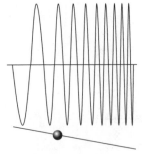

**Fig. 7.8** The general form of a wavefunction for a bead on a wire held at an angle to the horizontal, so that it has a lower potential energy towards the right. Note that the wavelength gets shorter as we go further to the right, which corresponds classically to the increasing kinetic energy of the particle as it slides down the wire.

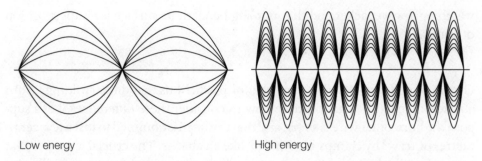

Low energy                                  High energy

**Fig. 7.9** A representation of the time dependence of wavefunctions. Wavefunctions oscillate in time with a rate that depends on their energy. We have tried to suggest how the two wavefunctions in Fig. 7.6 oscillate: the high-kinetic-energy wavefunction (right) oscillates more rapidly than the low-kinetic-energy wavefunction (left).

Next, we need to know something about how a wavefunction changes with time. The new point to keep in mind is that the wavefunction oscillates with a frequency proportional to the total energy of the particle. We can think of the wavefunction of the slowly moving (low-energy) particle as oscillating slowly and that of the rapidly moving (high-energy) particle as oscillating rapidly (Fig. 7.9).[11] The wavefunction in Fig. 7.8 behaves in exactly the same way, and oscillates at a rate determined by its energy.

Finally, suppose we don't know the energy of the bead exactly (our hands holding the wire might be trembling; air molecules are battering the bead). In this case, the wavefunction will not be pure like the one we have drawn, but instead will be a sum of a large number of similar wavefunctions with slightly different shapes. The resulting superposition is a wavepacket like that shown in Fig. 7.7. As we have just seen, each of the individual wavefunctions is oscillating in time as well as in space, so the shape they give when added together changes because at one instant the peaks might add together in one location but then a peak turns into a trough and the wavepacket takes on a different shape. When we examine the sum, it turns out that the region of constructive interference giving rise to the wavepacket moves from left to right. It also picks up speed where the waves have the shortest wavelengths, on the right. That is, the bead accelerates from left to right, just as we know from classical physics. So, when you watch everyday objects undergo their familiar motion—balls bouncing, aircraft flying, people walking—have in your mind's eye the thought that you are

[11] You might be wondering about the particle disappearing and reappearing as the wavefunction oscillates in time. I have simplified the discussion. What the wavefunction actually does is to oscillate from a real value to an imaginary value and then back to a real value, so its square remains the same. I don't want to get drawn into the discussion of that complication.

watching a wavepacket and that rippling beneath its surface is a superposition of waves.

Quantum mechanics makes a number of predictions that differ strikingly from classical mechanics, and it is time now to consider these differences. Let's suppose the horizontal wire is short, and that the bead is confined to just a few centimetres of travel by clamps at each end, like an abacus. The crucial feature is that *only the wavefunctions that fit between the ends are allowed*, just like the only allowed vibrations of a stretched violin string are the waves that fit between the stops. Because the curvature of the wavefunction determines the kinetic energy of the bead, and therefore its total energy (because the potential energy is constant), we conclude that in this arrangement the bead can possess only certain energies. In other words, the energy of the bead is *quantized*, meaning confined to discrete values, not continuously variable (Fig. 7.10). This is a general conclu-

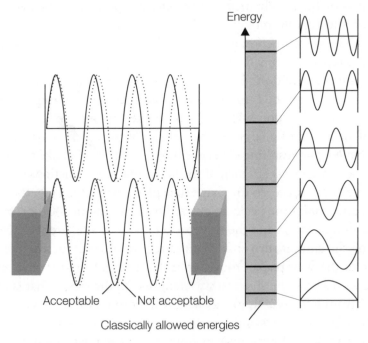

Energy

Acceptable    Not acceptable

Classically allowed energies

**Fig. 7.10** When a particle is confined to a definite region of space, only those wavefunctions that fit into the container, and the corresponding energies, are allowed. On the left, we see a perspective and a direct view of two wavefunctions: one fits the container and is allowed; the other (dotted) does not, and is not allowed. On the right, we see the effect on the energy: the grey bar shows the classically allowed energies, and the horizontal lines show the first six quantized, allowed energy levels. The corresponding wavefunctions are shown on the far right.

sion: *energy quantization*, such as that originally proposed by Planck and Einstein, *is a consequence of the Schrödinger equation and the requirement that the wavefunction must fit properly into the space where the particle can roam.* This is how energy quantization emerges automatically from the Schrödinger equation and the so-called 'boundary conditions' of the system.

Quantization shows up in an interesting way in a pendulum, with one extraordinary aspect. First, consider the wavefunction for the location of the bob that is swinging with an exactly defined energy (so it is in a definite quantum state). The potential energy rises as the bob swings to either side, so its kinetic energy falls to keep the total energy constant, and classically we can expect the wavefunctions to have their biggest amplitude at the ends of the swing where the bob lingers longest. We have already seen one such wavefunction (in Fig. 7.5). As for the bead on the clamped wire, the only allowed wavefunctions are those that fit into the range of values allowed by the swing, from turning point to turning point. Because only some of all conceivable wavefunctions behave properly, and each wavefunction corresponds to a different energy, it follows that only some energies are allowed. It turns out that these allowed energies form a uniform ladder of values with a separation between 'rungs' that we shall write $h \times$ *frequency*, where $h$ is Planck's constant and *frequency* (which we will say more about in a moment) is a parameter inversely proportional to the square-root of the length of the pendulum. For a pendulum of length 1 m at the surface of the Earth, *frequency* works out as 0.5 Hz, so the separation between allowed energy levels is a minuscule and wholly undetectable 300 trillion-trillion-trillionths of a joule ($3 \times 10^{-34}$ J); but it's there.[12] Some of these energies and the corresponding wavefunctions are shown in Fig. 7.11.

Now, here's the amazing feature. Suppose we pull the bob back and let it swing. It will swing with a range of energies, perhaps because of the impact of air molecules or a roughness in the support. Therefore, its actual wavefunction will be a wavepacket formed from a superposition of a large number of functions like those shown in the illustration. This wavepacket ripples from side to side, moving fastest when the pendulum is vertical and slowest at the ends of the swing, just like a classical pendulum. Moreover, and here is the extraordinary point, the frequency of the swing—the rate at which the bob swings from side to side—is exactly equal to the parameter *frequency* appearing in the expression for the separation of the quantized energy levels. So, when you watch a pendulum swing, not only are you seeing the motion of a wavepacket, you are

---

[12] For a pendulum, *frequency* = $(1/2\pi)\sqrt{(g/length)}$, where $g$ is the acceleration of free fall (at sea level on Earth, $g$ = 9.81 ms$^{-2}$).

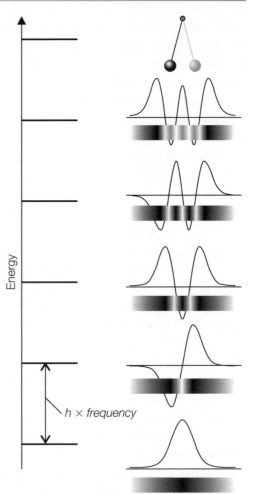

**Fig. 7.11** The first few energy levels and the corresponding wavefunctions of a pendulum. Note that the energy levels are equally spaced. You should also note that the lowest energy wavefunctions do not resemble what we suggested for the shape of high-energy wavefunctions (such as that in Fig. 7.5), because the pendulum is likely to be found close to zero displacement from the vertical, not at its turning points. We can use classical ideas to guide our thoughts about wavefunctions only at high energies.

also seeing, from its frequency, a direct portrayal of the exceedingly closely spaced energy levels. In other words, you are watching quantization directly. A pendulum is a powerful amplifier of the separation of its quantized energy levels, and when you watch a 1-m pendulum swing backwards and forwards, you are observing an energy separation of 300 trillion-trillion-trillionths of a joule directly. I think that's amazing.

The main messages to take from this discussion are that quantization follows naturally from the Schrödinger equation, and that classical behaviour emerges when the precise quantum level is unknown and we have to form a wavepacket.

I have sneaked into the discussion a word central to the problem of interpreting quantum mechanics, the word *probability*. For the remainder of this chapter, we explore the implications and ramifications of this weasel word, for it has profound significance for the way we think about the world. In fact, I want to return to several aspects of the discussion so far, and try to distil from it a number of philosophical issues. I hesitated then to write 'epistemological and ontological issues', that is, issues relating to the nature of knowledge and the fundamentals of reality. That is what they will prove to be, but I am not a philosopher, and I do not want to give the impression that my remarks have any pretensions to being philosophically technical. So, I decided to write simply 'issues' and leave it at that.

I would like to make another remark. The preceding material in this chapter is all you really need to know if you want to use quantum mechanics. Of course, I have left out the technical and mathematical details, but everything that has gone before is reasonably straightforward and uncontentious. The 30 per cent of the US economy that is based on quantum mechanics is a result of using that material and turning a blind eye to what is to come. Quantum mechanics gets philosophically interesting when we start to ask what it all means. That's the subject of the rest of this chapter. If you stop here, you will know the main principles of quantum mechanics and, in principle, can use it to do any calculation; if you go on, your ability to use it won't be helped, but you will know why people find it so deeply puzzling.

First, I shall deal with the uncertainty principle and attempt to justify the subtitle of this chapter: the simplification of understanding. Many people—among them, the fathers of the subject—consider that the uncertainty principle limits our understanding of the world in the sense that because we cannot know the position and momentum of a particle simultaneously, we are permitted only incomplete knowledge of its state. That pessimistic view, in my opinion, is a result of our cultural conditioning. We have been brought up, by our exposure to classical physics and our casual familiarity with the everyday world, to believe that a complete description of things in the world is in terms of both position and momentum. That is, to describe the path of a ball in flight—or just to anticipate when to kick it—we have to judge its position and its momentum at each instant. What quantum mechanics, and particularly the uncertainty principle, demonstrates is that such an expectation, a description in terms of both attributes, is *overcomplete*. The world is simply not like that. Quantum mechanics tells us that we have to choose. We have to choose between discussing the world in terms of the positions of particles and discussing the world in terms of the momenta of particles. In other words, we should speak only of the position of a ball, or speak only of its momentum. It is in that sense that the uncertainty principle is a major simplification of our description of the world, for it shows that our classical expectation is false: the world is simply not like the picture implied by classical physics and casual familiarity.

Let's take this further. There are, the uncertainty principle implies, two languages for discussing the world: the position language and the momentum language. If we try to use both languages simultaneously (as classical physics does and as those preconditioned by its principles still try to) we can expect to get into an awful muddle, just as we would if we tried to mingle English and Japanese in the same sentence. Heisenberg himself is reported as having remarked that 'in the statement that to predict the future we need to know the present, it is the premiss that is wrong'. It was he, though, who was wrong. The correct interpretation of the uncertainty principle is that it reveals that classical physics strives for an impermissible and misleading overcomplete knowledge of the present: momenta *alone* are adequate and, alternatively, positions alone are adequate as complete knowledge of the present.

This interpretation of the uncertainty principle is in line with the philosophical stance adopted by Niels Bohr in 1927 through his *principle of complementarity*, a term he appears to have picked up from William James's *The principles of psychology* and later incorporated into his coat of arms as the motto *Contraria sunt complementa*. Like so much of what Bohr wrote, this principle is not entirely clear, but broadly speaking it states that there are alternative ways of looking at the world, that we must choose one description or the other, and that we should not mix the descriptions. Bohr went on to apply this principle in literature and sociology in much the same way that the principle of relativity was hijacked and corrupted by irrelevant literary applications, but we shall stick with its more reliable relevance to quantum theory.

Bohr's principle is a central component of the *Copenhagen interpretation* of quantum mechanics that grew up around him. The Copenhagen interpretation is a web of attitudes built around Born's probabilistic interpretation of the wavefunction, the principle of complementarity as expressed quantitatively by the uncertainty principle, and—most centrally—a 'positivist' view of nature in which the only elements of reality are the outcomes of measurements made using apparatus that operates on classical principles. Measurement is our only window on to nature, and anything not perceived through this window is mere metaphysical speculation and not worthy of being considered real. Thus, if your laboratory apparatus is set up to examine the wave characteristics of a 'particle' (for instance, to demonstrate diffraction of an electron), then it is valid for you to speak in wave terms. On the other hand, if your apparatus is set up to examine the corpuscular properties of a 'particle' (for instance, to locate an electron's arrival on a photographic plate), then it is appropriate for you to use particle language. No instrument can bring out both wave and particle properties simultaneously, for these attributes are complementary. This essentially was Heisenberg's view, for he regarded quantum mechanics merely as a way of correlating different experimental observations and not revealing anything about an underlying reality: for him and other strict Church of Copenhageners, the outcome of observation is the only reality.

The aspect of the Copenhagen interpretation on which we shall concentrate is the act of measurement. Measurement is a crucial component of considerations of the interpretation of quantum mechanics, not just for those of a positivist disposition, and the generator of more paper, puzzlement, and grief than almost any other aspect of the theory. It is crucial to the Copenhagen interpretation because of that interpretation's insistence on the role of the measuring instrument in teasing out reality. But whatever interpretation one puts on quantum mechanics, there comes a point where we have to confront its predictions with observation, so understanding the interface between prediction and observation is of crucial importance and significance.

At this point, we come to what is perhaps the most difficult but central aspect of the interpretation of quantum mechanics. I have tried to simplify the issues as much as possible without losing the essence of the discussion. I am very sensitive to the subtlety of the argument, and have done my best to make it as transparent as possible. If the going gets too rough, don't hesitate to jump to the next chapter, for nothing that follows depends on the discussion here.

In the broadest terms, an act of measurement is the portrayal of a quantum mechanical property as the output of a macroscopic device. That output is commonly called a 'pointer reading', but the term can be taken to mean the output of any large-scale system, such as a number displayed on a monitor screen, a remark printed on paper, a click we hear in our ear, or even finding a dead cat inside a box. The Copenhagen interpretation insists that a measuring instrument acts classically, for it must portray the quantum world in terms of quantities to which we giants can relate. Although the Copenhagen interpretation was dominant for many years, not least on account of Bohr's influential role, it is by no means universally adopted. One Achilles heel in its fleshy under-side is this insistence on a particular type of measuring apparatus. The alternative is that the measuring device also acts on quantum principles; we explore that variation later.

Suppose we have a detector that turns on a red light if an electron is absent and a green light if an electron is present. An electron is described by a wave-function that spreads through space and tells us through its square, as we have seen, the probability that the electron will be found at each point in space. If we insert our detector into the region where we suspect the electron to be, we are more likely to get a green light where the wavefunction is large than where it is small, and the square of the wavefunction will tell us the probability (for instance, one time in ten tries) that we will get a green light.

If a green light does go on when we stick the detector in, then we know *with certainty* that the particle is at this position. Immediately before that detection event, we knew only the probability that the electron was there. So, in a very real sense, the wavefunction has collapsed from its spread-out form to a sharp peak located at the detector. This change in the wavefunction as a result of investigation using a classical apparatus is called the *collapse of the wavefunction*. Whenever we observers make an observation, the wavefunction collapses to a definite location corresponding to the pointer reading (in this case, the switches controlling the lights) that we observe. This intrusion into the system seeming-ly causing the collapse of the wavefunction at a particular point is the central concept and difficulty of the Copenhagen interpretation and the central dilem-ma of the connection between calculation and observation. It is also the source of the view that quantum mechanics eliminates *determinism*, the causal chain between the present and the future, as there is no way within quantum mech-anics, it is argued, for predicting, before we make the measurement, whether the wavefunction will or will not collapse at a particular point, for all it allows us to calculate is the probability that it will do so.

I need to introduce three technical details of quantum mechanics at this point, for they are central to the measurement problem and to its resolution. I will do so using the well-worn problem of *Schrödinger's cat*. In this quantum parable, Schrödinger imagined a cat confined within an opaque box together with a poisoning device triggered by radioactive decay. Radioactive decay is ran-dom, so for a given interval the decay is as equally likely to decay as not to decay. Correspondingly, according to quantum mechanics, the state of the cat is an equal mixture of its alive and dead states (Fig. 7.12), and we write[13]

*Cat's state = alive state + dead state*

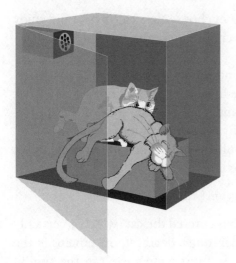

**Fig. 7.12** Schrödinger's cat. A living cat is confined to an opaque box, with a nefarious device that kills, or does not kill, the cat. Before we open the box, is the cat in a superposition of its dead and alive states? When does the wavefunction collapse into one state or the other?

This sum is the analogue of the superposition of wavefunctions that we used to construct a wavepacket, the only difference being that instead of the super-posed states being states of momentum, they are states of the cat. Writing the actual wavefunctions would be more than a little tricky, but we don't have to do that.

The description of states as superpositions is the root of all heartache in quantum mechanics, for in particular there seems to be no mechanism for predicting whether in a subsequent observation of the cat we will get the result 'it's alive!' or 'it's dead!'. As soon as we open the box, we immediately discover whether the cat is alive or dead, so in some sense the cat's wavefunction collapses into one or other of the two wavefunctions. But at what point does the cat's wavefunction collapse? Before we open the box? As we open the box? A fraction of a second later when our mind registers whether the cat is alive or dead? When the cat thinks it's dead? All that quantum mechanics does is to specify rules for predicting the probabilities that these states will be found. Thus, determinism appears to have leaked out of physics, and quantum mechanics appears to have surrendered to the lap of the gods. This feature was of deep concern to Einstein and stimulated his tiresomely often repeated objection that 'God does not play dice'. Bohr swept this criticism aside by remarking that causality is a classical concept anyway, and complementary (in a somewhat vague and ill-defined sense) to a spatial description of the location of the

---

[13] It might look naive to express states in this way, but quantum mechanics provides a set of rules which tell us how to manipulate expressions like this and how to draw precise, quantitative conclusions. Don't be fooled by the apparent triteness of these symbolic expressions.

particle. That is, according to Bohr, you either choose classical physics and enjoy the heady advantages of causality, or you choose quantum mechanics and pay the price with causality.

We can introduce another important notion by thinking about an aggressive modification of Schrödinger's parable in which the cat is not poisoned but shot. When the cat is first shut into the soundproof box, the state of the apparatus is *cat × bullet in gun*.[14] The gun is fired by the same random device as before, so there are equal probabilities that the bullet is in flight or still in the gun. At some stage, the state of the system has become

*State of system = cat × bullet in gun + cat × bullet in flight*

Immediately afterwards, when the bullet has entered the cat (which is certainly the case if the bullet was in flight), so generating a dead cat, or remains in the gun (if that was its state a moment ago), so preserving a live cat, the system becomes

*State of system = living cat × bullet in gun + dead cat × bullet in cat*

This is an example of an *entangled state*, in which the states of the cat and the bullet are inextricably entwined. If this is the true state of the system, we can expect there to be some very weird interference effects between the two states of the system. What on Earth is the interpretation of this description? What can it mean for there to be interference between the wavefunctions of the dead and living versions of the cat and the different locations of the bullet?

Let's deal first with the question of quantum mechanical interference between the various states. This will introduce the third important idea, *decoherence*. This is perhaps the most subtle part of the argument, and I will do my best to keep the concepts in sight of land. The cat is not a single, isolated particle. It is composed of trillions of atoms and its overall wavefunction is a very complex function of the locations of all these atoms. The two contributing states of the system (*living cat × bullet in gun* and *dead cat × bullet in cat*) evolve in time according to the Schrödinger equation, quite differently and extremely quickly. Within a tiny fraction of a second, the wavefunction of the dead cat becomes completely different from that of the living cat, and the interference between the wavefunctions of the living and dead cats is completely washed out. As a result, the system shows no quantum mechanical interference effects and we have either a dead cat or a living cat, not a funny superposition of the two states.

[14] This product—how do you multiply a cat by a bullet?—might look a bit odd. However, the product is well defined in quantum mechanics, and it really means that we should multiply the wavefunction of the cat by the wavefunction of the bullet. More formally, we might write the product as $\psi_{cat}\psi_{bullet}$, where $\psi$, psi, is a wavefunction.

But which state will we find? Is quantum mechanics silent on the prediction of the outcome of our experiment? The loss of causality and determinism, the scaffolding and underpinning of science and understanding, is thought by many to be a price too high to pay, especially when the argument against it is one of opinion and philosophical taste rather than of mathematics in alliance with experiment. One possible resolution grew out of a suggestion by Einstein that quantum mechanics is incomplete in the sense that there are *hidden variables*, or characteristics of particles (including cats) that are hidden from us but nevertheless influence their behaviour. Thus, a hidden variable might tell the particle to pop up at a particular location whereas all that quantum theory could do was to predict the probability of it appearing there and somehow couldn't get a grip on the hidden variable that controlled the actual outcome. Dealing with these hidden variables and making the precise prediction about the outcome of an observation rather than just its probability was then presumed to be the domain of an as yet undiscovered deeper theory underlying quantum mechanics.

The confirmation or otherwise of the existence of unknowable yet influential hidden variables might seem to be more a matter for inconclusive metaphysical debate than scientific resolution. However, in an extraordinary, simple, yet seminal article published in 1964, John Bell (1928–90) demonstrated that there is an experimental distinction between quantum mechanics and its modifications in which there are hidden variables, and therefore that the question can be resolved once and for all. To be precise, Bell showed that the predictions of quantum mechanics differ from those of *local* hidden variables theories. A *local* hidden variable is what its name implies: a hidden variable that can be identified with the current location of the particle, which seems to be a reasonable requirement for the possession of a property. *Bell's theorem* does not rule out *non-local* hidden variables, in which the behaviour of a particle here depends on a characteristic located elsewhere; that may seem a bizarre possibility, but quantum theory has taught us not to dismiss airily the bizarre from our armchairs. Bell's powerful theorem is a theoretical result, but it has been tested in a series of experiments of increasing sophistication. In every case, the results have been consistent with quantum mechanics and inconsistent with a local hidden variables theory of any kind.

So, if quantum mechanics really is complete, at least in terms of local properties, do we really have to give up causality? A number of alternatives have been suggested. One of the most radical suggestions, and therefore of great appeal to journalists if not scientists, was the ill-named *many-worlds interpreta-*

*tion* proposed in a somewhat obscure form by the chain-smoking, horned-Cadillac-driving, multimillionaire weapons research analyst Hugh Everitt (1930–82) in his doctoral dissertation in 1957. The central, straightforward, and seemingly innocuous idea of Everitt's proposal, scorned by Bohr, is that the Schrödinger equation is universally valid and controls the evolution of wavefunctions even when the particle is interacting with a measuring apparatus. A number of towering castles have been constructed by various commentators on this foundation and remarks that Everitt made concerning its apparent implications.[15]

The castle that has seized the popular imagination is that all the probabilities expressed by a wavefunction are actually realized (so the cat is in fact both dead and alive), but the realization splits the universe into any of an infinite number of parallel universes (one with a dead cat, the other with a live cat) as soon as a measurement is made and the state perceived. In essence, the interaction of the measuring device with the observer's brain selects a branch of the universe to follow. Every observation splits the universe, so there is an enormous and increasing multitude of parallel universes in which brains have followed different paths. It is hard to imagine a more profligate interpretation, but because distaste is not an instrument of scientific distinction, some people take this interpretation seriously. Unlike Bell's theorem, there appears to be no way of testing whether the mind is really involved in the detection step except for one experiment that has been proposed. Because that experiment requires the observer to commit suicide, it has not yet been executed.

We have to distinguish Everitt's seemingly unexceptionable (except to dyed-in-the-wool Copenhageners) basic idea that the Schrödinger equation is applicable to macroscopic objects from the interpretations built on this view, so you have to be very careful to define precisely what aspect of the many-worlds interpretation you mean when you ask someone to reveal whether or not they are a many-worlder. I think it fair to say that most physicists now accept the vanilla version of the many-worlds interpretation, that the Schrödinger equation is universal, but few subscribe to the more subjective flavours that have been added to the interpretation. The 'universal Schrödinger view' is in contrast to the Copenhagen interpretation, which asserts the distasteful thought that quantum mechanics is somehow invalid when it is applied to the macroscopic assemblies of atoms we call measuring instruments. This attitude seems unduly defeatist, and it is difficult to see how quantum mechanics could gradually blend or even sharply switch into another theory as the number of atoms involved in

---

[15] So much so, in fact, that it has been suggested that a better name might be the *many-words interpretation*.

the system increases. It is certainly the case that macroscopic objects behave to a very good approximation in accord with classical physics: but we know that that behaviour is simply a manifestation of quantum mechanics applied to large numbers of atoms.

Let's stick with the 'universal Schrödinger view' and look at its implications and problems. We are left with the possibility that the simplest scenario is adequate: quantum mechanics is complete, there are no local hidden variables, and it describes exhaustively bodies composed of any number of particles. Collapse of the wavefunction, a mysterious component of the Copenhagen interpretation, is also out of bounds, because a universal Schrödinger equation will, somehow, have to account for all the changes that a wavefunction undergoes, including the apparent collapse that occurs in the course of a measurement. How, then, under these constraints do we retain causality and determinism within the framework of quantum mechanics and in particular in the process of measurement?

The success of decoherence in eliminating quantum mechanical interference between living and dead versions of cats strongly suggests that decoherence is the white knight we need here too. A living or dead cat is an elaborate pointer reading. That being so, let's simplify the problem by thinking of a primitive measuring device consisting of a ball resting on the peak of a hump between two wells. The slightest nudge will send the ball into one of the two wells, and by observing which well the ball ends up in we can identify whether the ball got a tiny nudge to the left or the right (Fig. 7.13). The apparatus is a nudge amplifier. That indeed is the essential characteristic of all measuring devices: they are all nudge amplifiers. If we wish, we can label the left-hand well 'dead cat' and the right-hand well 'living cat'. The cat is then a bullet-position amplifier: I will leave it to you to translate between the Schrödinger cat indicator and its stylized ball-on-a-hump simplification.

As we have described it so far, the device is useless, because a ball that rolls down into the left-hand well will roll up the opposite face, back down, and over the hump. Only if there is friction to dissipate its energy will the ball come to rest in the well it rolled into initially. So friction traps the ball in its well, and enables us to inspect the output of the device at our leisure. We now have a viable measuring device, made viable by friction, the interaction of the system with its environment.

**Fig. 7.13** The 'nudge amplifier' that epitomizes the problem of measurement in quantum theory. The ball at the peak between the two wells is in its 'ready' state. If a nudge sends it to the right, then in the absence of friction it would roll back and forth between the two wells and we would find it in the left-hand well as often as it is found in the right-hand well. However, if friction is present (symbolizing decoherence, and indicated by the bars on the right), then the ball settles down into the right-hand well, and we have a viable measuring device.

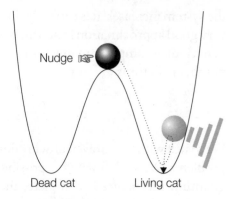

Friction is the analogue of decoherence. (That assertion involves an act of faith on your part: once again, I am trying to interpret the mathematical formulation rather than trying to justify every step.) We can think of the rolling ball as a pure Schrödinger particle, under the control of his equation. Initially, the state of the measuring device is with the ball balanced on the hump; we call this state *device ready*. Suppose the particle the detector is designed to detect is in a state that is in a superposition of travelling to the left, which we will call *particle travelling to left*, and the right, which we denote by *particle travelling to right*, then before the detection event the state of the system is

*Initial state = device ready × (particle travelling to left + particle travelling to right)*

When the particle hits the detector the ball moves into a superposition of being in the left and right wells, so it becomes

*Final state = ball on left + ball on right*

However, because the ball is connected to the environment by friction, there is very rapid decoherence of these two states and we never observe any interference between them: effectively the ball is on the left or it is on the right—the superposition has been resolved into two essentially classical states.

There is still the question about whether the ball is in fact in the left well or in the right well. We have to remember that in the *ready* state the ball is delicately poised on the top of the hump and poised to roll either way. This is just another way of saying that the detector is very sensitive and has no bias. Now we have to remember that even the ball is not completely detached from its environment and is subject to vibrations, impacts of air molecules, a brush with the odd passing photon, and so on. When the particle to be detected strikes the ball and triggers its roll to one side or the other, the combination of the impact, which triggers movement to both sides with equal probability, and the local disturbance which can be in either direction, triggers movement in a definite direction. As a result, the superposition evolves into the ball ending up in just one of the wells, where it is immediately trapped by decoherence.

The essential feature of the measurement device, therefore, is not the requirement that it is a classical device where the Schrödinger equation's writ does not run (as required by the Copenhagen interpretation) but that it is a macroscopic quantum mechanical device embedded in its environment.

I have touched only the surface aspects of quantum mechanics. There are several messages to take home from this account, and I will try to review them here.

First, we are no longer to think of waves and particles as distinct entities, for each can take on the characteristics of the other. If we think in terms of particles, then we are committing ourselves to thinking in terms of their locations. If we think in terms of waves, then we are committing ourselves to thinking in terms of wavelengths and therefore, through the de Broglie relation, momenta. The uncertainty principle expresses this essential complementarity by warning us that the determination of a particle-like property (location) interferes with the determination of a wave-like property (momentum). A simple, deterministic description of the world is obtained only if we discard one or other of these modes of thought.

The properties of particles (which we have agreed to call the entities that have this chameleon character) are calculated by solving the Schrödinger equation. The solutions of this equation contain all the dynamical information about the particle, such as where it is likely to be found or how fast it is likely to be travelling. The solutions also account for all the observations that led to the formulation of quantum mechanics in the first place, such as the diffraction of particles and the existence of quantized energy levels as introduced by Planck in connection with black-body radiation and by Einstein in connection with atoms in solids. The implementation of the Schrödinger equation—finding its solutions and thereby predicting the properties of objects—can be achieved almost automatically, and there is no question that quantum mechanics is a profoundly reliable theory.[16]

[16] It is incomplete as we have described it, because it lies outside special relativity. Special relativity was combined with quantum mechanics to give relativistic quantum mechanics by Paul Dirac (1902–84) in 1927. Its combination with general relativity has still to be achieved (see Chapter 9).

Where quantum mechanics is weird is in the interface between the microscopic and the macroscopic, for the outcome of measurements appears to suggest that quantum mechanics is entirely probabilistic and does away with determinism. That is not the case. Wavefunctions evolve fully deterministically in accord with the Schrödinger equation. Where determinism appears to be absent is in the prediction of the outcome of measurements. One solution, the incompleteness of quantum mechanics in the sense of there being local hidden variables that govern the actual outcome of an observation but lie invisible beneath the surface of the theory, is untenable as such a theory is incompatible with experiments that have been performed. The Copenhagen interpretation asserts that the Schrödinger equation should be replaced by a mysterious process called the collapse of the wavefunction. However, it is most unlikely that there is a domain of competence for quantum mechanics which fades away as the system becomes more elaborate. The modern view is that the Schrödinger equation is universally valid, and that subtle effects stemming from the involvement of the environment are sufficient to account for all observations. Some people, though, will disagree vehemently with that view. Richard Feynman's remark, quoted below the title of this chapter, is still mightily true.

# EIGHT

# COSMOLOGY

## THE GLOBALIZATION OF REALITY

### THE GREAT IDEA
*The universe is expanding*

*He gave man speech, and speech created thought, which is a
measure of the universe*[1]

SHELLEY

CIENCE is often considered to be arrogant in abrogating to itself, in the
eyes of some (my own included), the claim to be the sole route to true,
complete, and perfect knowledge. Yet, some of its greatest achievements are
extraordinarily humbling. Nowhere is its achievement so majestic and this
abject humbling appropriately so complete as in its role in putting Man in his
place in the world. The arrogance of majestic achievement is in science's ability
to apply itself to the greatest question of all: the origin of the universe. The
inescapable and ironical humiliation is that each astronomical and cosmological revolution has diminished the uniqueness of Man's position. Ptolemy had
us at the centre. Copernicus nudged us to a beautiful but nevertheless minor
planet in orbit around the Sun. Since then, the Sun has been elbowed out to an
insignificant location in an insignificant galaxy in an insignificant cluster in
what may prove to be an insignificant universe.

This chapter is the story of this successive humiliation in which from our
presumptiously claimed centrality our scientific discoveries have pushed us
ever more to one side and diminished our significance. Yet, at the same time as
we have been forced to realize our insignificance, we puny-brained little things
have grasped the extent of the universe, have put a measure on all there is, have

---

[1] *Prometheus unbound.*

identified what seems to be our origin, and have even found out the likely unfolding of our cosmic future. It is right that we be proud in the midst of our ceaselessly increasing humility.

In earlier chapters we have looked in; here we look out. Earlier we looked at the exceedingly small; here we look at the exceedingly large. Now we look out into the open spaces of the sky, see where our little arena lies, and ask what we can learn from the stars.

The stars did not escape the notice of the Greeks. At first, when in those darker days than now they looked up at night, they saw a shield pinpricked with holes through which shone the celestial light of an outer lustrous realm. This vision of the cosmos became a little more sophisticated when, in the opinion of the subtle Eudoxus of Cnidus (c.408–355 BCE), the shield gave way to twenty-seven concentric spheres.[2] There is still dispute about whether Eudoxus regarded the spheres as merely a computational device or, like Aristotle who elaborated his model and increased their number to an alarming fifty-four, as being real. In Aristotle's view, or at least in medieval elaborations of what he actually wrote, all except the outermost of these spheres were transparent; the outermost was black with pinpoints of light attached and rotating once a day. According to Aristotle, the heavenly bodies that inhabited the spheres were made of the fifth element, quintessence, that had no analogue on Earth. We might scoff now, but beware: quintessence will return at the end of this discussion. The shells were almost but not quite within reach, for height was then hard to fathom. Even Johannes Kepler (1571–1630) thought that all the stars lay in a thin shell only a few kilometres thick.

Our perception of the universe began to expand once Man put it under the convex lens and held it up to the parabolic mirror. By the time of Sir William Herschel (1738–1822), who started his career as an oboist with the band of the Hanoverian footguards but rose to prominence as an astronomer under the patronage of that other Hanoverian, George III, it had swollen to a grindstone-shaped cluster of myriad stars some six thousand light-years in diameter.[3] The perceived diameter grew once the Eiffel Tower had been built, not because

[2] For a biography of this clever man, see http://www-groups.dcs.st-and.ac.uk/~history/Mathematicians/Eudoxus.html

[3] A light-year is the distance that light, travelling at 300 000 kilometres a second, travels in one year; it is approximately a trillion kilometres (specifically, $9.54 \times 10^{11}$ km).

astronomers could then stand higher off the ground with their heads closer to the heavens,[4] but because the elevators in the tower were built by one William Hale, who thereby became wealthy enough to indulge the passion of his son, George Hale (1868–1938), for astronomy. Hale the younger was first the director of the Yerkes Observatory of the University of Chicago, named after Charles Yerkes, Chicago's ruthless streetcar tycoon, who, in the hope of recovering his place in society after he had been imprisoned for embezzlement, was cajoled into funding the construction of the biggest refracting telescope before or since (the lens was 1 metre in diameter). In 1904, Hale moved to the Mount Wilson Observatory just outside Los Angeles. He knew that by adding inches to mirrors he could reach more light-years into space. First his father helped provide a 60-inch (1.5-m) reflecting telescope there; then with the assistance of another businessman, John Hooker, the 100-inch Hooker telescope was built in 1918, and remained the largest in the world for thirty years.

In 1919, Hale persuaded Edwin Powell Hubble (1889–1953), a Rhodes scholar from Oxford who had studied law but grown tired of its demands, to join him. Hubble began his work by determining the distances of some hazy patches of stars—nebulae—that had long perplexed astronomers. The measurement of distances of objects far away is no easy task. When Hubble set to work there was only one way, which was to use a technique suggested by Henrietta Leavitt (1868–1921), who worked at Harvard College Observatory. She had noted a correlation between the brightness of a certain class of variable star, a *Cepheid variable*, which occur in the arms of spiral galaxies, and the period of its variation. The brightness an astronomer measures on Earth depends on the distance of the star, the greater the distance the dimmer the star appears. Therefore, by noting the period of the variable star we can judge its absolute brightness, and by measuring its apparent brightness, we can infer its distance. Hubble's conclusions were astonishing: whereas our own galaxy, the Milky Way, was known to have a diameter of about 25 thousand light-years, the closest of these nebulae, the Andromeda nebula, was 2 million light-years away. It had to be outside our galaxy; it was another galaxy.

Immediately, the perception of our universe became bigger than had previously been believed and the rack of our humiliation was stretched another notch. Not only did we have to accept that we are not central to our system of planets and hustled off to one side in the Milky Way, but it now became clear

---

[4] The Titans alluded to on the plinth of Galileo's finger's vessel piled Pelion upon Ossa, mountain upon mountain, to spy better on the stars; but mortal Galileo's 'fragile glass' brought them closer than these hooligan gods could achieve. Thus, Galileo's finger epitomizes the power of brain over brawn.

that our galaxy is but one of myriads. There was more and grander humiliation to come.

Hubble's next task was to determine the speeds with which these other galaxies were approaching us or receding from us and thereby to discover the dynamics of the universe. Was it, for instance, like a gas in which island galaxies were dashing about at random, or were they just hanging in the sky? That there was movement had in fact already been established in 1912 by Vesto Slipher (1875–1969) working at the Lowell Observatory in Arizona. He had measured the shifts in colours of galaxies caused by their motion and by 1924 had found thirty-six galaxies that were receding from us out of the forty-one that he had examined. Slipher had used the *Doppler effect*, the change in wavelength caused by the motion of the source: motion towards us decreases the perceived wavelength, giving white a hint of blue; motion away increases the perceived wavelength, giving white a hint of red. The effect is familiar in the context of sound, when a vehicle moving towards us sounds higher pitched than one moving away. The effect arises because the motion of the source helps to bunch together the peaks of the waves or to stretch them out (Fig. 8.1). The greater the velocity of the source, the greater the shift in wavelength, so by measuring shifts, the relative velocity can be determined. If the wavelength is increased, giving a so-called *red shift*, then the source is moving away from the observer. The light of most galaxies shows a red shift, so they are moving away from us.

Hubble went further. In the years 1923–29 he reached the astonishing conclusion that the speed of recession is proportional to the distance from us, and

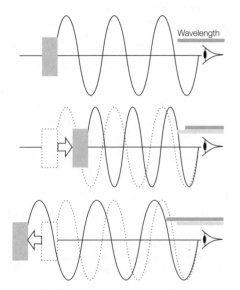

**Fig. 8.1** The Doppler effect is the modification of the wavelength of radiation (either light or sound) emitted by a moving source and received by a stationary observer. In the top diagram, the source is stationary and emitting radiation of a given wavelength. In the middle diagram, the source is moving towards the observer, and the train of waves is compressed, so the observer perceives a shorter wavelength or higher frequency (a shift to the blue, or to higher notes for sound). In the bottom diagram, the source is moving away from the observer, the waves are stretched out by the motion, and the observer perceives a longer wavelength and lower frequency (a shift to the red, or to lower notes for sound).

Wavelength

**Fig. 8.2** A model indicating how we can think of the expanding universe. The coins stuck on the surface of the sphere represent the galaxies. As the universe expands—represented by the sphere expanding—the galaxies move further apart but do not themselves expand. According to this model, an observer standing on any coin would see the other coins moving away: the recession of the galaxies does not imply that we are located at a special place in the universe.

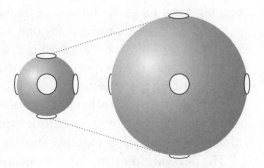

the further the galaxy, the greater its rate of recession. This observation is now expressed as a general law of the universe:

*Speed of recession = Hubble's constant × distance from us*

Hubble's constant is such that a galaxy 10 million light-years away seems to be receding from us at about 200 kilometres per second, one that is 20 million light-years away seems to be receding at about 400 kilometres per second, and so on.[5]

Hubble's conclusion, even though he forgot to mention it in his first paper, was that the universe is expanding. Each galaxy is like a point marking a location on a rubber sheet. For future reference, think of the galaxies as small coins stuck to the surface of a rubber balloon: as the rubber stretches, so the coins move apart but they do not themselves expand (Fig. 8.2). The implication of this expansion is awesome, for if we trace it back through time, then there must come a moment when all the coins are coincident and the universe itself is a single point. That is, *the universe seems to have had a beginning.* I have introduced the weasel words 'seems to have' because nothing in cosmology is ever quite straightforward, especially in curved spacetime, and I will need to elaborate on this conclusion later. At this stage, though, we can think of one of the consequences of the great idea that the universe is expanding as being that *there was a moment when it all began.* That really is breathtaking, and opens up all kinds of questions, some of which we explore as this chapter, like the universe, unfolds.[6]

There are several aspects of this description that we must lay to rest, some now, some later. Wherever we point our telescopes, we see the galaxies

---

[5] The determination of the Hubble constant is fraught with difficulty. Hubble himself greatly overestimated it and concluded rather awkwardly that the Earth was older than the universe. The currently accepted value is close to $71 \pm 7$ km s$^{-1}$ Mpc$^{-1}$, which corresponds to about 22 km s$^{-1}$ Mly$^{-1}$ (1 Mly = 1 mega-light-year, 1 million light-years) or $2.3 \times 10^{-15}$ s$^{-1}$.

[6] During the time it will take you to read this footnote—perhaps 10 seconds—the separation of two galaxies that are 1 million light-years apart will increase by about 200 km.

receding from us as the universe expands. Well, that is not quite true, some galaxies near us—Andromeda for one—is moving somewhat threateningly in our direction. This 'local' motion is a so-called *peculiar motion* of the galaxy (where 'peculiar' means specific rather than weird), a motion relative to a frame set in expanding space. We can think of the galaxies as wandering through space and responding to each other's gravitation.[7] For galaxies that are close together, this motion can overcome the cosmic expansion, just as two coins sliding over a rubber sheet might slide together even though the rubber is stretching.

The second point in this connection is that the expansion we observe appears to place us back at the centre of things, with every galaxy moving away from *us*. This uniqueness, though, is an illusion, for wherever we were to stand in the universe, we would also see expansion away from us. The coin-stuck-on-balloon analogy shows what is happening: on whichever coin we stand, we see neighbouring coins receding from us. This observation is the essence of that epitome of political correctness, the *cosmological principle*, which asserts that the universe looks the same wherever an observer is located. Humility is back in place again.

There is one more technical point before we get down to business. Hubble was not quite right in thinking that he was measuring the speed of recession of the galaxies. We can interpret the red shift as the Doppler effect, and therefore as an indication of the speed of the receding object, only for objects that are close to us. The light from very distant objects set out on its journey a long time ago; the universe has expanded since then and the waves of the light have been stretched. The correct interpretation of the red-shift, which is valid for both nearby and very distant galaxies, is that it is a measure of the *change in scale* of the universe between the time the light was emitted and the time it was detected. So, if the wavelength is shifted to the red by some factor, then it began its journey when the universe was that much smaller. It is an extraordinary fact that by looking out into the distance, we are seeing the universe as it was when its scale was smaller than it is now.

If the galaxies all moved at constant speeds, we could use the Hubble constant to work out when the entire visible universe was a single dot. We shall have to

[7] A large number of galaxies, including our local group (the Milky Way, Andromeda, and a few other bits and pieces) are all flowing towards a point in space known enigmatically as the Great Attractor, a region of space about 150 million light-years away with the mass of 50 thousand trillion Suns. It will take us time to get there, as we are drifting towards it at only 600 kilometres per second.

return to this point later, but it is a good place to start. On this basis we can put the start of the universe at about 15 billion years ago. The event that marked the beginning of the universe was called the *Big Bang* by the British astronomer Fred Hoyle (1915–2001) in the course of a radio programme in 1950. Hoyle used the term disparagingly,[8] for he favoured his own *steady-state theory* of the universe in which, as the universe expanded, matter popped into existence to ensure that the density was preserved. The known rate of expansion of the universe—which is accepted in the steady-state theory—needs only a few hydrogen atoms to pop into existence in each cubic metre of space each 10 billion years, so the demands on whoever is making the matter are not too great. Indeed, we could even think of the stress of stretching space as generating the atoms, so matter creation is not *a priori* absurd; but particle creation appears to be an abnegation of the law of conservation of energy, and therefore distasteful however discreet.

Hoyle was attracted to the steady-state theory because it avoided the problem of what happened at the beginning, for there was no beginning: the universe had always been here and has always been expanding. It also avoided the need to ask the even more puzzling question of what went on before the universe came into being. However, the avoidance of questions is not a justification for the introduction of any theory; indeed, it is only an apparent simplification, for it is arguable that it is more difficult to understand why the universe has always been here than it is to find a mechanism for a beginning. On the whole, to scientists chains of causality are more palatable than retrospects of eternity.

The steady-state model of the universe, which was developed independently by Hermann Bondi and Thomas Gold in papers published in 1948 and 1949, is now no longer accepted as plausible by the great majority of scientists and, like Hoyle himself, has been quietly laid to rest. However, we should not scoff at the discarded too soon: later we shall see that current thinking has returned to a more sophisticated version in which *whole universes* erupt into existence even more often than the steady-state theory required little hydrogen atoms to pop into being.

There is, in fact a great body of evidence that supports the Big Bang model, most impressively the existence and detailed properties of the cosmic background radiation that we describe shortly. Few cosmologists now doubt that the early universe went through a stage when it was very dense and very hot. Indeed, through an extraordinary combination of theory and observation, and

---

[8] Hoyle said 'This big bang idea seemed to me to be unsatisfactory ... for when we look at our own Galaxy there is not the smallest sign that such an explosion ever occurred.'

drawing on our knowledge of the very small to explain features of the very large, we can now with reasonable confidence trace the story of the universe back to within a tiny fraction of a second after its birth. Hubble's astronomical legacy is the experimental discovery of the expansion of the universe; his intellectual legacy, however, is greater, for it is nothing less than the realization that we midgets can trace our history back almost to the beginning of time. It is his intellectual legacy that we explore in the remainder of this chapter and see that the scientific ideas generated in our Lilliputian laboratories have the reach to encompass the cosmos.

A hypersharp intellect can see at a glance that the universe has expanded. In 1826, the German astronomer and physician Heinrich Wilhelm Olbers (1758–1840) saw with half an eye that the universe was expanding, but did not realize that that was what he had seen. He publicized a question now known as *Olbers' paradox*, even though the problem had been known since Kepler suggested a resolution in 1610. Olbers pointed out that it was right to be puzzled by the fact that the sky is dark at night. You and I, with our untutored minds, might think the answer obvious: the Sun has set. But Olbers reminded his public that if the universe is infinite and eternal, then wherever you draw a line from your eye far enough out into the sky, it will end up at a star. So the sky at night should be as bright as the surface of the Sun, for the sky is effectively a sheet of Sun covering the heavens. Although our Sun may set, the myriad of others' suns do not.

There are two points to consider. The first, the simpler, is that if the universe was formed a finite time ago, then Olbers' argument fails because there has not been time for the light from very distant stars to reach us. So, instead of the sky being a sheet of sunlight, the sheet has gaps where stars are too far away to have contributed to our night sky.

The second point is rather more subtle and reduces further the intensity of the light that even with a finite universe we expect to reach our eyes. When we look into the distance, we are looking back in time, for light takes time to reach us. We see what was there when the light left, not as it was when the light arrives at our eyes. Even reading this page is a part of history, for you are looking at it as it was about a billionth of a second ago ($10^{-9}$ s, 1 nanosecond), not as it is at this instant. Most spectators of sports see them as they were yesteryear, or more precisely yestermicrosecond, not at the very instant a goal is scored but nearly a microsecond later. Astronomically distant objects emitted the light that now reaches us billions of years ago when the temperature of the universe was so high that the entire sky shone with the intensity of the Sun. Looking that far away and that far backwards in time we would expect, like Olbers, to see the sky awash with light. But since then, the universe has expanded and the waves of light that were typical of an object at about 10 thousand degrees ($10^4$ K) have been stretched out by an enormous factor. Instead of wavelengths measured in nanometres and visible they have become wavelengths measured in millimetres and invisible. Those waves are now characteristic of a much cooler body, one close to 3 degrees above absolute zero (3 K). The sky at night, then, does shine with something approaching the intensity of the surface of a star, but it shines with starlight so ancient and stretched that we consider the sky to be dark.

Scientists stumbled on to this explanation once the hot Big Bang model of the universe had become established as a theoretical possibility. On the basis of this model, it had also been predicted that the temperature of the universe should fall as it expanded because the wavelengths of the radiation filling all space are stretched. As a result, what once was short becomes long and the density of energy in the universe falls. The temperature turns out to be inversely proportional to the scale of the universe, so when the universe doubles in size its temperature falls to half its former value. Efforts were set in train to discover the remnant radiation of the Big Bang, but they were forestalled by two post-doctoral students, Arno Penzias (b. 1933) and Robert Wilson (b. 1936), whose job it was to clear pigeon droppings from a large microwave antenna. Well, that was not their sole employment: they were radio-astronomers who had taken over an antenna made redundant once the primitive Echo satellite transmission system had been replaced by Telstar, in the hope of using it for more fundamental radio-astronomy and were looking for the origin of a background hiss that plagued reception. After eliminating all terrestrial sources, which included scraping out pigeon droppings and turning their backs on Manhattan, they were left with the conclusion that the radiation is cosmic in origin. They had stumbled on the remnants of the fireball, Its brilliant radiation stretched out into microwaves and its electric thunder quenched to an almost silent electronic hiss.

Detailed study of the *microwave background radiation* over the following years has shown that it is exactly that expected to be emitted by a body at a temperature of 2.728 degrees above absolute zero (that is, at about minus 270 degrees Celsius, Fig. 8.3). Once allowance is made for our motion round the Sun, the motion of the Sun around the centre of our galaxy, and the wholesale drift of our local group of galaxies towards the Great Attractor, the radiation is

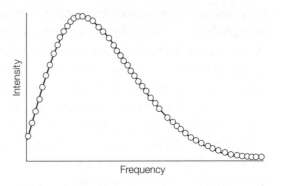

**Fig. 8.3** The intensity of radiation filling empty space can be measured at each wavelength, and the points show the values obtained. The solid curve is the intensity predicted by Planck's black-body radiation law (Chapter 7) for a body at 2.728 K.

the same in whichever direction we look. It is uniform to within one part in a hundred thousand and has characteristics that rule out a variety of other suggestions that have been made to account for its origin by those for whom a hot Big Bang is distasteful. There is no doubt that the universe was once exceedingly hot and exceedingly dense.

We can bring observation and theory together at this point and devise a little history. We know (by solving Einstein's field equations, his mathematical description of the gravitational field in the presence of massive bodies, Chapter 9) how the scale of the universe will change with time subject to certain assumptions about how much matter it contains (of which more anon). We know the present rate of expansion, from our determination of Hubble's constant, and we know how the temperature of the universe is related to its scale. How do we know that? The intensity of radiation at different wavelengths depends on the temperature (remember our discussion of black-body radiation in Chapter 7, and Fig. 8.3), and the wavelengths are stretched as the universe expands, so there is a relation between temperature and scale. By combining the relation of temperature to scale and of scale to time, we can work out how the temperature of the universe changes with time.

We can take this connection further because we know from experience in our laboratories how temperature brings about change. We know how the temperature of the universe, the cosmic furnace, then oven, and later refrigerator, has changed over time, so we have a means of inferring how the properties of the universe have changed from shortly after its beginning. Broadly speaking, the effect of elevated temperature is to shake things apart, with particles that are tightly bound able to survive at high temperatures but particles that are only weakly bound able to survive only at low temperatures. We use that principle in the kitchen, where roasting and boiling helps to break matter down into smaller, more digestible, more aromatic molecules, and freezing helps preserve them by slowing the reactions that result in decay. The temperature of the cosmos has a similar culinary function, but the material we are cooking in the cosmic oven is the stuff of matter itself.

'Shortly after' in the last paragraph is a weasel phrase that needs interpretation. When the currently observable universe was packed into a region with a diameter equal to a quantity called the *Planck length*, of just under 200-billion-trillion-trillionths of a metre (that is, $1.6 \times 10^{-35}$ m, a fundamental quantity we shall encounter again in Chapter 9), our current physics falters. For events when the universe was this compact, we need a quantum theory of gravity. That theory is starting to emerge, but currently we have so little confidence in it that I will detach this quantum-palaeolithic era from our history and look at it separately later. Our burrowing back in time emerges from this fog of ignorance at the *Planck time*, about $5.4 \times 10^{-44}$ s after the beginning, when the temperature had its Planck value of about $1.4 \times 10^{32}$ degrees. That was about 15 billion years ago: not within living memory, but not so awesomely distant that we cannot imagine it. It is really quite remarkable that so much has happened in such a short time. We cannot, like Bishop Ussher and his minute analysis of the Bible, give an exact date like his 23 October 4004 BCE, at noon, just in time for lunch,[9] but the precision of our identification of the beginning is growing as we get to understand the dynamics of the evolution of our universe, and soon we can hope to pin the beginning down to within a billion years or so.

There is one further initial point we need to address. It is sometimes asked *where* the Big Bang took place. The answer is extremely simple and precise (as good answers always are): it took place *everywhere*. The universe did not explode *into* anything, and in so far as 'big bang' gives the impression of an explosion, it is an unfortunate name. The Big Bang filled the whole of space: it occurred everywhere.[10] Nor, necessarily, has the universe ever been a point. If the universe is set to expand for ever (rather than crunch back down again), then there has *always* been more mass outside any given region than there is inside that region, even at the instant of creation. That is, if the universe is 'open' and set to expand for ever, then it has always *already* been infinite. So even if the *visible universe*, the universe with which we can currently interact—which extends to about 15 billion light-years from us in all directions, the light of creation that far

---

[9] Often misquoted as 26th at 9 a.m. See http://www.merlyn.demon.co.uk/critdate.htm.

[10] Perhaps the word 'plosion' would convey a better image than 'explosion'.

distant just having had time to reach us now—was once compacted inside an infinitesimal point, there would still be an infinite region outside that point. Only if the universe were 'closed', in the sense of undergoing a Big Crunch at some distant time in the future—an event increasingly thought unlikely as experimental evidence relating to the rate of expansion accumulates—would it be correct to think of the entire universe as initially packed into a point.

We also need to understand how the expansion of the universe is expressed. In what follows, I will refer not to the size of the universe, which is probably infinite at all times, nor to the size of the visible universe, which corresponds to a radius of about 15 billion light-years now but was smaller earlier, but to its *scale*. By its 'scale' I mean the factor relating the distance between two points that are currently separated by 1 metre. So, when the scale is 100, those two points will be 100 metres apart; when the scale was one-billionth ($10^{-9}$), the two points were one-billionth of a metre apart ($10^{-9}$ m). Einstein's field equations can be adapted to calculate the time dependence of this scale factor for various models of the universe. The first reasonably realistic solutions were found by the Russian mathematician, aviator, balloonist, and meteorologist Aleksandr Aleksandrovich Friedmann (1888–1925), who proposed them in 1922 shortly before he died of typhoid. They are known as the *Friedmann equations* (Fig. 8.4). The same equations were found by the Belgian cleric Abbé Georges Lemaître (1894–1966) in 1925; he was the first to trace them back in time and identify what he termed the 'cosmic egg' and we now refer to as the Big Bang.

Cosmologists currently believe that the universe is neither open nor closed but 'flat'. A flat universe is like an open universe insofar as its scale will expand for ever, but its expansion gradually slows and becomes infinitely slow as its

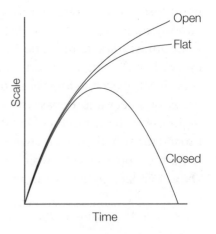

**Fig. 8.4** The history of a Friedmann universe. If the universe has less than a certain density, then it will be 'open' and expand for ever. If the universe has more than a certain density, then it will be 'closed', and after an initial expansion phase it will contract back to a 'Big Crunch'. If the universe has exactly the critical density, it will expand for ever, but glide to a halt as time approaches infinity. Current measurements suggest that the universe is not closed. There is new observational evidence that suggests that the universe is open and might recently have entered into an accelerating phase.

Open

Flat

Scale

Closed

Time

scale approaches infinity. In a flat universe, as in an open universe, there is no limit to the ultimate separation of two points that are currently 1 metre apart. An implication of being flat, like being open, is that the universe has always been infinite in extent, and therefore that the Big Bang occurred everywhere over an infinite volume of space. When people say that the universe was initially very small, they mean—they should mean—that the scale was initially extremely tiny and that two points that are now 1 metre apart were then a tiny fraction of a metre apart. With a huge amount of matter stuffed into a tiny region, you can appreciate that it was very dense; in fact, it was about $10^{97}$ times more dense than water. And it was that dense everywhere, everywhere over an infinite region. There was, there always has been, and there always will be, an awful lot of universe.

The final preliminary point that sometimes gives people pause is to understand that although the scale of the universe is increasing with time, that does not mean that the objects it contains are getting bigger. Ourselves and our measuring rods do not expand with time, nor do the distances between the stars within a galaxy. There are several ways of understanding this sometimes perplexing point. The simplest is to accept the view that the Friedmann equations describing the expansion are based on a model in which the matter is regarded as averaged out over the entire universe, with the galaxies best regarded simply as notional points indicating locations in space. The expansion of the scale refers only to this 'average universe' and is silent on the behaviour of tiny systems inhabiting the space. Another way of coming to the same conclusion is to note that if two points, such as two stars in a galaxy, are bound together by an attractive force, then that force is not overcome by the expansion of the universe and the points remain at the same separation however long we wait.

The purist's more subtle way of thinking about this tricky but important point is that the Friedmann equations tell us how two points will move apart *given that they are moving apart initially*. That's a bit like Newton's equations of motion, which tell us how to calculate the distance a ball will travel once we know how fast it was travelling initially. If the ball is stationary, then however long we wait it will remain at the same place. Likewise, if two points in space— your head and your heels—are not moving apart initially, then however long we wait they will always remain in the same relative positions. We don't get stretched by the expansion of the universe any more than in classical physics a stationary ball moves to another location.

With these remarks in mind, it is time to begin to come to terms with our history. At the Planck time, it is presumed that all the forces holding matter together (the gravitational, electroweak, and strong forces discussed in Chapter 6) had the same strength, but as the universe cooled to below the Planck temperature, the gravitational force separated from the other two. These two continued to have identical strengths and to be propagated by massless bosons. Then nothing much happened for ages. To be precise, the electroweak and strong forces maintained their equal strengths for 10 billion ticks of the Planck clock, until what we would call 1 billion-trillion-trillionth of a second ($10^{-33}$ s) after the Bang. Using the ticks of our own ponderous clocks is misleading, for our clocks have been devised to suit human convenience, and the ticks of clocks on town halls are unsuited to the discussion of events when the universe was very young, very hot, and very dense. The early expansion of the universe was extraordinarily slow—like slimemould on Valium—when measured in natural units, Planck ticks; from that point of view, it is easy to see how so much change could come about in what we ponderous, lethargic giants would call the twinkling of an eye.

After this immense time (10 billion Planck ticks, what you and I would call 1 billion-trillion-trillionth of a second) had passed, the temperature had fallen far enough for the strong force to separate from the electroweak force, so from now on in this ever colder universe they would seem to be unrelated. Once again, events in the universe came to a virtual standstill. The universe expanded and its temperature fell, but we have to wait almost an eternity—to be precise, until Planck's clock has ticked $10^{30}$ times—before anything discernible occurs in this extraordinarily lazy world. You might be tempted to think of the wait as just another twinkle of an eye, as one ten-trillionth of a second ($10^{-13}$ s), but that would give you a false sense of the awesome slowness of events in the early universe and you might wonder how anything had time to happen at all. By now, the scale of the universe has expanded to an enormous $10^{15}$ Planck lengths. Of course, when measured in units more suitable to later epochs we think it very small, with what will become a separation of 1 metre being only $10^{-20}$ m then, but farmyard units are not at all appropriate and very misleading. It has cooled to 10 thousand trillion degrees ($10^{16}$ K), which is cold enough for scalar particles (perhaps they are Higgs bosons) to stick to the W and Z gauge bosons to give them mass, thereby limiting their range and distinguishing the weak force from the electromagnetic force for the rest of time. The universe is now so cold that the forces have acquired the separate identities that will distinguish them for ever.

There is still nothing that we would identify as matter: the temperature is still enormously high, and thermal motion shakes anything apart that might, under the influences of the forces, start to coalesce. The first forms of matter to crystallize out of the inferno as its temperature falls are the nucleons (protons and neutrons), formed as quarks are gripped together by the strong force. This coalescence can occur only when the temperature has fallen to an extraordinarily cold 10 trillion degrees ($10^{13}$ K). Cold? Well, it is bitterly cold on a Planck scale, for it lies only $10^{-19}$ Planck degrees above absolute zero. It is hugely hot, of course, on our everyday scale of temperature, but that scale was invented for reporting our terrestrial weather and is not in the least fundamental.

I will now relax my insistence on talking in terms of fundamental units and resort to farmyard units, for at this stage of the evolution of the universe they are much more convenient than Planck's natural units. However, you must keep in mind that twinkles of the eye in conventional units are in fact almost immeasurably lengthy epochs. What seems brief to us can be a concatenation of countless events in natural, fundamental units. A bullet travelling at the speed of sound takes almost an eternity, a hundred trillion trillion ($10^{26}$) Planck ticks, to travel the width of an atomic nucleus.

At one second after the beginning, neutrinos decoupled—shook themselves free—from matter. Never again would they interact appreciably with it and from that moment on they would fly almost unimpeded through the universe, speeding freely through space and penetrating planets as though they were almost perfectly transparent crystal spheres. If we had eyes to see with neutrinos, the almost massless particles of spinning flavour, then we would consider the world to be almost empty, with just a ghost of a shadow here and there.

At first thought, we might expect the neutrino sky to be brighter than the photon sky, as neutrinos kept the stamp of the universe, in the form of its temperature, when they first decoupled and the continuing expansion of the universe would have cooled them less. But in fact the neutrino background is cooler than the microwave background, being slightly under 2 degrees above absolute zero.[11] The reason for this cooler neutrino sky is that various events, specifically the collision of electrons with their antiparticles the positrons, have augmented the number of photons and have increased the brightness, and therefore the temperature, of the microwave sky.

At three minutes after the beginning, the temperature has fallen to 1 billion degrees. It is so cold (at only $10^{-23}$ Planck degrees) that under these Arctic

---

[11]  The temperature of the neutrino background is expected to differ from that of the photon background by a factor of $(4/11)^{1/3}$, giving 1.95 K for its temperature compared with 2.73 K for the photon background.

conditions even nucleons can stick together, forming deuterium (heavy hydrogen, a nucleus consisting of a neutron stuck to a proton) and helium (two protons and two neutrons stuck together). Calculations show that as the temperature continues to fall, this epoch of the universe results in about 23 per cent helium, 77 per cent remaining as hydrogen (uncombined protons), and just a hint of slightly heavier elements (lithium and beryllium, for instance, with three protons and four protons, respectively, and a few neutrons managing to hang on and help to keep the protons close together). The abundance of helium depends critically on the number of types of neutrino, and is incompatible with any number greater than four. As we saw in Chapter 6, there are three known flavours of neutrino, which is in line with this constraint. More important perhaps, is that we see how the very large—in this case, the abundance of helium throughout the universe—is a consequence of ideas emerging from the study of the very small. It is this mutual compatibility of knowledge emerging from the immense and the minute that inspires confidence in the reach of science.

Now nothing significant happens for ages. Even in farmyard units, the composition of the universe remains much the same for a hundred thousand years. The universe continues to expand for all that time and to cool, but it remains a plasma, a swarm of nuclei bathed in a sea of electrons. As such, the universe is brilliantly hot but opaque, rather like the Sun we see today, for light can travel only short distances through such a medium. For the same reason, the Sun is opaque to us, not a transparent ball.[12] A photon has a tiresome journey of about 10 million years from the centre of the Sun to its freedom at the surface. Every fraction of a second it is absorbed and re-emitted, travelling first one way and then another. Only when light decouples from this boggy plasma and enters empty space does it fly off exuberantly at the speed of light. If the centre of the Sun died today, its light wouldn't falter for another 10 million years. Much the same conditions prevailed in the early universe, with light inching its way through almost impenetrable, brilliant plasma.

Suddenly, at a hundred thousand years into its unwinding, the skies clear as though on a cloudy summer's day: the universe becomes transparent and light is free to travel. There isn't much to see when the sky clears; indeed, there is nothing to see, for the stars have not yet formed, but it is a crucial moment in our history. At this clearing of the skies, the Arctic cold has fallen to a mere 10 thousand degrees ($10^4$ K), and under these freezing conditions electrons are at last able to stick to the nuclei. The plasma condenses to neutral atoms, the elec-

---

[12]  Metals are opaque for the same reason. They too consist of nuclei surrounded by an electron sea, the difference from the early universe being that the nuclei form an orderly array.

trons, once free but now captured, are no longer able to scatter radiation so effectively, and light can pass freely through the void.

The electromagnetic radiation—the light—released from its bondage to matter is now brilliantly hot, at 10 thousand degrees, not unlike the surface of the Sun today, and all around us is this searing brilliance. All is photosphere; Kepler's messenger Olbers would be pleased, for this is the origin of his darkless night. As the universe continues to expand, this light is stretched into the microwave background that surrounds us today. As we have already seen, our present sky is still a flaming fiery furnace, but its temperature has fallen to 2.7 degrees above absolute zero. The *cosmic background radiation* peaks in the microwave region: it is invisible to us unless we extend our eyes with radio telescopes and listen to the gentle hiss of the waves as they brush against our detectors.

At last there are atoms in the universe. They are not particularly abundant and there is hardly any variety in them. Were we to average out the matter today and spread it over the entire universe, then we would find only about one hydrogen atom in any cubic metre. The only elements emerging from this immediate era of the Big Bang are hydrogen (lots of it), helium (lots of it, but less than hydrogen), and a relatively tiny sprinkling of lithium and beryllium. The universe, at three minutes old, is an incredibly lonely and primitive place.

So it would stay for a billion years. However, the universe had potential for extraordinary diversity, and that potential slowly began to be realized. For reasons to which we must remember to return, the primeval universe was not absolutely smooth. In some regions the primeval gas of hydrogen atoms, helium atoms, and the enigmatic 'dark matter' of the universe that we come to later, was a little denser than it was elsewhere: there were slight ripples in its distribution. As the universe aged, the gas in the denser regions began to condense under the influence of gravity. As these local globular regions formed and the gas became compressed, they became hot. In due course, they became so hot that the nuclei of the hydrogen atoms collided with such force that they fused together, releasing energy. *Nuclear fusion* had begun, the star had begun to shine, and the clusters of stars we call galaxies had burst into life. The distribution of the galaxies is far from random, since they mark out the denser regions of the ripples: there are clusters and great voids on a scale of hundreds of millions of light-years in extent (Fig. 8.5). This enormous pattern is a magnification of the ripples that marked the inception of the universe, when densities varied on a

**Fig. 8.5** The distribution of galaxies as seen from Earth. Each dot represents the location of one galaxy. The feature to note is that the distribution is not uniform: there are long fibres of galaxies and huge regions with a lower than average number of galaxies. These inhomogeneities are the colossally amplified remains of density fluctuations in the primordial universe.

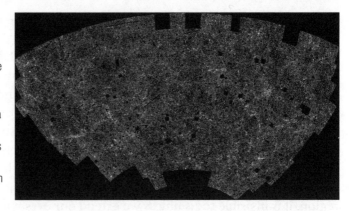

scale of a few Planck lengths but have been stretched into their current enormities. It has taken the universe 15 billion years to reach this stage, but that relatively brief period in terms of our irrelevant human units (for why does it matter how many times a minor planet has swung round a minor star?) is an immense stretch of time on Planck's more fundamental clock and amounts to no less than about $10^{61}$ ticks (Fig. 8.6).

The ancient stars are formed from hydrogen, but as they consume hydrogen in the process of nuclear fusion, they form new elements. *Nucleosynthesis*, the

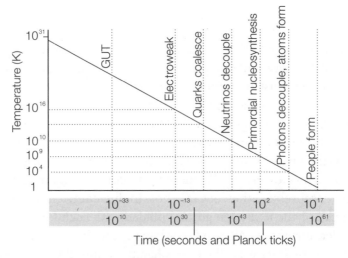

**Fig. 8.6** The time scale of events during the lifetime of the universe. The temperature during the inflationary era is still a subject of speculation, and the linear dependence on time suggested by the graph should not be interpreted literally. After the GUT era, the strong force separates from the electroweak; after the electroweak era, the weak and electromagnetic forces separate. The temperature quoted is that of the electromagnetic field, and people form when the local environment is close to 300 K, even though the electromagnetic field is much colder.

formation of the elements, has begun and the universe has started to become more diverse. The formation of elements in the very early life of the universe, before any stars were formed, is called *primordial nucleosynthesis*. It doesn't get very far, largely because nuclei are built up by successive addition of nucleons to protons, giving deuterium (one neutron hanging on, by its strong-force finger-nails, to a proton), helium (two protons and two neutrons in a reasonably stable arrangement), and so on. However, there are no stable nuclei with five and eight nucleons, so there is a bottleneck at this stage and it is hard for heavier nuclei to grow as a result of collisions. The most abundant element formed at this stage is helium, which then and still makes up about 23 per cent of the universe, almost all the rest being hydrogen. This abundance of helium can be predicted from the Big Bang theory, and its experimental confirmation is powerful support for the theory.

Almost all the other elements in the universe had to await the formation of stars before they saw the light of day. This is not the place to go into this branch of nuclear physics any more than to say that the fact that the stars are shining, including the Sun, is a sign that elements are still being made (or, at least, they were being made about eight minutes ago). The astronomer Arthur Stanley Eddington (1882–1944) was the first to suggest that the fuel of stars is the energy released as hydrogen nuclei collide and fuse together into helium.

Stars are very dangerous objects, as may be suspected of such huge globules of fiercely hot matter hanging in the sky and undergoing unconstrained nuclear fusion. They do not burn smoothly like a firelighter and then slowly fizzle out. Theirs is a violent history, with nuclear reactions taking place in shells deep within the star, these shells growing, shrinking, collapsing, and giving off pulses of energy that can ablate the outer layers of the star and puff them off into space.

The tumultuous life story of a star begins with a cloud of gas. Whether or not such a cloud will draw itself together under the influence of gravity depends on a variety of factors, including its density, temperature, and mass. The minimum mass of a cloud of given temperature and density that can form a star is called the *Jeans mass*, after the astrophysicist James Jeans (1877–1946) who studied and built theories of the formation of stars. Diffuse clouds, with their low density, are typically stable against gravitational collapse and do not form stars. A dense cloud, however, will collapse, and for a typical cloud composed of hydrogen and helium the Jeans mass is equivalent to about seventeen Suns. However, as the cloud collapses in on itself, its density increases, the Jeans mass decreases, and instead of forming just one huge star, smaller regions of the cloud can themselves undergo gravitational collapse, so the cloud fragments and forms clusters of smaller stars. Potential stars about one tenth as massive as our Sun don't get hot enough for nuclear reactions to begin and are stillborn: they never shine. Potential stars about ninety times more massive than the Sun are unstable: they break into oscillation and fall apart. So, all stars have masses between these two limits.

The gas destined to form a star—which is mostly hydrogen and helium—is in free fall towards a common centre. As the atoms fall, they jostle together, and these collisions cause the temperature to rise. There comes a stage when the temperature in the collapsing cloud is so great that the nuclei collide with such violence that they meld together and form helium, and helium nuclei smash together and form heavier elements. For stars about 20 per cent more massive than the Sun, the temperature can rise even higher. Above about 20 million degrees particles are moving so fast that protons can crash successfully into more highly charged nuclei, such as carbon, nitrogen, and oxygen, and release energy as they are captured.

Stars bigger than about eight Suns have a violent future. The temperature in these giants can rise so much, to around 3 billion degrees, that 'silicon burning' takes place, in which helium nuclei can merge with nuclei close to silicon and gradually build heavier elements, stepping through the periodic table and finally forming iron and nickel. These two elements have the most stable nuclei of all, and no further nuclear fusion releases energy. At this stage, the star has an onion-like structure with the heaviest elements forming an iron core and the lighter elements in successive shells around it (Fig. 8.7). The duration of each of these episodes depends critically on the mass of the star. For a star twenty times as massive as the Sun, the hydrogen-burning epoch lasts 10 million years, helium burning in the deep core then takes over and lasts a million years. Then fuels get burned seriously fast in the core. There, carbon burning is complete in 300

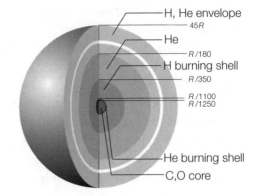

**Fig. 8.7** The internal structure of a typical star of about five solar masses as it approaches its red giant phase with a carbon–oxygen core. For clarity, the radii of the inner shells have been increased relative to the surface (the white band indicates the change of scale).

H, He envelope
45R
He
R /180
H burning shell
R /350
R /1100
R /1250
He burning shell
C,O core

**Fig. 8.8** The remnants of a Type II supernova (the Vela remnant). This supernova occurred about 11 thousand years ago, and we can see the manner in which matter—elements formed in the interior of the star— is spread through the universe. Vela, the Sails, is a bright constellation of the southern Milky Way; it was once considered part of the constellation of Argo Navis, Jason's ship. It is very hard to distinguish between the various types of supernova.

years, oxygen is gone in 200 days, and the silicon-burning phase that leads to iron is over in a weekend.

The temperature is now so high in the core, about 8 billion degrees, that the photons of radiation are sufficiently energetic and numerous that they can blast iron nuclei apart into protons and neutrons, so undoing the work of nucleosynthesis that has taken billions of years to achieve. This step removes energy from the core, which suddenly cools. Now there is little to maintain the structure of the core, and it collapses. The outer parts of the core are in free fall and their speed of collapse can reach nearly 70 thousand kilometres a second. Within a second, a volume the size of the Earth collapses to the size of London. That fantastically rapid collapse is too fast for the outer regions of the star to follow, so briefly the star is a hollow shell with the outer regions suspended high over the tiny collapsed core.

The collapsing inner core shrinks, then bounces out and sends a shockwave of neutrinos through the outer part of the core that is following it down. That shock heats the outer part of the core and loses energy by producing more shattering of the heavy nuclei that it passes through. Provided the outer core is not too thick, within 20 milliseconds of its beginning, the shock escapes to the outer parts of the star hanging in a great arch above the core, and drives the stellar material before it like a great spherical tsunami. As it reaches the surface the star shines with the brilliance of a billion Suns, outshining its galaxy as a *Type II supernova* (Fig. 8.8),[13] and stellar material is blasted off into space.

[13]  We meet Type I supernovae later.

The death of a star brings life to the universe. The explosion of the star leaves the compressed core as a *neutron star*, a tiny, immensely dense, smooth body composed of neutrons, or, if the initial mass of the star was greater than about twenty-five Suns, even a *black hole*, a region with so fierce a pull of gravity that even light cannot escape. Far more important, though, in the short term at least, is the shrapnel, for in this way the elements cooked and baked in the star from the primordial hydrogen and helium are scattered through the galaxy. Those elements might become incorporated in new generation of stars. Some though, will aggregate into dust, dust will aggregate into rock, rock into boulder, and boulder into planet. The planets, should they form around a hospitable star, as the Earth did around the Sun, are now rich in the building blocks of life, life that in at least one place and almost certainly myriads more, became capable of discovering its own grand cosmic and petty local history (Chapter 1). We are the creatures of starlight:[14] from cosmic violence slowly emerged science, art, and joy.

Let's go back to the beginning for a moment. The Big Bang account of our history has major successes. Predictions based on it are quantitatively in agreement with observation, where observations can be made, and there is little doubt that in general terms the history is correct. However, there are several difficulties with the theory, and it cannot be the last word about the First Word.

First, we have seen that 'expansion of the universe' actually means that two points moving relative to each other will become further apart as time goes on. That is, all the theory is saying is that if two points are moving now, then they will be moving later. There is no explanation, within the theory, of why they were moving in the first place!

Second, the universe is extraordinarily uniform, despite the fact that different parts of the universe had no time to communicate with each other. To understand this point, think of two points 15 billion years away from us in opposite directions across the visible universe, with us in the middle. Light has just had time to reach us from each point, but light has not had time to travel between those two points, for they are separated by 30 billion years. If we do the calculation carefully, it turns out that the sky can be thought of as divided into a hundred thousand little patches, each about 1° on a side, that have never had time to exchange signals at the speed of light. Why, then, is the sky so uniform,

---

[14]   I am tempted to say that 'human flesh is stardust'. But it has been said so many times elsewhere, that I will confine this wonderful remark to a footnote. I think Nigel Calder said it first.

with almost exactly the same temperature (2.7 degrees) wherever we look? This is called the *horizon problem* because each part of the universe needs to be able to communicate, somehow, with regions that are in some sense over its immediate horizon for otherwise the currently observed universe would not be so uniform, just as two blocks of hot iron wouldn't have the same temperature unless they had once been in contact.[15]

Thirdly, there is something very odd about the shape of the universe. In fact, the shape is doubly odd. One oddity is that the universe has *almost* the right amount of matter to indicate that it will *just* expand for ever. This criterion is normally expressed by saying that the universe has almost its *critical density*. The other oddity is that the universe doesn't *quite* seem to have the right amount of matter: current determinations of the amount of matter in the universe are levelling out at only about 1 per cent of the critical density.[16] There are very good theoretical reasons for believing that the difference between the observed density and the critical density increases as the universe expands and that by now, 15 billion years after the beginning, this difference will have been magnified by a huge factor. For instance, if the difference was only one part in 10 thousand trillion (1 in $10^{16}$) when the universe was 1 second old, then the difference would be enormous now, not just a factor between ten and a hundred. The requirement is even more stringent the further back we go. For the density to be anywhere near its critical value now, after one tick of the Planck clock the difference could not have differed by more than 1 part in $10^{60}$! These figures strongly suggest that the density had exactly its critical value when the universe was born and has retained that value ever since. This awesome requirement is called the *flatness problem* and is part of the more general *fine-tuning problem*. The fine-tuning problem continues to puzzle cosmologists and suggests to those of a more sentimental disposition that someone had to make sure that the density was exactly critical initially and that various other parameters must have been ascribed particular, special, and ultimately (for us) benign values in the original specification of the universe.

The related problem is that it would be exceedingly surprising to find that we were alive at just the epoch when the critical density has arrived close to its critical value. It is far more likely that the density has always had and now

---

[15]  More specifically, the horizon of a point lies at the distance light can have travelled to it by the current age of the universe. For a universe that is $10^{-8}$ s old, the horizon of a point lies 3 metres away.

[16]  The density is expressed in terms of the parameter $\Omega$ (omega), with $\Omega = 1$ the critical density. For $\Omega > 1$ the universe is closed; for $\Omega < 1$ it is open. For $\Omega = 1$ it is flat: its rate of expansion slowing to zero as its scale approaches infinity.

continues to have exactly its critical value.[17] If that is so, then because the meas-ured density is seriously less than the critical density, it follows that we have not identified all the matter in the universe. There is further evidence for that con-clusion from the rate at which galaxies rotate, which suggests that they contain more matter than we see by counting the stars, and current estimates of the density put it at at least 20 per cent of its critical value. Where and what is this *dark matter*? The simplest answer is that it consists of the corpses of old, dead stars. If this were the form of the dark matter, there would have to be a thousand or more Jupiter-sized bodies for every star about the size of the Sun. Surely we would have seen this buzzing hive by now? At least these bodies have a name, which is often the first step towards existence: each one is a MACHO, a *massive astrophysical compact halo object*. The alternative explanation, inevitably, is the existence of a WIMP, a *weakly interacting massive particle*. The latter are particles that interact so weakly with matter that we might be able to detect them only by their gravitational pull or weak interaction. One candidate was once thought to be the neutrino, provided it has mass, but that suggestion is now thought unlike-ly as neutrinos travel almost freely on the scale of galaxies and give rise to too much structure on much larger scales. A more exotic candidate is one of a num-ber of varieties of sparticles, the undiscovered, supposed, speculative, super-symmetric partners of the known particles (Chapter 6). Whatever the solution, it is a sobering thought for scientists that they have not yet identified the most abundant form of matter in the universe.

The fourth problem with the Big Bang is that there don't seem to be any 'magnetic monopoles' around today. We are all familiar with a bar magnet, with its north and south poles. A *magnetic monopole* is one of these poles without the other, the magnetic equivalent of an electric charge. If electricity and mag-netism are two faces of a single force, why do magnetic monopoles always go round in pairs and are never, like electric monopoles ('charges'), found alone? In the Big Bang model, such is the stress of the initial tumultuous event that lots of defects—nicks, tears, creases, badly aligned bits—were introduced into space-time, the point-like nicks being the magnetic monopoles. According to the Big Bang theory, there are predicted to be more monopoles than ordinary matter; but not one—not one—has been found.

The fifth problem is one we have mentioned already: the large-scale struc-ture of the universe as represented by Fig. 8.5, with galaxies clustered around voids on a scale of a hundred million light-years. There we saw that this struc-

---

[17] If the density has its critical value initially, then $\Omega - 1 = 0$ initially, and multiplication of 0 by any factor, however enormous, leaves $\Omega - 1$ equal to zero at all later times and therefore $\Omega = 1$ always, including now.

ture is a greatly magnified version of the lumpiness of the primordial universe, when its scale such that today's observable universe was little more than an infinitesimal point. But why was the point lumpy in the first place, and why did it have the lumpiness that, in due course, became what we find today? This problem is completely outside the reach of the Big Bang theory. We cannot claim to understand our universe if we have no idea about the origin of the biggest objects in it!

These five problems—the origin of the expansion, the horizon problem, the flatness problem, the missing monopole problem, and the presence of large-scale structure—are very serious. The Big Bang theory, though, is so successful in other respects that it would be hard to give it up. Indeed, experiments virtually confirm that the universe went through a very hot phase and has been expanding ever since. The answer must lie in events taking place in the earliest moments of the Big Bang, events preceding those considered so far (but still on this side of the absolute beginning). The currently favoured theory is a variety of *inflation*.

Inflation is no ordinary expansion. Inflation is *very* rapid expansion. You will have realized by now that I do not use 'very' lightly, and that I use '*very*' even less lightly. In this instance I mean expansion at greater than the speed of light. In fact, I mean at *very* much more than that. Don't worry about something happening faster than light can travel: there is no particular difficulty in the concept of superluminal expansion, for it is the scale of space that is expanding; we are not considering the propagation of signals through that space. In the inflationary scenarios (there are several versions, each revolving around one central axis of an idea), something—we'll come back to what—switches on at about $10^{-35}$ s after the beginning. Then the action begins. Each $10^{-35}$ s thereafter, the universe more than doubles in size,[18] and it continues to more than double in size in each subsequent $10^{-35}$ s until inflation switches off at about $10^{-32}$ s: that is time for a 100 more-than-doublings. Think what that means in more human terms. Let's take the initial size to be 1 cm. One more-than-doubling brings us to 2.7 cm. Two doublings to 7.4 cm. Three to 20 cm. By ten we have reached 220 metres. By twenty we have got to 4852 kilometres. By fifty we are at 5480 light-years (achieved, remember in $5 \times 10^{-34}$ s). Two more more-than-doublings embraces the galaxy. A few more the local group. After 100 more-than-doublings, the original object has grown by a factor of $10^{43}$. In some versions of inflation, the expansion is even greater, such as by a factor of 10 multiplied by itself a trillion

---

[18]  By a 'more-than-doubling' I mean an e-folding, an increase in size by the factor e = 2.718 . . .

times, or $10^{1\ 000\ 000\ 000\ 000\ 000}$. That is an enormous, really enormous, magnification to have taken place in $10^{-32}$ s!

We should stand back a little from that remark. I have purposefully dramatized inflation by talking in terms of human units. By now, though, you will realize that a better way of thinking is in terms of fundamental units. From this more fundamental viewpoint, inflation is as leisurely as it comes. First, that $10^{-35}$ s to get under way. That initiation period is actually a very long time indeed, for it corresponds to a hundred million Planck ticks (there are nearly a hundred million seconds in three years, so to make the time digestible, think of it as three years). Whatever was getting ready to switch on had plenty of time to gather its resources. Then the more-than-doubling period: that takes another leisurely hundred million ticks—another 'three years'—for each episode, which is hardly frantic.

Let's see how inflation solves the problems with the Big Bang model. The horizon problem is solved because all the points that today are far too far apart to be in contact at the speed of light were in fact extremely close together initially and had plenty of time to communicate with one another. In other words, all our current visible universe was once packed together into such a small region that signals had time to travel across it and bring about its homogenization. The flatness problem is solved because inflation flattens any initial curvature, just as blowing up a crinkly balloon smooths its surface. The monopole problem is solved because even if monopoles were present initially, there would now be only about one in our region of the universe and it is not surprising that it has not been spotted. The reason why there is matter here is that it forms after inflation, whereas monopoles formed before and were inflated away. The final point to emphasize is that, if inflation is true, then the universe is very much bigger than we ever thought, and what we can see—what we can ever see—is only a very tiny fraction of all there is. Humiliation too has inflated, and there is more to come.

We still have the problem of how inflation began. We also have a new problem: how did it end? Why did inflation run out of steam after $10^{-32}$ s? The inflationary idea was first introduced by the Dutch astronomer and mathematician Willem de Sitter (1872–34) in 1917. He realized that if the vacuum has an energy, then inflation would occur. The possession of energy by the vacuum should not be thought to be too troublesome: what we regard as 'vacuum' is arbitrary anyway, and empty space should not be thought of as absolutely nothing. We shall suppose that the vacuum is filled with a field, which is called the *inflaton field*. A very primitive way of imagining the inflaton field is to think of the universe as being connected to one electrode of a battery, then it would have a uniform voltage, of say 12 volts, throughout. That voltage would not be detectable by any experiment that we could do, and we could call it a *false vacuum*. We could then imagine disconnecting the battery and discharging the universe, in which case the 12-volt vacuum would change into a zero-volt *true vacuum*. The two versions of the vacuum would look the same to us, but they are different.

Because these ideas are rather odd, it might help to see them in a broader context. First, it is notable that chemists didn't think that the air was anything to study until later in the development of their subject, for how could the seemingly insubstantial have chemical properties? We might think the same about the vacuum. The history of science appears to be heading along the track of studying more and more about what is less and less: air is old hat, various different vacua are now at the centre of physicists' attention, and presumably, when they come to build theories of the actual moment of inception of the universe, then they will have to study absolutely nothing at all. Maybe we shall discover that absolutely nothing has properties and can also take on different forms![19]

The question we have to tackle is how the possession of energy by a vacuum results in rapid inflation. The mechanism is a kind of positive feedback. First, we note that as more vacuum is created as the universe expands, then if the vacuum possesses energy the total energy of the universe increases. Next, the Friedmann equations show that the rate of expansion of the universe increases with the energy it contains, so the rate of expansion increases as it expands. Because the rate of expansion turns out to be proportional to the scale of the universe, so the scale increases exponentially with time. Exponential changes grow very rapidly, so ever more rapid inflation occurs so long as the inflaton field is present (Fig. 8.9).

The problem with de Sitter's model, though, was that there was no method to halt inflation. Inflation continued for ever, with the result that all matter and radiation was quickly diluted to zero, leaving an empty universe. That is contrary to experience, so his inflationary model was discarded and largely forgotten. However, in the late twentieth century the concept of inflation was revived in two isolated islands of intellectual activity, each over the horizon of communication of the other. One centre of activity was in the then Soviet Union, where in 1979 Alexei Starobinsky used ideas drawn from general relativity to

---

[19] I am not entirely joking. If the universe did emerge from absolutely nothing, then presumably there will come a stage when we will have to examine how something can come from absolutely nothing, *ex nihilo*. One day scientists will have to study absolutely nothing.

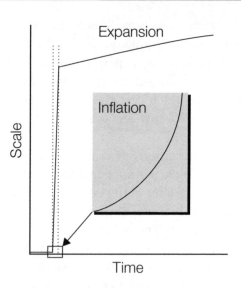

**Fig. 8.9** The inflationary universe. A short time after its inception, the scale of the universe begins to increase at a colossal rate, more than doubling its radius every $10^{-35}$ seconds. The inflationary era sees an exponential growth in size, but comes to an end after $10^{-32}$ seconds. From now on, the expansion is much more leisurely and corresponds to one of the scenarios shown in Fig. 8.4.

develop an earlier idea that another Russian, Erast Gliner, had suggested in 1965. Over the horizon in the United States, Alan Guth was considering the problem of the unwanted creation of magnetic monopoles as a problem in particle physics, and in 1981 arrived at a similar idea.

The central idea of this early version of inflation was to regard it as taking place like a 'phase transition', a change of state analogous to the freezing of water to form ice. As the universe inflates, it cools, and may cool to almost absolute zero, so colossal is the expansion. There comes a moment, though, when the false vacuum collapses into the true vacuum and releases a huge amount of energy. To picture this transition, think of the false vacuum as like liquid water, a transparent medium that looks as though nothing is there. The inflated state of the universe is like supercooled water: it has a temperature far lower than its freezing point, but has remained liquid. When the water suddenly freezes, it releases its 'latent heat' as the water molecules relax into an arrangement with lower energy, ice. Similarly, the supercooled false vacuum suddenly collapses into the true vacuum, releasing all the energy of the inflaton field, raising the temperature of the universe, and bringing inflation to an end. From that point on, the Big Bang scenario takes over, and the universe expands in a much more leisurely way.

That is the essence of the 'old' inflationary scenario. As you can suspect from the name, there is a newer version. The problem with the old model is that the release of energy reheats the universe so much that many defects—monopoles—appear at the end of the inflationary era, too late to be wiped out by inflation. Other problems also emerged with the rate at which inflation took place and came to an end. For instance, in its early form, a universe would collapse before inflation had time to get under way. One of the 'new' inflationary scenarios solves these problems.

One promising scenario is *chaotic inflation*, introduced by Andrei Linde in 1982 and developed in detail by him and others since then. No high-temperature phase transition of the inflaton field is required. Instead, a cold universe came into existence with an arbitrary value of the inflaton field (in terms of our earlier analogy: an arbitrary voltage), and if the field was large enough, inflation occurred. In due course the field rolled slowly back down into the state corresponding to the true vacuum (the voltage slowly discharged), and inflation comes to a graceful end. The rise in temperature accompanying this so-called *graceful exit* from the inflationary era is much less than in the phase-transition model. Consequently, far fewer defects—meaning no monopoles—are produced, but the temperature produced is high enough to initiate the standard Big Bang epoch that we still inhabit. The fluctuations in density that emerge from this scenario appear to be just right for accounting for the distribution of galaxies and also for the tiny fluctuations in the cosmic background radiation that have now been observed. Although inflationary theories are highly speculative, and when written about qualitatively might sound no better than primitive creation myths, they are rigidly constrained by mathematics and make predictions that are experimentally testable in this our current era. Cosmogenesis is one pinnacle of the imaginative application of science, but it is science, for it remains testable.

One entertaining outcome of chaotic inflation theory is the realization that far from eliminating the point defects we have called magnetic monopoles, inflation actually inflates them, and these point defects continue to expand even when inflation has ceased in their neighbourhood. Point defects can act as the seeds for the emergence of a new universe. That, of course, is the final humiliation, this one of hypercosmic proportions: this universe may be but one of countless others. Not only are we the inhabitants of an insignificant (but wonderful) planet near an insignificant (but wonderful) star at an insignificant location in an insignificant (but wonderful) galaxy in an insignificant location in the visible (and wonderful) universe, but that universe is perhaps an insignificant universe among innumerable others, a 'multiverse', each of them infinite in extent.

Nor need our universe be close to the beginning of time, for it might be a descendant of a branching tree of countless trillions (Fig. 8.10). Although we

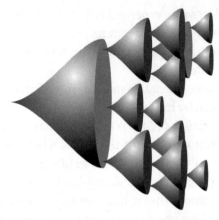

**Fig. 8.10** In one version of inflation, an existing universe can form buds of new universes, which immediately inflate, just as our own seems to have done. This view of the universe puts the actual moment of creation of the entire system of universes far back in time, for we might be inhabiting an inconsequential universe that has descended from myriad others, with the ur-universe, the truly initial universe, formed trillions and trillions of years ago—if times in these other universes are additive. One possible answer to 'where are these other universes' is that they are among us: if we think of spacetime as made up of points that we regard as close to each other, then it is speculatively conceivable that other universes make use of the same points, but regard them as related in a different way, so the points that make up a cubic millimetre in this universe might be distributed over the entire spacetime of another universe.

can say that our Big Bang took place 15 billion years ago, the actual inception of the original universe was probably far, far further back in time, almost, but let's hope not, beyond the reach of scientific imagination.

There are two further big questions to which we should turn. One is why is the universe (our particular universe, we might now add, in the multiverse) so lopsided. The other is why the universe is three-dimensional.

First, why is there more matter than antimatter? One possibility is that there are antimatter galaxies out there somewhere. The fact that we haven't seen any is not an inhibition in other areas of cosmology, as we have seen, but there is no evidence for such galaxies. Indeed, because intergalactic space is full of hydrogen atoms—well, full is an exaggeration, but there are a lot of them—we should expect to see intense radiation from the annihilation of these atoms when antimatter galaxies drifts into them. No such radiation has been observed, so it looks as though there really is more matter than antimatter. To be more precise, if initially there were equal abundances of matter and antimatter, then they would have annihilated each other and all we would have left would be photons from the radiation the annihilation developed. In fact, there is one particle for every billion photons, so there must have been a slight preponderance of particles over antiparticles initially. How could it arise?

The Russian physicist and dissident Andrei Sakharov (1921–98) came up with the ground rules in 1965, but was stumped for a mechanism for implementing them. He argued that three conditions needed to be fulfilled. One was that there should be processes that did not preserve the number of hadrons in the sense that hadrons (protons, for instance) could turn into leptons (positrons for instance).[20] The second was that CP symmetry should be violated (C denotes charge conjugation, P parity; see Chapter 6). The third was that events should take place slowly enough for equilibrium to be avoided: any imbalance, should it occur at some instant, needs to be left frozen in as the universe evolves rapidly into its future.

We now know that putative grand-unified theories (GUTs, as discussed in Chapter 6) eliminate the distinction between hadrons and leptons, so at high enough temperatures (before symmetry breaking has brought about distinctions between the particles), hadrons and leptons can change into each other. We can think of the transformation as being brought about by some kind of force that drives a hadron into becoming a lepton. These transformations are mediated—like any force—by the exchange of gauge bosons. Because a fully fledged GUT has not yet been formulated, we don't know much about the properties of these force-carrying particles, and currently they are called simply *X gauge bosons*. However, we do know that because an X achieves the transition between a hadron and a lepton, it will be able to decay into a positron and an antidown quark. Similarly, the antiparticle of an X, an anti-X, can decay into an electron (the antiparticle of a positron) and a down quark. If the rates of these decays were slightly different, then a small imbalance of matter and antimatter would result even if there were equal numbers of X and anti-X present initially. This is where CP violation steps in, as it can slightly upset the rates of such processes. We saw that CP violation is equivalent to the breakdown of time-reversal invariance, in the sense that a process run backwards is distinguishable from a process running forwards in time, and that this lopsidedness of the universe in time has in fact been detected. It is now believed that the excess of matter over antimatter is a manifestation of this lopsidedness of the universe. Why the universe is lopsided, no one knows. Perhaps only our universe is lopsided: the multiverse as a whole—if there is one—might be symmetrical overall.

[20] We saw in Chapter 6 that hadrons are particles that interact by the strong force whereas leptons do not. Hadrons include the quarks and the particles built from them; leptons include electrons and neutrinos.

The remaining problem is why there are three dimensions of space. The first glimmerings of a possible explanation are starting to emerge from string theory. We have been suspiciously silent about string theory in this chapter, except for a glimmer of its presence in a footnote, largely because it is still so speculative. However, there are some indications that string theory is relevant to the very early stages of the emergence of the universe—as must be the case if it is a fundamental theory of matter—and that the earliest moment of the universe was not an explosion of particles but an explosion of strings: a burst of spaghetti rather than semolina. For instance, we have seen that at very early times and therefore at very high temperatures, before symmetry breaking occurred, all the forces had the same strength. That is not quite true, for it turns out that if the calculations are done carefully, then the gravitational, strong, and electroweak forces do not quite coincide in strength in the very early universe, at the first tick of the Planck clock (Fig. 8.11). However, when string theory is brought into play, that small discrepancy is erased, and the forces are exactly equal the instant they are born.

We have seen that one fascinating feature of string theory is that it suggests that the universe has ten spatial dimensions (eleven including time), with six of them furled up into a Calabi–Yau space, with strings threading through the multidimensional holes in these spaces. We can think of strings as winding one way and of anti-strings as winding the opposite way. When a string and an anti-string meet, they annihilate, so we can picture a ten-dimensional space as writhing with strings and anti-strings, annihilating where they meet. Where they don't meet, the strings keep the space from unfurling, just like real string wrapped round a coil of paper.

**Fig. 8.11** In Chapter 6 we saw that the fundamental forces converge to a common value as we approach the instant (and temperature) of the Big Bang. That is not entirely true, for there is a tiny difference between them at very short times. This discrepancy seems to disappear when string theory is invoked.

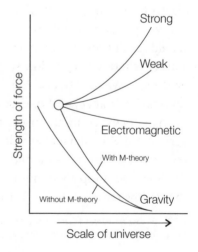

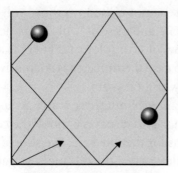

**Fig. 8.12** Two particles confined to one dimension—like two beads on a wire (upper illustration)—are certain to meet if they are moving unless they have exactly the same speeds. In two dimensions, though—like billiard balls on a billiard table (lower diagram)—the chance of their meeting is greatly reduced.

**Fig. 8.13** Two strings, a string, and an anti-string moving on a furled-up dimension will meet, annihilate, and leave the dimension free to unfurl. According to string theory, there are hints that strings are likely to meet in three dimensions, like point particles in one dimension. The remaining dimensions are trapped, so only three dimensions unfurl to give the dimensionality of our familiar universe.

Now we need a further fact. In one-dimensional space, like an abacus wire, a point and its antimatter counterpart, another point, are virtually certain to meet and annihilate, providing they are not travelling at exactly the same speed in the same direction. In two dimensions, like on a billiard table, it is much less likely that two points will meet (Fig. 8.12). When instead of points we try to think of strings and anti-strings meeting, it turns out that they are likely to meet *provided the dimensionality is no greater than three*. This suggests—and it is no more than an intriguing suggestion at this stage—that the strings and anti-strings that can be thought of as keeping up to three dimensions coiled are likely to annihilate each other and release the corresponding dimensions, enabling them to unfurl (Fig. 8.13). That is, three dimensions unfurl, and before the rest have time to do so, the universe moves on to its next phase, perhaps inflation, leaving six dimensions trapped for all time.

So much for the past; what of the future? I will deal with our presumably infinite future more briefly than with our ostensibly finite past. The general consensus is that we do have a future; a rather long one if we are prepared to move. I will take my starting point that the universe is not closed, so there will be no future crunch: the universe is currently infinite and its scale will expand forever. This seems to be the generally accepted view of cosmologists. There is always the possibility that they are wrong, in which case the universe is currently finite and will end in a Big Crunch in perhaps a few trillion years.

Not only is the universe likely to expand forever, but there is also accumulating evidence that its expansion is accelerating. This discovery shook the cosmological world, as it has profound implications for the universe. We should recall that Hubble used the Cepheid variables to determine the distances of galaxies. An alternative approach is to use a *Type Ia supernova* as a standard candle. A Type Ia supernova forms when a white dwarf, a star with approximately the mass of the Sun but the size of the Earth, in a close binary system accretes enough material from its neighbour to undergo a runaway nuclear reaction. Unlike the Type II (core collapse) supernovae that we discussed earlier, Type Ia supernovae are highly uniform in their intensity. Therefore, just as for the Cepheid variables, they act as standard candles and we can use their perceived intensities to judge their distance from us. Another advantage is that a supernova is very much brighter than a Cepheid, so they can be used to study objects much further away.

In 1998 it was discovered that a number of distant Type Ia supernovae were fainter than they should be if the expansion of the universe was slowing down or even just coasting at constant speed. Provided the evidence is sustained, there must be a contribution to the energy that can be ascribed to the vacuum, rather like in the inflationary era but currently much smaller in value. This contribution is the so-called *cosmological constant* initially introduced by Einstein to balance the gravitational pull and stop the universe collapsing and then discarded by him as 'his greatest blunder' when he learned of Hubble's results. Now it is beginning to look as though Einstein's recognition and acceptance of 'his greatest blunder' was in fact an even greater blunder.[21] The mysterious energy responsible for the acceleration is called dark energy, or more imaginatively, harking back ironically to Aristotle, *quintessence*. One possible scenario that stems from the existence of a non-zero cosmological constant is that a new inflationary era has already begun and that the acceleration of the universe will

---

[21]  In speaking of Einstein's blunders I do not wish to denigrate his wonderful contributions: these blunders were themselves wonderful, and I wish I had had the intellectual power to have made them myself.

in due course—in about a million trillion trillion years ($10^{30}$ years) or so—rise to colossal rates. If so, we shall experience the sudden onset of almost absolute loneliness, with only the merged remnants of our galaxy and Andromeda in view. I will assume that this exponentially fast expansion phase does not take over before other events have occurred, but that is by no means certain.

The Sun will go out quite soon, in about 10 billion years. It will swell up and become a Red Giant, with a radius reaching out well beyond the orbit of the Earth, so on a simple view of the matter we can expect the Earth to become a cinder in orbit. The Earth will experience drag as it hurtles through the very dilute solar matter in its vicinity and within about fifty years will spiral to its death into the Sun. All that will be left of our achievements will be a slight con-tamination of the Sun: we will become just one more contribution to pollution. I say 'on a simple view'. There is the possibility that in the process of puffing up into a Red Giant and becoming a hundred times brighter than now, the Sun will push a lot of material off into space and so become less massive. As a result of the lower gravitational pull on the planets of the leaner Sun, the Earth's orbit will expand and may move so far out that we avoid incineration; Venus our inner neighbour might escape, too. The Sun, meanwhile, will be left as a *white dwarf* with a mass about one-half its present value. Bigger stars, which have shorter lives than smaller stars, will also end dramatically, forming either neutron stars or black holes.

Galaxies can live only as long as their stars, just as human societies can live only as long as humanity. Based on the dynamics of star formation and evolu-tion and the way they recycle matter into their galaxy, the era of star formation is likely to come to an end in about one hundred trillion years ($10^{14}$ y). Long before that, in about 6 billion years time, there will be a little local difficulty when Andromeda smashes into the Milky Way or at least scores a near miss, but on the cosmic scale that is an event of little significance. When star formation ceases, we can expect the galaxies to consist of an almost equal mixture of white dwarfs and *brown dwarfs* (cold failed stars insufficiently massive to ignite; their masses must be less than about eighty times that of Jupiter), with a smattering of black holes. Very slow star formation might in fact continue as these brown dwarfs collide, merge, and become sufficiently massive to ignite. White dwarfs will also collide, and coalesce into bigger dwarfs. Black holes will also accrete stars, and in about a hundred trillion trillion years ($10^{26}$ years) the black holes presumed to exist at the centres of galaxies will have gobbled up their stars. These huge black holes, with masses of about 10 billion Suns, will wander through the universe like sharks, feeding off the sprats of lone stars that

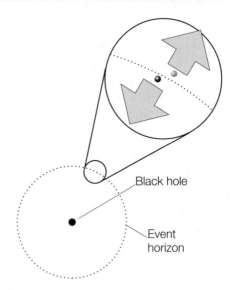

**Fig. 8.14** A pictorial illustration of the formation of Hawking radiation, by which black holes lose mass and shrink. A black hole is surrounded by a horizon at the Schwartzchild radius, from within which nothing, not even light, can escape. However, if a particle–antiparticle pair (such as an electron and a positron) is formed at the horizon, the antiparticle might be found inside the horizon and the particle might be formed outside the horizon. That allows the particle to escape and thereby reduce the mass of the black hole. It turns out that the intensity of this radiation has the characteristics of black-body radiation, with a temperature inversely proportional to the mass of the black hole.

Black hole

Event horizon

evaporated from galaxies in earlier eras. If these stars are white dwarfs, their nuclear reactions will be long gone, but they will glow dimly with the radiation as their protons decay with lifetimes of the order of $10^{35}$ years. The intensity of this radiation will be so low that you will have to be very alert to notice it: a typical white dwarf fuelled by proton decay will have the luminosity of a 400-watt lamp.

Black holes die. *Hawking radiation*, predicted by the cosmologist Stephen Hawking in 1974, can be thought of as follows. The vacuum (and we have learned to wonder what we mean by that) is a seething foam of particles and antiparticles coming into fleeting existence. If we think of a particle–antiparticle pair coming into existence at the horizon of a black hole, the surface surrounding the hole bounding the region of space from which nothing can escape, then one particle might find itself formed inside the horizon and its partner is formed outside (Fig. 8.14). As a result, one particle is captured and its partner escapes. The escaping particle carries energy away from the region of the hole, so the mass of the hole is reduced. This process is very slow. For a black hole with the mass of a galaxy, we can expect it to take about $10^{98}$ years. We can conclude, therefore, that after about $10^{100}$ years, the universe will consist of electromagnetic radiation, electrons, and positrons. In due course, the electrons and positrons will meet, annihilate, and decay into electromagnetic radiation. The wavelengths of the radiation in the universe will be stretched as the universe continues to ex-pand, just as the brilliance of the Big Bang has been stretched into the micro-wave background radiation of the cosmos.

As the universe becomes infinite, the wavelengths will become infinite too. We shall be left with dead flat spacetime with all trace of our achievements, aspirations, and existence wiped clean. Our end, though, is not the same as our beginning. In the beginning, there was nothing; absolutely nothing. In the end, in contrast, there will be completely empty space. How happy we should be, therefore, to be alive in the interval of exuberant activity between two boundaries of bleakness.

# SPACETIME

## THE ARENA OF ACTION

**THE GREAT IDEA**
*Spacetime is curved by matter*

*Time and space are modes by which we think, and not conditions
in which we live*

ALBERT EINSTEIN

**W**HERE is it that anything happens? When we look around us the answer seems obvious. We exist in space and we act in time. But what is space, and what is time? Here too, our intuition appears to suggest a ready answer. We think of space as a stage—an immaterial stage, perhaps, but some kind of stage nevertheless. And time? Time distinguishes successive actions, time is a feature of the universe that permits us to recognize the present as an ever-shifting frontier between the past and the future. In other words, space disentangles simultaneous events; time distinguishes the unforeseeable future from the unchangeable past. Space and time jointly spread events out over locations and in an orderly sequence, so rendering them comprehensible. Science springs from the existence of time, for science's mainspring, *causality*—the effect of an event on its successors—is essentially the systematic succession of events through time: if *this* now, then *that* then.

Such interpretations of space and time, though, are akin to sentiment rather than true knowledge. They are, perhaps, a philosopher's starting point rather than a scientist's ending point. As we shall see in this chapter, the development of our current view of space and time has been a refinement of the intuitive view that the world is a stage, an arena of action; but in its more recent developments that view has started to dissolve. Some scientists currently consider that they are on the edge of the discovery of an even greater idea than the target of this chapter.

Our story begins when Man (as it happens, it was almost certainly men) began to measure the surface of the Earth, then perceived as the arena of action. Actually, they began to measure not the Earth but the earth, which proved to be sensible, given what was to follow. Indeed, one aspect of the scientific method is to limit ambition to what is plausibly achievable: science nibbles at buns, it doesn't try to ingest great questions in a single gulp.

The key to understanding anything is a combination of observation, especially the quantitative kind of observation we call measurement, and the systematic way of thinking we call logic. In the first steps that ultimately led to our present understanding of our arena of action, the Babylonians and the Egyptians provided measurements and the Greeks provided logic. The Babylonians had procedures, but lacked proofs; the Greeks brought proofs. The Babylonians, for instance, knew for a thousand years before the Greeks that the sum of the squares of the sides of a right-angled triangle is equal to the square of the hypotenuse, but it was left to the Greeks, possibly in the form of the mathematical collective we know as Pythagoras, to prove that the relation is true for every conceivable right-angled triangle. Procedures are the basis of knowledge and the root of application, but proofs sharpen insight and lead us to deep comprehension.

I shall dwell a little on Pythagoras' theorem, because it teaches a number of important lessons. Indeed, we shall see that a number of features of our current understanding of space and time were anticipated in the works of Pythagoras (*c.*500 BCE), Euclid (*c.*300 BCE), and Apollonius of Perga (*c.*200 BCE). Of these individuals we know virtually nothing, and because most of the anecdotes about them were written down centuries after their deaths, we cannot rely on anything we are told. However, much of their extraordinary thought survives, and is a golden treasury of proof and insight into the properties of empty space.

We shall begin with a parable, and fantasize about the approach that the early conqueror of Mesopotamia, Hammurapi, might have made as he considered his prize about 3500 years ago. We shall pretend that Hammurapi lived in a world of multiple inconveniences, not the least of which was the convention that distances north–south were measured in metres and distances east–west were measured in yards.

As Hammurapi's surveyors surveyed his newly conquered fields, they measured the lengths of their sides and, for an arcane reason related to taxation, the lengths of their diagonals, reporting the latter in either metres or yards as the fancy took them. As you might suspect, Hammurapi found little rhyme or reason in the numbers his surveyors collected. For instance, one north–south aligned field had sides of 120 metres and 130 yards with a diagonal of 169 metres whereas another east–west aligned field had sides of 131 yards and 119 metres with a diagonal of 185 yards. Yet Hammurapi was puzzled, for the two fields looked the same.

One day, he had a brainwave. He decided to override the ancient conventions about units, and ordained that henceforth all distances be reported in metres (m). After a great deal of diligent work, his surveyors presented him with a new list of sides and diagonals. At a stroke, he saw that their measurements were now much more helpful. The sides of the two fields that looked the same were 120 m and 119 m and their diagonals were 169 m.[1] By bringing all his measurements into line by using the same units, Hammurapi had detached shape from orientation: all objects with the same shape had the same dimensions regardless of their orientation.

Hammurapi had further to go in tidying up the mensuration of his kingdom. Not all the fields in his kingdom had the same size, and his surveyors presented lists of sides and diagonals that even when expressed in metres seemed to their eyes to be little more than random. One rich farmer, for instance, owned a field that had sides of 960 m and 799 m with a diagonal of 1249 m; another poorer farmer had a field of sides 60 m and 45 m with a diagonal of 75 m. But then our fictional but brilliant Hammurapi suddenly had his *eureka*. He saw that, if for each one of his fields, regardless of its size, he squared the lengths of its sides and added the two squares together, then the answer was equal to the square of the length of the diagonal. That is, all the measurements his surveyors had collected fitted the formula

$$Distance^2 = side_1^2 + side_2^2$$

where *distance* is the length of the diagonal. Being a parsimonious ruler, he could now order his surveyors to save time and wages by measuring only the sides of the fields, for he could work out the lengths of the diagonals for himself. Indeed, he realized that even if they insisted in using the quaint units of the kingdom, he could still work out the length of the diagonal by writing

---

[1] I am taking these figures (but not the units, of course), and the ones below, from the Babylonian tablet *Plimpton 322* (approximately 1700 BCE).

$$Distance^2 = (C \times side_1)^2 + side_2^2$$

where $C$ is the factor needed to convert from yards to metres, one of the highly revered fundamental constants of the kingdom.[2]

There we can step back from our mythical version of Hammurapi, with his formula, his efficiency, and his taxes. More important than the secular use to which he put his formula is the fact that he had identified an expression that somehow summarizes the properties of space in Mesopotamia. The unknown Indian who wrote the *Salvasutras*, an account of the intricate sacrificial procedures of priests of the Vedic era (in about 500 BCE), also knew the formula, for the Brahmin needed reliably designed and constructed rectangular altars. The Chinese who wrote the *Jiuzhang suanshu*, compiled during the early Han period (starting in about 200 BCE), also knew it.

As we shall see, the existence of a particular formula for the distance between two points corresponds to the existence of a *geometry*, a description of space in terms of the points, lines, planes, and volumes that can exist in it. To identify the geometry of the space we inhabit, we have to identify the formula. Hammurapi's identification of the geometry of Mesopotamian space required two steps. First, he had to unify the units along the various coordinate axes; then he had to find the formula that gave the distance between two points. Because the same value of $C$ and the same formula apply in India and China, it follows that space in India and China has the same geometry as space in Mesopotamia The *proof* that Hammurapi's formula applies to any field anywhere in the universe (or so they thought), not only in Mesopotamia, may have been provided by Pythagoras and his school, but there is no firm evidence that they did anything other than just use it. To find the proof of the theorem we have to turn to Euclid's *Elements*, written about 2300 years ago and in print ever since, but there is no reason to suppose that Euclid's proof is original.

Euclid found that he could summarize the characteristics of space, including deducing Hammurapi's formula, from five simple and apparently obvious statements, his 'axioms'. That was truly a wonderful achievement. Had I been writing 2000 years ago, I would certainly have included Euclid's axioms as a great idea of science, for apart from one little defect, they fulfil the criteria of an idea being great: they are simple, but have consequences of unbounded richness. The defect, of course, is that they are wrong (in the sense that they are an

---

[2]   For the record, $C = 0.9144$ m yard$^{-1}$. Hammurapi's scientists used to spend a lot of time determining $C$ by diligently comparing metre rods and yards. Hammurapi thought this a waste of time, and ordered them henceforth to *define* $C$ with that value, in effect defining the yard in terms of the metre (as we do today).

inaccurate description of the space we inhabit); but we can ignore that detail for the moment and give Euclid the respect he truly deserves.

Euclid compressed his description of space into the following five remarks:

> **1.** *A straight line can be drawn between any two points.*
>
> **2.** *A straight line can be extended in either direction without limit.*
>
> **3.** *A circle can be drawn with any centre and any radius.*
>
> **4.** *All right angles are equal to one another.*
>
> **5.** *Given any straight line and any point not on that line, we can draw through that point **one and only one** straight line parallel to the original line.*

(I have simplified the statements somewhat but have preserved the essentials.) The fifth axiom is known as the *parallel postulate*. It is responsible for more grief than almost any other statement in mathematics, for it has a complex look out of line with the others, hinting seductively that it should be provable from the four simpler axioms. Lifetimes, however, have been wasted on trying but failing to derive this axiom from the others. We now know that it is independent of the other axioms and that is it possible to devise perfectly acceptable geometries in which the parallel postulate is replaced by others, such as

> **5'.** *Given any straight line and any point not on that line, we can draw through that point **no** straight line parallel to the original line.*

Or even

> **5".** *Given any straight line and any point not on that line, we can draw through that point **an infinite number** of straight lines parallel to the original line.*

The description of space that uses Euclid's parallel postulate is called *Euclidian geometry*; descriptions based on the alternative postulates are called *non-Euclidean geometries*.

For the time being, we shall stick with Euclidean geometry, for it certainly seems to fit the space we inhabit. The thirteen books of Euclid show that an enormous number of properties can be derived from the five axioms, and these properties are found to be valid when checked against actual measurements. One consequence of the axioms, and in particular the parallel postulate, is Pythagoras' theorem, which is proved at the end of Book I. Therefore, the existence of our mythical Hammurapi's formula for *distance* is implied by Euclid's five axioms, and Hammurapi's geometry is Euclid's too.

So far, we have expressed the Euclidean geometry of a plane, a flat two-dimensional region, like the surface of a sheet of paper. We all know, however, or think we know, that we inhabit a space of three dimensions, with up-and-down as well as freedom in the plane. Pythagoras' theorem is easy to extend to three dimensions by including the length of the additional side and writing

$$Distance^2 = side_1^2 + side_2^2 + side_3^2$$

We don't have to stop there. Mathematicians live by an insatiable passion to generalize, and Euclid's geometry is a fertile ground for generalization. Although most of us can't imagine anything beyond our homely three dimensions, it is easy to express the properties of such spaces using formulas. Thus, a four-dimensional Pythagoras, a 4-Pythagoras, would write

$$Distance^2 = side_1^2 + side_2^2 + side_3^2 + side_4^2$$

You might think that there is little point in thinking about spaces of dimension higher than three, except as an intellectual amusement, but you would be wrong. We shall see, for instance, that the ability to step between dimensions is a valuable way of appreciating the structure of our world. Moreover, can we be sure that there are only three dimensions in this our actual world, or are there several—even many—other dimensions that are somehow hidden from us? We saw in Chapter 8 that we cannot be sure, for we might inhabit ten dimensions of space together with one of time.

I have asserted that we cannot imagine beyond three dimensions. That is not quite true. Some people who have spent a lifetime studying the geometries of higher dimension claim to have an inkling of the relationships that exist in four but nine in three dimensions, and stunning computer images have been constructed portraying three-dimensional slices through four-dimensional worlds (Fig. 9.1)[3]. I will not ask you to bend your mind in that way, but in preparation for what is to come, we do need some familiarity with four-dimensional landscapes. To achieve it, we have to retread fragments of the path of the intellectual revolution initiated by the Italian painters of the late thirteenth and early fourteenth century, such as Giotto di Bondone and Piero della Francesca, who began to express three dimensions on two, with perspective put on a more precise mathematical basis by Gaspard Monge, Conte de Pélouse (1746–1818), at the end of the eighteenth century in his *Géométrie descriptive* (1798). Then we have to go beyond, and see a little of how four-dimensional scenes can be repre-

---

[3] An animated, stereoscopic image of a rotating hypercube will be found at http://dogfeathers.com/java/hyprcube.html.

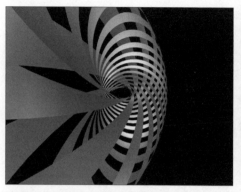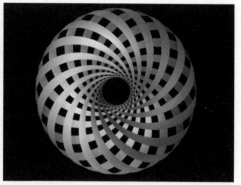

**Fig. 9.1** Some inkling of the shapes of objects in hyperspace can be obtained by exposure to graphical images and animations. Here are two scenes from an animation of the rotation of a flat torus in four dimensions, projected out into three dimensions and then rendered in two.

sented by two-dimensional portrayals of three-dimensional projections. It all sounds rather complicated, for it is like asking an ant that has forever been confined to a flat world to use its imagination to think in up-and-down too. But we are better equipped intellectually than ants, so we can expect to make some progress.

A zero-dimensional cube (a 0-cube) is a point. Think of the 0-cube as a pencil point, then a one-dimensional cube (a 1-cube) is the line drawn by the pencil as it is moved along a straight-edge (Fig. 9.2). A two-dimensional cube (a 2-cube) is a plane generated by dragging the 1-cube (the line) in the new dimension lying perpendicular to the first. All this is easily within the compass of our imagination, and that of an intelligent ant, and is easily performed on a sheet of two-dimensional paper. A three-dimensional cube (a 3-cube), a common-or-garden, everyday cube, is generated by dragging a 2-cube, a plane, in a perpendicular direction. There should be no problem with imagining that step, although the ant will be puzzled because it can't see how there can be a third perpendicular direction. There is also no problem with representing a 3-cube on a 2-page, an ordinary sheet of paper, because we are now so familiar with two-dimensional representations in art that we can decode the representation effortlessly. To help the puzzled ant, we can do the following. We cut carefully along some of the edges of the 3-cube, unfold it, lie it flat (Fig. 9.3), and tell the ant how the edges are joined together to form the 3-cube. The ant will be puzzled how the edges I have marked with a heavy line can ever be contiguous, but it will at least have an inkling of what is meant by a 3-cube and will perhaps begin to be able to interpret our two-dimensional representations of a 3-cube, including funny views in which the ant would swear we were showing it a hexagon.

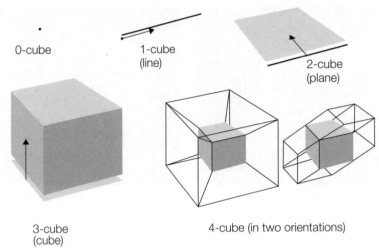

0-cube

1-cube
(line)

2-cube
(plane)

3-cube
(cube)

4-cube (in two orientations)

**Fig. 9.2** Cubes of different dimensionality can be constructed by moving the preceding cube in a new, perpendicular direction. Here we see a family of cubes built up from a 0-cube (a point). A line (a 1-cube) is obtained by dragging the point in one direction, a plane (a 2-cube) by dragging that line in a perpendicular direction, an ordinary cube (a 3-cube) by dragging that plane in a new perpendicular direction. We have learned to interpret the resulting two-dimensional representation of the cube. Finally, a four-dimensional hypercube (a 4-cube) is constructed by dragging that 3-cube in yet another perpendicular direction. We humans haven't yet learned how to interpret the resulting diagram: I show two views obtained by rotating the hypercube into different orientations.

**Fig. 9.3** An ordinary cube in three-dimensional space can be constructed from the cross-like shape consisting of six squares by gluing together adjacent sides and folding the long strip over and joining the edges marked with a bold line. It is easy for us to see that the dimension perpendicular to the page can be used to join the bold edges, but hard for a being confined to two dimensions.

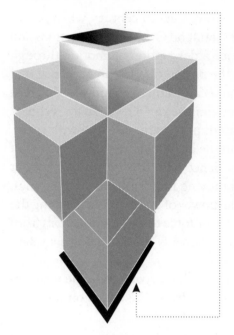

**Fig. 9.4** We now step up a dimension and construct a hypercube from this collection of eight three-dimensional cubes (one is hidden inside the crossing point), by gluing together adjacent faces. We also have to glue together the two faces indicated by the dark planes and dotted lines. As beings confined to three dimensions, it is hard to see how that can be accomplished in three dimensions, but it is easy to see in four.

We now know enough to build a four-dimensional hypercube (a 4-cube). A lot of mathematics proceeds by analogy. So, just as we dragged a 0-cube to build a 1-cube, and so on, we build a 4-cube by dragging a 3-cube (an ordinary cube) in a direction perpendicular to the first three dimensions. Now *we* are puzzled ants, because we can't conceive of a direction perpendicular to our three dimensions. Still, just as an ant can't conceive of a third dimension, we can take an intellectual leap and accept there is one, and—just like an ant—try to comprehend it by analogy. To help us understand the two-dimensional image of the 4-cube shown in Fig. 9.2, we could get a hyperbeing to cut along some of the faces of the cubes and then unfold them in three dimensions (Fig. 9.4). Just as a 3-cube unfolds into six 2-cubes, a 4-cube unfolds into eight 3-cubes (one 3-cube is hidden in the centre of the upper cross).[4] To imagine how the 4-cube is constructed from the eight 3-cubes that make up its face, we think of the as glued together. We 3-readers, the analogues of 2-ants, find it impossible to conceive how the two marked faces, for instance, could be joined, just as a 2-ant had a similar problem with three dimensions. A 4-reader would have no difficulty.

---

4  Salvador Dali knew what was involved. The difference between Piero della Francesca's *Flagellation of Christ* (c.1460) and Dali's *The Crucifixion: Corpus hypercubicus* (1954) represents the progress we are trying to make.

Euclidean geometry was completed in the seventeenth century when, as we saw in Chapter 3, Isaac Newton (1643–1727) built on Galileo's observations and added to Euclid's static description of space a description of motion through that space. To do so, Newton introduced the notion of *force*, an influence that pushes particles away from straight lines and causes them to move at different speeds. In the context of our current discussion, we can view Newton's contribution as the first reasonably successful attempt to combine time with space. Aristotle had tried, but was unsuccessful, for he did not appreciate the power of geometry in dictating paths: from his experience with the Earthly, he thought that forces were needed to keep particles moving uniformly on straight lines. Newton, on the other hand, did perceive the power of geometry in dictating the paths of particles. He introduced the concept of force to express *deviations* from the natural travel, which he took to be that of steady motion along a straight line.

For Newton, like Aristotle 2000 years before him, space and time were absolutes, with space a stationary platform shared by all the actors on the stage and time a ticking parameter common to them all:

Absolute space, in its own nature, without relation to anything external, remains always similar and immovable . . . Absolute, true and mathematical time, of itself, and from its own nature, flows equably without relation to anything external.[5]

If for me it is Tuesday, then for everyone it is Tuesday, and if for me if took an hour, then for everyone it took an hour. If for one observer the separation of two points is 1 km, then for all observers the separation of the two points is 1 km. In other words, space is a fixed, absolute stage and time has a universal tick.

The concept of action at a distance, that a star could wrap the path of a distant planet into something akin to a circle, was deeply puzzling. Newton himself saw that it was a defect of his theory, but he had a pragmatic vision of his abilities and was content to leave that puzzle to later heads: he was a nibbler, not a gulper. The head that almost single-handedly solved the puzzle belonged to Einstein, and in the remainder of this chapter we shall see what immense understanding he achieved.

Albert Einstein (1879–1955) moved civilization forward in two steps. In one step he bonded space to time in a much more intimate manner than Newton had

---

[5] The *Principia*, more formally *Philosphiae naturalis principia mathematica* (1687).

attempted. In doing so, he destroyed Newton's perception of absolute space and time and erased the universal tick. In the second step, he eliminated one of Newton's greatest achievements, the concept of universal gravity as a force. Great puzzles are often resolved by elimination, and scientists should relish the overthrow of the major concepts, including their own.[6] We shall join Einstein in these two steps. The second, the greater, could not have been taken without the former, and we need to emulate his progress if we are to understand truly and deeply what, where, and when we inhabit.

Einstein's first achievement was the *special theory* of relativity. The special theory of relativity is a description of the observations that people make when they are in uniform unaccelerated relative motion. Einstein's central notion was that it is impossible for anyone in uniform motion to detect, without looking out of the window, whether or not they are moving. Einstein expressed this conclusion succinctly by the remark that *inertial frames are equivalent*. An 'inertial frame' is simply a platform moving at a uniform speed in a straight line. Galileo had had the same idea in the early seventeenth century, when he imagined travelling in the sealed, windowless cabin of a boat on a calm sea: no experiment he could imagine could detect whether the boat was in motion. To find a modern example of an inertial frame we might think of experiments being performed inside the cabin of an aircraft in smooth flight: provided we could make no reference to the outside world, we would be unable to detect motion. The crucial difference between Galileo and Einstein, and the two centuries that separated them, is that Einstein had available to him information about electricity and magnetism as well as the dynamics of moving bodies (pendulums, and so on).

To see the significance of Einstein's idea about the equivalence of inertial frames, we could suppose that you and I are both authors of textbooks. I regard myself as stationary in a laboratory where I make a series of measurements; you think of yourself in a laboratory moving relative to me in a straight line at 1 000 000 000 kilometres an hour (kph; this speed corresponds to nearly 93 per cent of the speed of light, and encircling the Earth would take about 0.14 s). Unlike most authors, who rely on the work of others to compile their texts, we have decided to carry out all the classic experiments—Galileo dropping balls from Pisa's leaning tower, Faraday discovering electromagnetic induction, Michelson and Morley fruitlessly searching for evidence of motion through the ether, and so on. Einstein's view was that we would each write essentially the

---

[6] I speak of an ideal world. I am not at all sure that Newton would have taken kindly to Einstein. Newton took kindly to few of his contemporaries, and to any scientist, even the apparently modest, intellectual overthrows are unwelcome while one is still alive.

same textbook despite the fact that you are travelling at 1 000 000 000 kph relative to me. Our words would of course be different, but the physics we teach would be indistinguishable. If we exchanged textbooks, I could use yours just as you could use mine. The equivalence of our textbooks extends to the whole of physics, not just to moving particles (Galileo) but also to electricity and magnetism (Einstein).

Now, we get to the central point. Many equations in physics, particularly those describing electricity and magnetism, depend on the speed of light, $c$.[7] The problem is this. In my chapters on electromagnetism, the expressions I use require a particular value of $c$, which I have measured in my laboratory. The expressions you use in your chapter also use a certain value of $c$, and for me to be able to teach physics from your textbook, the value of $c$ you have measured must be exactly the same as the value I have measured. In other words, when you measure $c$, you measure exactly the same value as I do, even though you are moving at 1 000 000 000 kph relative to me. Only in that way can your textbook be consistent with mine.

The fact that observers in different inertial frames travelling at different speeds (you and I) measure the same speed of light has profound implications for our understanding of space and time. It does away with the concept of universal simultaneity, for instance, and eliminates the concept of space alone as an arena. Because these remarks do away with everything we have been brought up to believe, this is a crucial moment in revising our understanding of nature. We need to see more precisely what is entailed.

How can it be that you measure the same value of $c$ even though you are travelling so fast relative to me? One solution is that your measurements of distance and time are different from mine. For instance, if your measuring rods are shorter than mine and your clocks run more slowly, then you will report values that differ from mine even though we may be observing the same phenomenon. Thus, it could be that the 'boost' you are giving to a beam of light by emitting it from a lamp travelling 1 000 000 000 kph faster than mine, is cancelled by these modifications of your perception of space and time. That is, the boost your motion gives to the motion of light might be cancelled completely by this change of perception. Such modifications had already been suggested

---

[7]   Scientists used to spend a lot of time measuring the value of $c$. Now though, like Hammurapi's $C$, it is *defined* to have the value $c = 299\ 792\ 458$ m s$^{-1}$; the speed of light is no longer a quantity we measure.

independently by the Irish physicist George Fitzgerald (1851–1901) and the Dutch physicist Hendrik Lorentz (1853–1928), and were jointly known as the *Fitzgerald–Lorentz contraction*. Einstein's achievement was to put their *ad hoc* suggestions on to a firmer, deeper theoretical foundation by proposing that they were consequences of the geometry of space and time.

Einstein cut to the heart of the matter. Although he didn't express it this way, he might have imagined Hammurapi's surveyors as being pressed for time and making their measurements as they ran across the fields. But surveyors moving at different speeds over the same fields would have reported different lengths of the diagonals, and Hammurapi's formula for *distance* would no longer work, with different surveyors reporting different values for it, depending on how fast they were moving and in which direction. In a leap of insight, our fictional Hammurapi and our actual Einstein said that reporting the location of a point in space was no longer sufficient: surveyors must from now on report both the location of the point and the time at which, according to their clocks, it was recorded. We call this joint measurement an *event*. Einstein proposed that the true 'invariant', the figure everyone could agree on regardless of their speed, was the *interval* between two events. The interval between two events separated in space by *distance* (as measured by a particular surveyor) and separated in time by *time* (as measured by the same surveyor) is defined as

$$Interval^2 = (c \times time)^2 - distance^2$$

where *distance* is calculated by using the same expression as we had before. Because the distance that you measure between the spatial locations of the two events is smaller than the distance I measure, but the interval is the same, the time between the two events must be smaller too, to preserve the value of the difference $(c \times time)^2 - distance^2$. In other words, time runs more slowly for you than it does for me.[8] The time that we each measure is called our *proper time*: I consider that your proper time ticks away more slowly than my proper time. Because you consider that I am moving relative to you, you also consider that my proper time ticks more slowly than yours.

Einstein's proposal requires a remarkable revision of our perception of space and time. First, it eliminates the concept of universal simultaneity: it is no longer possible for observers in different inertial frames to agree about whether

---

[8] We do not need the details, but for completeness, if you know that the length of your spaceship is *length*, then the length that I measure is *length* $\times (1 - speed^2)^{1/2}$, where *speed* is your speed relative to me expressed as a multiple of the speed of light. At 100 kph (about 60 mph, well beyond the capacity of the fleetest bullock cart), the factor $(1 - speed^2)^{1/2}$ differs from 1 by about one part in 100 trillion, so Hammurapi would have been totally oblivious to the need to worry about the speed at which his scurrying surveyors took their measurements.

two events are simultaneous. To appreciate this conclusion, suppose you are in a spaceship which you know to be 100 m long. You hurtle past me at 1 000 000 000 kph. I note the location of the two ends of the spaceship at a given instant, and find that they are separated by only 38 m. The separation in time between my two events (the two measurements) is zero because they are simultaneous, so the interval between them is the same as the spatial separation I measure, which is 38 m. You know that your spaceship is 100 m long, so for the interval to be the same, the time that *you* measure between the two events cannot be zero. In fact, you think the time between my two measurements is 0.31 microseconds (0.31μs)! In short, you do not consider the two events to be simultaneous. The reliability of the concept of simultaneity has vanished, as no two observers in uniform relative motion can agree on what events are simultaneous. Gone, in other words, is Newton's perception of absolute space and time.

The second revision of common sense it the fusion of space and time. First, let's clean up the expression for interval. Just as Hammurapi helped to simplify the description of Mesopotamia by converting measurements east–west and north–south to the same units, so we can simplify the description of space and time by converting measurements of time and space to the same units. It's a matter of convenience which we convert, but let's choose to express measurements of time in 'metres of light travel', the distance light travels in that time, by multiplication by *c*. Thus, 1 s will be reported as 300 000 km, because that's how far light can travel in 1 s, and '1 m of time' is equivalent to $3 \times 10^{-11}$ s (30 picoseconds, 30 trillionths of a second) in traditional units. When you look at the hand of your watch ticking off the seconds, think of each tick as indicating another 300 000 km. This conversion is just housekeeping, but it simplifies the definition of interval to

$$Interval^2 = time^2 - distance^2$$

just as Hammurapi simplified the definition of *distance* from $(C \times side_1)^2 + side_2^2$ to $side_1^2 + side_2^2$ by insisting that his surveyors report $side_1$ in metres.

Now we come to a very important point. Just as the formula for distance given by Pythagoras' theorem for distance summarizes the geometry of two-dimensional space, and can be generalized to three and more dimensions, so Einstein's formula for *interval* strongly suggests that time should be regarded as a fourth dimension perpendicular to the three dimensions of space. This realization is the origin of Herman Minkowski's (1864–1909) remark, made in 1907, that:

Henceforth, space by itself, and time by itself, are doomed to fade away into mere shadows, and only a kind of union of the two will preserve an independent reality.

The union of space and time is now called *spacetime*.

We must not confuse four-dimensional spacetime with four-dimensional space, because their geometries are quite different: *distance* in 4-space is given by $t^2 + x^2 + y^2 + z^2$ whereas its analogue *interval* in 4-spacetime is given by $t^2 - (x^2 + y^2 + z^2)$, or $t^2 - x^2 - y^2 - z^2$. We say that 4-space and 4-spacetime have different *metric signatures*, the metric signature of 4-space (the pattern of signs in the expression for *distance*) being $(+,+,+,+)$ whereas that of an *interval* in spacetime is $(+,-,-,-)$. You can perhaps begin to get a hint of the essence, or at least the definition, of time: it is the coordinate corresponding to the uniquely different sign in the metric signature, the $+$ rather than the $-$. A world in which the metric signature of spacetime is $(+,+,-,-)$ would have two dimensions of time, so 'today' would have to be distinguished by two dates. If we ever had to envisage higher dimensional spacetimes, such as a five-dimensional spacetime with the metric signature $(+,-,-,-,-)$, we could immediately identify the first coordinate as time; we encountered higher dimensional spacetimes in Chapter 8, and this is the basis of deciding whether one of the additional dimensions was space or time. Throughout this chapter, by spacetime we shall mean the four-dimensional version with metric signature $(+,-,-,-)$.

I have to admit that the geometry of spacetime, which is called *Minkowskian geometry*, is harder to grasp than the geometry of space alone. However, the following remarks should give you an impression of some of its features and its differences from space itself. The material is not central to what comes later, so if it seems a bit bewildering, don't worry: just move on. To build up your confidence for thinking about this sort of thing, I shall use the same device as before: just as we found that we could get a glimmering of understanding of four-dimensional space by gradually increasing the number of dimensions, so we can edge towards understanding four-dimensional spacetime by starting out with a smaller number of dimensions.

There is no such thing as zero-dimensional or one-dimensional spacetime. The distinction between space and time (as expressed by the metric signature) is significant only when we get to two-dimensional spacetime (2-spacetime), with one dimension of space and one of time. Moreover, 2-spacetime can be represented by a flat drawing, with one axis denoting space and the other time (Fig. 9.5). The lines in the diagram show various paths of the particle through the world, and are what Minkowski called *worldlines*. A vertical worldline is the history of a stationary particle: the particle stays at the same point in space as

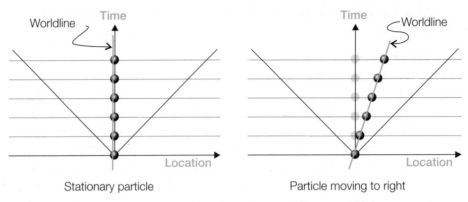

Stationary particle                    Particle moving to right

**Fig. 9.5** The worldline of a particle is simply the trace of its position as time increases. The illustration on the left shows a stationary particle. It remains at the same location as time increases, so its worldline is vertical. The illustration on the right show that same particle moving at uniform speed to the right, so it is located further to the right as time increases. Its worldline therefore slopes to the right. The lines at 45° in each illustration are the worldline of light, which can travel 1 m in 1 m of light-travel time. Nothing can travel faster than light, so nothing has a worldline that slopes at more than this angle.

time increases. A worldline slanted slightly to the right corresponds to a particle moving slowly to the right, because the location of the particle moves to the right as time increases. A worldline sloping at 45° corresponds to a particle that is moving to the right at the speed of light, travelling 1 m of distance in 1 m of light travel time (30 trillionths of a second in conventional units). This line represents the fastest possible motion of a particle, because nothing can travel faster than light, and only massless particles (such as photons) can reach that speed. All possible worldlines lie between the left-sloping line at 45° (a particle moving to the left at the speed of light) and the right-sloping line at 45° (a particle moving to the right at the speed of light).

Now we step into 3-spacetime, with two dimensions of space and one of time (Fig. 9.6), with the particle free to move in two spatial dimensions—anywhere in a plane—as time increases. Because no particle can move faster than the speed of light, all possible worldlines lie within the cone of half-angle 45°. This cone is called the *light-cone* of the event at its apex, because the worldlines of light, which does travel at the speed of light, lie on the surfaces of these cones. We could imagine a circular pulse of light starting out from a point: it spreads with time as shown by the rings on the plane and depicted at different times on the light-cone.

Once we turn to 4-spacetime, we have to think of a four-dimensional version of a cone starting out from the event, with slices through the cone at any

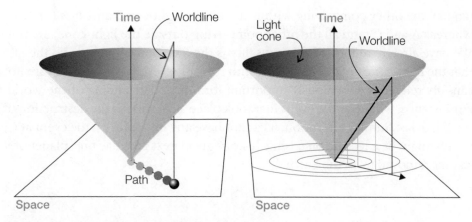

**Fig. 9.6** In two-dimensional space, when the particle is free to move over a plane, the worldline lies somewhere within the cone shown in the illustration on the left. The cone itself is the light cone, the worldlines of a pulse of light starting from the origin. No worldlines lie outside the cone because that would correspond to motion faster than light.

instant being a three-dimensional sphere (representing the propagation of a spherical pulse of radiation). Imagining that is completely beyond me, and I won't even pretend that I have a way of representing it on paper. Fortunately, the illustration of the light-cone for pulses in two spatial dimensions in Fig. 9.6 is all we really need to understand.

A light-cone divides events into two classes. Consider, for instance, the events A and B in Fig. 9.7. Because B lies within the light-cone based on A, signals from A can reach B in plenty of time for them to influence the event B. Now consider the two events A and C. The event at A cannot influence the event at C because the latter point lies outside the light-cone based on A, so no signal from A can reach C in time for it to have any effect. We say that A and B

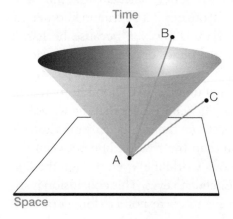

**Fig. 9.7** The light cone divides events into those that are causally related and those that are not. If an event occurs at A, it can influence events within the light cone, such as B, but it cannot influence events located outside the light cone, such as C, because a signal cannot travel from A to C.

(and all the other points lying within and on the light-cone) are *causally related*, whereas A and C (and all the other points lying outside the light-cone) are not. We have already remarked that causality is the lifeblood of science, so the fact that the light-cone divides spacetime into regions of events that are and are not causally related is enormously important for our understanding of the world. For instance, whatever the event that took place at A, such as the destruction of the Earth last Sunday afternoon, it cannot have any influence on the event at C, which might be a lecture on cosmic history given next Monday on a planet of a star far removed from the Earth.

All the preceding might feel reasonably familiar, because the lines and cones we have drawn echo the properties of ordinary space. Now we come to the principal difference between Euclidean space and Minkowskian spacetime, and the feature hardest to grasp intuitively. In space, a straight line is the shortest distance between two points. In spacetime, with its funny Minkowskian geometry, we have to get used to the idea that a straight line corresponds to the *longest* interval between two events. The following parable about Castor and Pollux will help to explain this point.

Let's imagine Castor remaining at home. His worldline is vertical and stretches from his twentieth birthday to his twenty-first birthday. Pollux, celebrates his twentieth birthday with his brother, and immediately embarks on a journey that Castor considers to last for 12 months, travelling at 1 000 000 000 kph out to a distant point in interstellar space and back again, arriving on his, Castor's, twenty-first birthday. According to Castor, Pollux travels 8.8 trillion kilometres. Castor uses the time his brother is absent to calculate that the interval between the start and end of his brother's journey is 3.30 trillion kilometres. Pollux agrees, because interval is an invariant. However, because he hasn't stepped outside his spaceship and has kept the blinds drawn, Pollux considers that he hasn't been anywhere, so he ascribes the entire interval to the passage of time, not relocation in space. In conventional units, 3.30 trillion kilometres of light-travel time corresponds to 4.6 months (Fig. 9.8). We see that the worldline representing the journey made by Pollux between the events marking Castor's birthdays corresponds to a *shorter* interval than the straight line between the two birthdays (which corresponds to Castor's worldline) even though the line depicting his journey looks longer to our Euclidean eyes. This conclusion justifies our remark that straight lines between events correspond to longer (actual-

Time

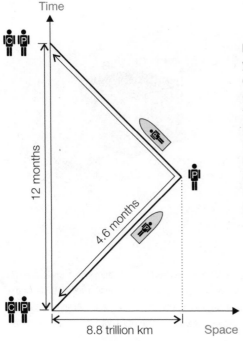

12 months

4.6 months

8.8 trillion km      Space

**Fig. 9.8** Castor stays at home for a year: he ages a year, but travels nowhere. The interval between the two events he calls his birthday is 3.30 trillion kilometres (one year). Pollux sets off on a journey at 93 per cent of the speed of light and goes to a point 8.8 trillion kilometres away, turns round, and arrives back on Castor's birthday. Pollux doesn't think he has been anywhere, but he agrees with Castor that the interval between his departure and arrival is 3.30 trillion kilometres. However, he regards that as assigned to the passage of time, which he thinks is 4.6 months.

ly longest) intervals than indirect paths. That is true in general. When you look at a spacetime diagram, don't be fooled into thinking like Euclid.

The second point to note is that Pollux has aged less than Castor. Pollux, whose metabolism keeps in step with the passage of time that he experiences, has aged only 4.6 months while Castor has aged a year.[9] So, to avoid aging, we should travel fast.

Another feature of spacetime that distinguishes it from space is the significance of volume. At some stage we won't be able to avoid thinking about four dimensions, but we can build up to that step by thinking in a smaller number of dimensions and then arguing by analogy. Let's consider a cubic box in three-dimensional spacetime, two dimensions of space and one of time. Like an ordinary cubic box in three-dimensional space, such a box has six square faces (Fig. 9.9). The face marked A in the illustration lies wholly in the two dimensions of space, and corresponds to an ordinary spatial plane at a given time: think of it as a flat sheet of paper at a particular instant. The face marked B

[9] The term 'twin paradox', which is commonly ascribed to this description, stems from some people's inability to see that Pollux has a different history from Castor. I have simplified the discussion by ignoring the effect on Pollux of the deceleration and subsequent acceleration when Pollux turns round to come home. When these effects are taken into account, the conclusion is the same.

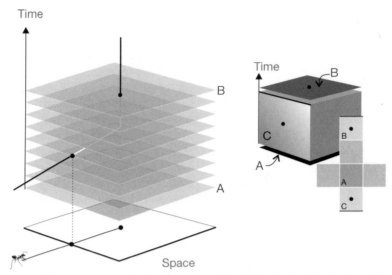

**Fig. 9.9** An ant ambles on to a rectangular sheet of paper, and stops in the middle. The bottom face of the 3-cube (A) is empty initially, because the ant is not on the sheet, but when we inspect the sheet later, we find the ant there, and note its presence with a dot on the corresponding plane (B), which is the upper face of the 3-cube. At some instant, the ant must have crossed the left-hand edge of the sheet, and we mark that position by a dot, which appears on the corresponding face of the 3-cube (C).

is the same plane at a later time: think of it as the same sheet of paper 5 minutes later, and lying in the same place. The face marked C is made up of the vertical worldlines of all the points on one edge of the stationary sheet of paper (one edge of face A); similarly each of the other vertical faces consists of the vertical worldlines of the points on each of the other three edges of the sheet of paper.

The four vertical faces summarize all the events that take place on each edge of the sheet of paper over the 5 minutes we are considering. For instance, suppose an ant runs on to the paper from the left after 2 minutes. Initially, the sheet of paper is empty, so plane A is empty too. The ant runs in from the left and traces out its worldline. It passes across the left-hand edge, so we see a point appear there. Then (we suppose) the ant stops in the middle of the sheet and stays there. Its worldline is now vertical, and after a further 3 minutes a dot appears on face B. Notice that the difference between planes A (no dot) and B (one dot) is matched by a dot somewhere on one of the vertical planes (on C in this instance). Providing a particle cannot simply appear out of nothing, the change between the two horizontal 'space-like' planes, A and B, must be matched by an event on one of the vertical 'time-like' planes (C, for this ant).

Now we fasten our intellectual seat belts and take off into four-dimensional spacetime. Here is your comfort blanket: hold on to the fact that four dimensions is exactly analogous to three, but spatial planes (sheets of paper) are replaced by spatial volumes (rooms). Ants walking on to paper are replaced by people walking into rooms.

As we have seen, the walls of a 4-cube are formed from eight 3-cubes (recall Fig. 9.4). In spacetime, two of these 3-cubes, we'll denote them X and Y, are purely spatial, and correspond to three-dimensional regions of space—actual rooms—at the initial and final times of interest (Fig. 9.10, where the events I am about to describe are spelled out in a bit more detail), just like the planes A and B in three-dimensional spacetime correspond to the actual sheet of paper at two different times. We call these ordinary cubes 'space-like'. What is the significance of the other six 3-cubes? Each one has sides made up of two spatial dimensions and one time dimension, so they summarize the history of what takes place at each two-dimensional face of the actual box, just as side C

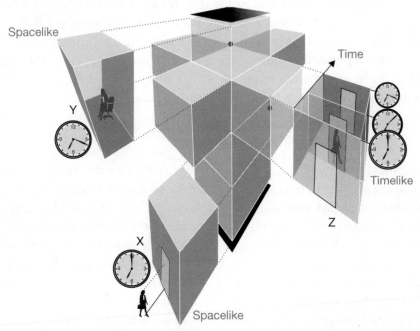

**Fig. 9.10** A 4-cube shows the history of occupation of a three-dimensional region (a room) just like a 3-cube shows the history of the presence of an ant on a two-dimensional sheet of paper. The 3-cube X is the original empty room at 7:00 p.m. Twenty minutes later, if we inspect the room, corresponding to the 3-cube Y, we find it occupied, and mark the position of the occupant's head with a dot. If at intermediate times we monitor the door, we can show what is happening there by the sequence of images that make up the time-like cube Z. For a fleeting moment at 7:10 p.m. the occupant appears on the plane as she enters the room, so we mark the position of her head with a dot in the corresponding cube. The hypercube is a record of the history of the room between the initial and final times.

summarized what happened at the edge of the sheet of paper. We call these cubes 'time-like'. To see their significance, suppose our space-like cube represents the room you are currently in. Before you came into the room, it was empty, so the space-like cube X is empty. When you entered the room, you passed through a door in the wall, so a dot marking your entry point and time appears in the corresponding time-like cube, cube Z, for instance (the dot might mark the position of your nose). If we inspect the room some time later, while you are still in it, we will find your position marked by a dot in the space-like cube Y. As in 3-spacetime, any difference between cubes X and Y must be matched by a point inside one of the other six cubes: where that latter point lies depends on where and when you entered the room.

Now, to round out this discussion and prepare you for what will come later, I need to press you to think a bit more generally. When we come to talk about the energy or mass in a region of space, we will be able to build on this picture. The total energy (or mass, through the relation $E = mc^2$) in cube X will be the total energy in a region of space initially, the total energy in cube Y will be the energy in that region after a given time has passed. The total energy in the time-like cubes will represent the flow of energy into or out of the region through its bounding walls, and the net influx of energy must account for the difference between the quantity of energy in the space-like cubes at X and Y.

That is probably enough about spacetime hypercubes for the time being. With luck, you are starting to begin to appreciate the structure of spacetime and the significance of points and volumes within it. Before we take the next step together, I need to introduce you to one more aspect of special relativity. This important step will reveal the origin of the most famous expression in the whole of physics, $E = mc^2$, or *energy = mass* $\times c^2$, and will show us that this intellectually, economically, commercially, militarily, and politically important expression, which has already cropped up in this and earlier chapters, is yet another aspect of the geometry of spacetime. In units in which time is expressed as a length, $c = 1$, because light travels 1 m in 1 m of light-travel time, Einstein's equation takes on the less familiar but far simpler and revealing form *energy = mass*. In other words, there is no distinction between energy and mass.

I can't escape from using a tiny amount of mathematics, but it will be trifling and the consequences breathtaking, and I hope you will see it through (or skip it: the result is the important point). We know the relation between interval, time, and distance:

$$Interval^2 = time^2 - distance^2$$

It's easy to rearrange this relation by dividing both sides by $interval^2$ to give

$$1 = \frac{time^2}{interval^2} - \frac{distance^2}{interval^2}$$

Next, we multiplying both sides by $mass^2$, where *mass* is the mass of any particle we happen to be thinking about (a uranium atom, a frog, Jupiter). These steps give

$$Mass^2 = \left( mass \times \frac{time}{interval} \right)^2 - \left( mass \times \frac{distance}{interval} \right)^2$$

Because *distance/interval* resembles the normal expression for speed, and mass multiplied by speed is the definition of linear momentum (Chapter 3), we can suspect that the second term on the right is the relativistic expression for the square of the momentum. I won't go into details, but this suspicion is confirmed by thinking about the collision of two particles and finding that the total value of *mass × distance/interval* is left unchanged by the collision. One of the central tenets of physics, as we saw in Chapter 3, is the 'conservation of linear momentum', the principle that although all manner of complicated events may occur in a collision between two bodies, the total momentum is left unchanged, so this identification is valid.

But what is the first term on the right? If we set up the equations for the collision between two particles, then we find that the quantity *mass × time/interval* is also left unchanged in the collision even though a lot of complicated individual events take place. Another great principle of physics, as we saw at length in Chapter 3, is that *energy* is conserved. This observation strongly suggests that we should identify *mass × time/interval* with energy, and that we should write the last equation as

$$Mass^2 = energy^2 - momentum^2$$

The identification of *mass × time/interval* with energy is also justified by the demonstration that, like the momentum, it too is conserved in a collision. One implication of this expression, which resembles the expression for *interval*, is that just as space and time should be thought of as unified into spacetime, so linear momentum and energy should be thought of as two faces of a combination that could be, but rarely is, called by the ungainly name *momentumenergy*. Mass, calculated according to this expression from energy and momentum, just as *interval* is calculated from time and distance, is an invariant, a property found to be the same for all observers however fast they are travelling.[10]

We can now make a run to our final conclusion. Let's suppose the particle is stationary in our inertial frame—it could be a lump of iron we are holding. Because the particle is stationary, it has zero momentum; so $mass^2 = energy^2 - momentum^2$ becomes $mass^2 = energy^2$ and we can immediately conclude that $mass = energy$, which is what we wanted to derive. You should notice how this extraordinarily important expression is a direct consequence of the geometry of spacetime in combination with two of the conservation laws of physics, which enabled us to identify the terms.[11]

Our investigation of the geometry of spacetime has led us to the conclusion that mass and energy are equivalent. We have to conclude that if energy vanishes from a region, then the mass of that region diminishes. If energy flows into a region, then the mass of the region increases. In practice, the difference in mass is totally negligible for normal objects. For instance, the difference in mass of a 10 kg cannon ball between room temperature and 1000 K is only 50 picograms (50 million-millionths of a gram), which is completely undetectable (with current technology).[12] Changes in the energy accompanying rearrangements of the subatomic particles, the protons and neutrons, that make up atomic nuclei are much greater than those obtained simply by heating cannon balls. *Nuclear fission* is a process in which the nucleus of an atom falls apart into two smaller nuclei, which enables the protons and neutrons in the nucleus to settle into energetically more favourable arrangements and hence to release the excess energy. When 10 kg of uranium-235 undergoes fission, the energy released corresponds to a mass loss of as much as 10 g. That is equivalent to the energy released by burning 30 thousand tonnes of coal. Geometry is astonishingly potent.

A lot of the chapter up to this point has sprung from the elimination of one fundamental constant, the speed of light, and the simplicity of the expressions that

---

[10]  In older presentations of special relativity, mass is presented as a quantity that increases with velocity. That is an old-fashioned view, and mass is now regarded as an invariant.

[11]  In Chapter 6, we saw that these conservation laws are also aspects of the symmetry of spacetime, so nuclear power—and thermonuclear warfare—is nothing but an expression of the potency of geometry. Joseph Conrad sensed that in *The secret agent*, where anarchists bombed an observatory in an act of aggression against geometrical abstraction.

[12]  Nevertheless, expressed as a number of iron atoms, it is equivalent to the addition of 540 billion iron atoms to the ball.

result. We will now move on to the elimination of another fundamental constant, and thereby achieve an even deeper understanding of Nature (we saw this process in action in Chapter 3, when we eliminated the mechanical equivalent of heat and were rewarded with thermodynamic insight). I suspect that if we were to eliminate all fundamental constants, we would understand Nature perfectly! This is where we move on to the Great Idea that is the real heart of this chapter. It took Einstein about a decade to move on from special relativity to a more general theory, which is commonly called *general relativity, Einstein's theory of gravitation*, or just plain 'Einstein's theory'.

Newton's theory of gravitation, by him regarded as a force acting across empty space, is characterized by a universal fundamental constant, the *gravitational constant, G*.[13] According to Newton, the force of gravity exerted by a body is proportional to the product of $G$ and the mass of the body. This proportionality means that, at a given distance from their centres (and outside them), the force of gravity of the Sun, which has the mass of 336 thousand Earths, is 336 thousand times greater than that of the Earth.

First, some housekeeping. From now on, we shall absorb $G$ into the mass of the object and thereby express mass as a length.[14] Expressed as a length, the mass of the Earth is 4.41 mm and that of the Sun is 336 thousand times greater, at 1.48 km. You should notice that we have now expressed mass, length, and time all in units of length; we could have expressed them all in units of time, by dividing by $c$, but then the numbers so obtained would be more ungainly and would have less direct significance.[15] You should also notice that we have eliminated the mysterious constant $G$ from Newton's description of gravity, which suggests that gravity is in some way an artificial concept and that $G$ appears in physics not because it has any deep fundamental significance but because our ancestors adopted a bizarre unit (for instance, kilograms) for expressing mass rather than the natural unit, that of length (such as the metre). But, the deepest remark that I can make in this connection, and which you should bear in mind as I develop the thought, is that by expressing all quantities in terms of length, we move towards a description in which the effect of mass on spacetime becomes a branch of geometry. Euclid would have been enthralled to know the reach of his considerations.

Now we get down to business. General relativity sprang from a remarkable

---

[13] The currently accepted value of $G$ is $6.673 \times 10^{-11}$ m$^3$ kg$^{-1}$ s$^{-2}$.    [14] To be specific, we replace $m$ by $Gm/c^2$.

[15] Black holes will not feature in this discussion. If they did, we would find that the radius of the event horizon around a black hole of mass $m$, the location of the boundary from which there is no escape, is $2m$ (in length units). The horizon of a black hole with the mass of the Earth lies at 8.8 mm from its centre.

coincidence that Einstein noticed. Remarkable coincidences are always as suspicious in science as in ordinary life. In ordinary life they commonly point towards deception; in science, they commonly point towards revelation. The coincidence in question is that the mass used to express the resistance of a body to a force, what in Chapter 3 we called its 'inertial mass', is the same as the mass used to express the ability of the body to generate gravitational attraction, its 'gravitational mass'. This equivalence has been confirmed experimentally to a precision of about one part in a trillion, which strongly suggests that inertial mass and gravitational mass are exactly the same. That should strike you as very peculiar. There is no immediately apparent reason why the resistance of a cannon ball to my kick should be identical to the strength of the gravitational field that the cannon ball generates.

Einstein built on this coincidence to identify another. Suppose you and I are travelling in the same elevator but something goes amiss. First, we find ourselves stranded on the 100th floor of a building. To pass the time before our rescue, we throw a ball to each other. Being observant, we notice that the path of the ball is curved (Fig. 9.11). If our elevator car had been in deepest space, outside the gravitational pull of any star or planet, the path of a ball would be a straight line. With that in mind, we ascribe the curvature of the ball's path to gravity. Being scientists, and having done a quick calculation, we also happen to know that the path of the ball is a parabola, the shape of the section obtained by slicing a cone through a plane parallel to one of its sides.[16]

Suddenly, calamity strikes. Our incompetent rescuers inadvertently slice through the cable holding up the elevator car and simultaneously disable all its safety features. We plunge downwards in free fall. Being scientists, we coolly exploit the unique opportunity provided by our fatal predicament, and continue tossing the ball to each other. Much to our amazement we find that the ball now flies in a straight line between us, just as it would in gravity-free space. Falling freely has eliminated the effects of gravity! If our elevator had been on the surface of the Sun, the parabolic path of the ball would have been more sharply curved, but when the elevator car went into free fall it would have accelerated more rapidly and that motion would also have ironed out the parabola into a straight line. The lesson to learn is that wherever we are, we can eliminate the effects of gravity by stepping on to a platform in free fall. If everyone who had ever lived had been forever confined to freely falling elevator cars, the artificial concept of gravity would never have been invented.

[16]  To understand parabolas fully we could turn to Apollonius's book, *Conics*, for the Greeks had worked out their properties long before they were found to be relevant not just to the description of the trajectories of balls but also to the shape of spacetime. Conic sections are classified as parabolas, hyperbolas, and ellipses: see Fig. 10.4.

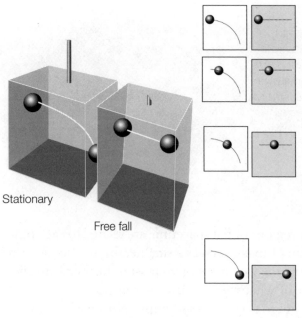

**Fig. 9.11** In a stationary elevator (left), the path of a ball projected horizontally is a parabola curving down towards the floor. In free space far from any gravitating mass, the path of the ball is straight (middle). In a freely falling elevator the path is also straight (middle). The sequences of images on the right show what is happening. The white boxes show in an exaggerated way the parabolic path of the ball in the stationary elevator. The grey boxes show that the elevator is changing its vertical position at an accelerating rate, and that the change of position just cancels the fall of the ball.

Stationary

Free fall

Einstein identified and built on this striking observation. First, he proposed, in effect, that all observers inhabiting freely falling elevator cars would write the same textbooks of physics. That is the essential content of the *principle of equivalence*. In particular, as the observers moved around in the cars, making measurements and sharing their results, they would experience the same contraction of space and time predicted by his special theory of relativity. We can put that proposition into more geometrical terms: the geometry of spacetime is the same (and Minkowskian) in any freely falling elevator car. So, everything that we discussed earlier in relation to special relativity applies in any such freely falling car.

Einstein's greater achievement, however, was to think about how the geometry in our falling car relates to another car, which might be falling with a different acceleration. For instance, your skyscraper might be built on an asteroid, and your fall would be accelerating very, very slowly. Mine might be located on Earth, and is accelerating at about 10 m s$^{-2}$ (so that after 1 s it is falling at 10 m s$^{-1}$, after 2 s it is falling at 20 m s$^{-1}$, and so on). The geometry of spacetime is 'flat'— Minkowskian—in each of our cars, but my little piece of flat geometry is twisted and rotated relative to yours. You could think of trying to cover a ball with coins (Fig. 9.12): each little region is flat, but one region is at an angle to another region. The question that Einstein tackled, and after years of hard

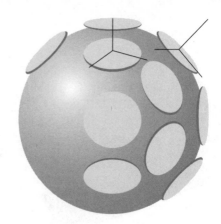

**Fig. 9.12** The local geometry at any point in space is Euclidean (represented by the flat circles attached to different points of the spheres). However, near a heavy body, such as a star or a planet, space is curved and a local Euclidean region is twisted and rotated away from another local Euclidean region. Einstein's general theory of relativity shows how to relate the different local coordinate systems to each other.

thought finally solved, is how regions of flat spacetime are related to each other when there is an accumulation of mass, such as a star, nearby. If I can describe my Earth-based spacetime from the viewpoint of your asteroid, then I am effectively describing the influence of what scientists used to call gravity.

Earlier in this chapter we got to grips with spacetime. Now we have to take that mindbending experience a stage further, to spacetimebending, and get to grips with *curved* spacetime. This is not as fearsome as it might sound, for we can put Minkowskian geometry into the back of our minds and pretend to forget its complications. In fact, many people consider the *qualitative* ideas of general relativity much easier than those of special relativity, because they can think about curved space (which is easy) rather than curved spacetime (which is not). That's a delusion, because general relativity is all about curved spacetime, but it's an agreeable delusion, because it makes the concepts accessible, so we will go along with it.

First, therefore, we concentrate on curved space, as the concepts are fairly straightforward. As before, it is conceptually easier to cut down the number of dimensions we have to try to imagine, and then build the number up again later. However, to think even of a two-dimensional curved surface, we already seem to need a third dimension to think of the surface as curved 'in', so you can see that to think of four-dimensional curved spacetime we would have to think in five dimensions! I won't be asking you to do that, because that is beyond me (and anyone I know) but if you want to visualize curved spacetime fully, then that is what you would have to try to do. The technical term for thinking about a curved space in one higher dimension is to 'embed' it in a space of one higher dimension. To think about curved four-dimensional spacetime, we would have to embed it in a space of five dimensions.

Let's stick with two-dimensional curved space (not spacetime) for the moment. To think of it as curved, we imagine the 2-space, a surface, embedded in 3-space, a volume. Let's think of the 2-space as the surface of a 3-sphere (an ordinary sphere, like an idealized Earth). Now think of a scene in which I am standing on the Equator at longitude 0° (that puts me somewhere inconveniently wet off the west coast of Africa) and you are standing on the Equator at longitude 90° (that puts you off the coast of Ecuador). A whistle blows, and we both begin to walk due north, ensuring at every step that we deviate to neither right nor left for the whole journey. Being theoretical physicists, we ignore the inconvenience of intervening deserts, oceans, and icecaps. Eventually, when we reach the North Pole, we find ourselves nose-to-nose (Fig. 9.13). We have to conclude that apparently parallel lines *do* meet in a space of this geometry. A space in which all apparently parallel lines meet when extended far enough— or, eguivalently, a space in which there are no truly parallel lines—is said to have *positive curvature*. This space is an example of one of the non-Euclidean geometries I mentioned earlier.

An immediate implication of the existence of non-Euclidean geometries is that geometry is an experimental science, not (as Immanuel Kant thought, as we see in Chapter 10) something that can be shown to be valid by introspection alone. Introspection alone is never a guide to truth, as Aristotle so wonderfully illustrates; introspection allied with experiment, of course—being the theme of this book—is an extraordinarily wonderful and reliable guide, as Galileo so wonderfully illustrates. We are confronted with the prospect of the geometry of space being either Euclidean, as Euclid and his followers thought, from their armchairs, for 2000 years, or non-Euclidean. To decide the question, we have to

**Fig. 9.13** You start off on the Equator and walk doggedly up the Greenwich meridian (at 0° longitude), always facing forward. I do the same from a point on the Equator at 90°W. When we get to the North Pole, our noses touch. These two lines of longitude are therefore not parallel: there are no parallel lines in this geometry. This illustration also shows how we imagine a two-dimensional surface of uniform positive curvature as the surface of a three-dimensional sphere. We say that the two-dimensional surface is 'embedded' in three-dimensional space.

resort to experiment and see, for instance, whether we do or do not come nose-to-nose if we walk along parallel paths for far enough. Carl Friedrich Gauss (1777–1855), one of the greatest of all mathematicians, had an inkling that Euclidean geometry might have competitors:

Indeed, I have therefore from time to time in jest expressed the desire that Euclidean geometry would not be correct.

Once the conceptual log-jam had been broken, largely by the tragically short-lived German mathematician Bernhard Riemann (1826–66) in an extraordinary lecture delivered in 1854 in pursuit of what we would now call tenure, people's minds were liberated to imagine non-Euclidean spaces with negative geometry too. Figure 9.14 shows a negative two-dimensional surface embedded in a three-dimensional space. When you sit on a saddle, you are supported by a two-dimensional surface with negative curvature. In this space, there is an infinity of parallel lines through a given point.

Once we pass over the intellectual hump and accept that there are different kinds of non-Euclidean geometry, we can start to imagine that the geometry of space can change from place to place. That is, different regions of space may have different curvatures. For instance, we could think of a dumbbell-like space obtained by squeezing a sphere round its equator and giving it a waist. This space would have positive curvature near its poles and a saddle-like negative curvature near the equator. We could go on to envisage more complex spaces by poking our fingers into the surface to give little craters pockmarking the domain, so that the curvature varies from place to place. You might like to think of various everyday objects that have surfaces with curvatures that vary from place to place (yourself, for instance).

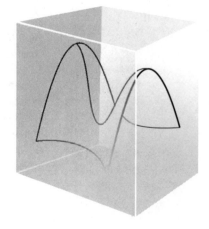

**Fig. 9.14** A two-dimensional surface with negative curvature is like a saddle-shaped surface embedded in three-dimensional space.

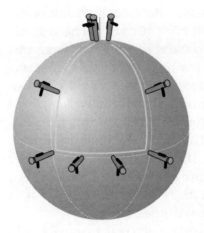

**Fig. 9.15** The curvature of a surface can be measured without thinking of it as being embedded in a space of higher dimension. One approach is to make a circuit around the point of interest and to measure the change in angle of a directed line. For instance, if as shown here we stand at the North Pole with our arms pointing south, and you walk down to the Equator along meridian 90°W, then along the Equator to the Greenwich meridian, then back up to the North Pole, all the time with your arm pointing south, when you arrive, we find that your arm points at 90° to mine. From this observation, we can deduce that the curvature of the surface is equal to $1/radius^2$, where *radius* is the radius of the sphere.

When we think of spaces embedded in spaces of higher dimension, we are adopting the point of view of a lofty hyperbeing who can inspect the actual world and judge at a glance whether it is curved. Suppose, though, that we are as an ant, and confined in our imagination to the actual space that we inhabit: could an ant determine whether the Earth is curved; can *we* determine whether our spacetime is curved? The answer is already implied by the discussion, because the journeys that you and I took, and either always came nose-to-nose or never did, can be thought of as taking place on a surface regardless of whether we think of it as embedded or not. Thus, if you and I set off on apparently parallel paths and come nose-to-nose, then we know that the space we inhabit has positive curvature. That conclusion is independent of whether or not we can *imagine* the space as embedded in a space of higher dimension.

We can develop this thought further and arrive at a quantitative measure of the curvature of space. Come with me to the North Pole (Fig. 9.15). Now we are there, let's both hold out one arm pointing directly south down longitude 0°, towards Greenwich. A whistle blows, and you walk due south until you reach the Equator. Still with your arm pointing south, you walk along the Equator until you reach 90° E. At that point, still keeping your arm pointing due south, you walk back to the North Pole. In due course, I see you coming over the horizon. Much to our mutual surprise, however, we find that your arm now points at 90° to mine despite the fact that you have meticulously kept it pointing south for your whole journey! In a flat space, our arms would coincide, so we have to conclude that the actual surface of the Earth isn't flat. Moreover, we can report the quantitative measure of 'curvature' as the change in angle of your arm divided by the area your journey enclosed, which works out as

$1/radius^2$, where *radius* is the radius of the Earth.[17] Because the radius of the Earth is 6400 km, the curvature of its surface is $2.4 \times 10^{-8}$ km$^{-2}$. This is a very dilute curvature, and indicates that we have to travel round a very large area before the effect be-comes appreciable. That is why Hammurapi's surveyors didn't notice it: the fields they measured in Mesopotamia were only a few thousand square metres in extent, and the curvature of the Earth simply didn't show up. The curvature of a soccer ball, of radius about 10 cm, is 0.01 m$^{-2}$, so its curvature is discernible for regions of its surface that cover quite small areas. For a sphere, the curvature is the same wherever we start our journey and whatever area we encompass. The curvature is also positive everywhere. A chicken's egg also has positive curvature everywhere, but it ranges from about 0.2 cm$^{-2}$ at its big end to about 0.4 cm$^{-2}$ at its more sharply curved little end.

We don't have to make the journey on the surface of a real, material Earth, soccer ball, or egg to discern curvature. If I stayed still and you travelled through empty space around a closed loop and at the end of your journey we found our arms pointing in the same direction, we would be able to conclude that that region of space is flat and Euclidean. If we found that our arms were at an angle, then we would have to conclude that the region of space is curved and therefore non-Euclidean. In that case, the relative direction of our arms would show the sign and magnitude of the curvature of that region of space. In general, journeys through different regions of space might give different results. We might even find that different orientations of a loop-like journey around the same point gave different outcomes. That is the kind of experiment that we could do to determine what kind of geometry prevailed in each region of space.

We need one more concept before we can appreciate fully the properties of curved space. A *geodesic* is a path through space that diverges to neither right nor left. In flat space, a geodesic is a straight line. Much of Euclidean geometry is about the properties of figures (such as triangles and rectangles) made up of geodesics—straight lines—in flat space. In any kind of space, the shortest distance between two points is along the geodesic that joins the two points. On the surface of a sphere, a geodesic lies along a great circle. For instance, if we travel along a line of longitude (such as the Greenwich meridian), then we are tracing out a geodesic between two locations with the same longitude. If the two points lie at different latitudes and longitudes, like London and New York, the shortest distance between them is along the shorter arc of the great circle that passes through them. Broadly speaking, commercial aircraft follow geodesics to their destinations.

[17]   The change in angle of your arm is $\pi/2$ radians and the area of the octant of the globe of radius $r$ is $\frac{1}{8} \times (4\pi r^2)$, or $\pi r^2/2$; so the curvature is $(\pi/2)/(\frac{1}{2}\pi r^2) = 1/r^2$.

Now it is time to step from curved space to curved spacetime. The step is not as traumatic as you might expect, because most of the concepts we need can be imported from our discussion of curved space. To imagine curved spacetime, we could think of a two-dimensional surface, one dimension being of space and the other of time, embedded in a three-dimensional space, just like the way we imagined a two-dimensional space. If the spacetime is flat, geodesics are straight lines in the surface. However, the funny geometry of spacetime implies that the geodesic joining any two points corresponds to the longest interval between them (remember Castor and Pollux). Curved two-dimensional space-time can be represented by a buckled sheet in three dimensions. Just as in flat spacetime, the geodesics—which now might meander across the surface depending on its local curvature—correspond to the longest intervals between the points they join.

Now we get to the crux of the whole discussion. This is the point where we bring together all the previous concepts. The Great Idea introduced by Einstein in 1915 was that *mass curves spacetime*. His extraordinary achievement was to find the precise relation between the detailed curvature of spacetime and the distribution of mass. I cannot possibly give you the precise relation, which is one of the most elegant yet complicated relations in the whole of science. However, it would be wrong of me, having made you work so hard to come to this point, to leave you stranded. So, I will do two things. First, I will give you a hint about the form of Einstein's result. Then I will outline some of its consequences.

At this point, I need to ask you to imagine a cube with slightly curved sides, a bit as though you had taken a cube made of rubber and stood on it so that the edges bulge out. The additional feature I need to get you to think about, though, is a cube in spacetime, not just space. To be perfectly honest, thinking about an ordinary spatial cube is almost good enough for conveying the essence of what I want to say, so don't hesitate to fall back on that image. However, bear in mind that we should really be talking in terms of spacetime, not space.

Recall the four-dimensional hypercube that we discussed earlier (and drew in Fig. 9.4). From now on, we should think of the edges of the constituent cubes as lying along geodesics in the region of spacetime of interest. That means we should think of the edges as twisted and tilted slightly but in such a way as to match each other properly when folded up to form the hypercube. Think of our carefully pasted together hypercube as having been sat on by the mass in its vicinity. The significance of the cubes remains the same, with the contents of the time-like cubes (those representing the history of the comings and goings through a face of the actual box) representing the mass flowing into and out of the box-like region through different faces and the two purely space-like cubes (the box at the beginning and end of the time period in question) representing the total mass in the box initially and finally. All that Einstein's 'field equation' does, is to state that the twisting and tilting of the faces of the eight cubes making up the hypercube is proportional to the total mass inside each one of them.[18] That is general relativity in a nutshell (admittedly a four-dimensional spacetime nutshell).

Einstein's field equation is simple to write down (provided the symbolism is sufficiently rich) but enormously difficult to solve. Nevertheless, one solution was found within a few months of its first being proposed. In one of the few positive acts of World War I, while serving in Russia the German mathematician Karl Schwarzschild (1873–1916) found a solution for the region of space outside a spherical region of mass, such as the empty space around a star or planet, and a solution for the interior of a sphere of uniform mass. He was dead a few months later, invalided home and felled by a rare skin disease, but the terms *Schwarzschild solution* and *Schwarzschild radius* have given him a pure kind of immortality. Another solution was found in 1934 by H. P. Robertson and A. G. Walker for spacetimes of *all* isotropic, homogeneous, uniformly expanding models of the universe.

Let's imagine moving out from the centre of a uniform Earth into the space outside and think about the shape of spacetime. To do so, we will think of an arrangement of six points arranged at the corners of an octahedron in space (Fig. 9.16). Inside the Earth the curvature of spacetime is entirely 'contractile' in the sense that the six points lie closer together than in empty space. It is as though spacetime itself is squashed together inside the Earth. This behaviour is a manifestation of Schwarzschild's interior solution of Einstein's equation for a spherical region of uniform mass. We can think of the lines of free fall as lying closer together inside the Earth, and of four-dimensional spacetime as having a positive curvature—like a sphere—with the same value in each two-dimensional time and space plane. The curvature in each plane is constant within the region of uniform density, and to some extent we can think of the curvature as like that of a rubber sheet in the region where a heavy ball rests on it (Fig. 9.17).

Once the array of six points bursts out through the surface of the Earth and enters the empty space outside, Schwarzschild's interior solution gives way to

---

[18] So that you don't feel entirely cheated, here is the Einstein field equation:

*Sum of the moments of rotation for the faces of each 3-cube* = $8\pi \times$ (*momentumenergy within the 3 cube*)

Crudely speaking, a 'moment of rotation' is the twist of a face of one of the 3-cubes multiplied by the distance of the face from the centre.

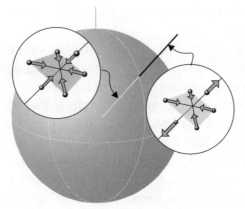

**Fig. 9.16** The tidal force of gravity can be detected by considering the forces on six test masses arranged in an octahedron. The two masses along the direction away from the centre of the Earth (or other massive body) are pulled apart, but the four masses in the grey-tinted plane are pulled together. This is a characteristic of the Schwarzschild exterior solution. Inside the Earth, where the geometry is given by the Schwarzschild interior solution, all six masses are pulled together.

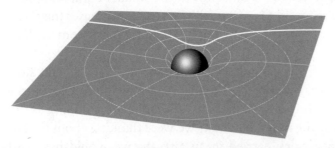

**Fig. 9.17** The effect of a massive body is to distort space rather like the effect of a heavy ball resting on a rubber sheet. Particles travel along geodesics (one of which is shown by the heavy white line). Because geodesics snake through the curved spacetime, steady motion along them may look to an observer like a path that is attracted to the heavy body. If we could show the time dimension too, then we would also observe what we would interpret as accelerations and decelerations as the body approaches and then leaves the region of the heavy mass.

his exterior solution. Now the geometry of spacetime is 'tidal' in the sense that the two points perpendicular to the surface move apart twice as fast as the four points in a plane parallel to the surface move together, the volume they surround remaining constant. We can think of the effect on space as stretching it in one direction (along the direction pointing to the distorting mass) and squashing it in the two perpendicular directions. This tidal effect is by no means negligible: the tidal effect of the Earth is enough to distort the sphericity of the reasonably rigid Moon by about 1 km. The tides of our oceans are just manifestations of the influence of the Moon on the geometry of spacetime at the surface of the Earth, with the fact that there are two high tides a day a manifestation of the bulge of geometry along the Earth–Moon direction. So, as you stand at the shore and watch the ebb and flow of tides, you are watching the shadow of Schwarzschild geometry pass over the surface of the Earth. Canute could not keep geometry at bay.

We can put numbers to the curvature. The radial curvature (the curvature of a plane with one side along the radial direction and the other along the time axis) is $-2 \times mass/radius^3$, where *radius* is the distance of the point of interest from the centre of the spherical concentration of mass (the star or planet, Fig. 9.17). Note that this curvature is negative (saddle-shaped), just like on the rubber sheet in the region outside the zone where the ball is resting. Each of the two planes that have one side perpendicular to the radial direction and one side along the time direction has a curvature equal to $mass/radius^3$. This curvature is positive, so we can think of each of these two-dimensional surfaces as like the surface of a sphere. These curvatures preserve the volume of a 3-cube, because the stretch in one direction is cancelled by the smaller compressions in the two perpendicular directions. Moreover, the curvature decreases the further we get away from the centre of the Earth, and at great distances from the Earth, space-time is flat.

One feature of Schwarzschild geometry is that clocks run slower when located close to a massive object. The fractional slowing compared to a clock far from the massive body is equal to $mass/distance$, where *distance* is the distance from the centre of the massive body. If we were thinking about the effect of the Earth's mass on a clock carried in an aircraft, we would have to take into account that it runs faster relative to one at sea level (because an aircraft is slightly further away from the centre of the Earth and in a region of spacetime with slightly less curvature) but time runs more slowly because the aircraft is in motion. The Earth is small (its mass is only 4.4 mm), so the effect of motion of a commercial airliner is small. Nevertheless, on a round-the-world trip at 10 000 m and 850 kph, the gravitational effect speeds up the clock by about 0.2 microseconds whereas the effect of speed slows it down by only about 0.05 microseconds. Tests of general relativity made in this way actually take into account the effects of landing and taking off as well as the varying speed of an aircraft in flight.

Why have we paid such a lot of attention to geodesics in spacetime? In empty space, particles travel in straight lines. In other words, they travel along the geodesics of flat spacetime.[19] This observation emphasizes the importance of geometry for the determination of paths. As spacetime becomes distorted by

---

[19] Why do they do that? Because the wavefunctions cancel for all other paths, as we saw in Chapter 7.

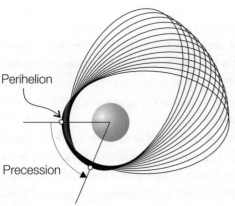

**Fig. 9.18** According to Einstein's theory, the path of a planet (particularly a planet close to its parent star, like Mercury) is not a perfect ellipse, but more like a rosette. The point of closest approach to the star rotates round the star. This motion is called the *precession* of the perihelion. Classical (Newtonian) mechanics also predicts precession, but accounts for only half the observed value of 43 seconds of arc per century (0.12 thousandths of a degree per year). General relativity predicts the correct value. The precessions of the orbits of binary star systems—the advance of the periastron—are much larger, as much as several degrees per year, and easier to measure.

Perihelion

Precession

the presence of mass—as we get closer to a star—particles continue to travel along geodesics, but now these geodesics are curved. In fact, the curvature of spacetime may be so great in the region of a massive body like a star that the geodesics are twisted into a spiral. In other words, as time progresses, a planet appears to move almost repetitively in a nearly closed path, nearly an ellipse, around the star. That is, a planet moving along a geodesic in spacetime describes an almost closed orbit in space. Far from its star—at Pluto rather than at Mercury—spacetime is less curved, and the planet needs to sample spacetime for a longer period before its path is almost closed. In other words, a distant planet orbits more slowly than a planet close to a star. In fact, the paths of planets are not perfect ellipses: they trace out slightly different paths on each revolution and to a space-bound observer describe a rosette around the central star. Accounting for the precise shape of Mercury's rosette-like path—the so-called *precession of its perihelion*—was an early success of general relativity (Fig. 9.18).

We have eliminated gravity. We now see that the motion of planets is not the response to a force called gravity, but simply the natural motion of a body along a geodesic in spacetime. In other words, motion is a manifestation of geometry.

There is a major problem with the description of spacetime given so far: on a small enough scale, geometry probably doesn't exist. One of the great outstanding problems of modern physics is the unification of the theories of general relativity and quantum theory (Chapter 7) into *quantum gravity*. Despite a huge effort, and despite making a lot of progress, scientists do not yet have a unified theory. Currently, there is no such theory as 'quantum gravity': in its place there are many speculations, most of them highly contentious, expressed with a variety of degrees of mathematical sophistication. When unification is achieved, however, everyone expects there to be a revolution in the way we think about space and time that will probably be greater in its impact than that caused by the introduction of relativity and quantum theory themselves. However, despite the currently foggy nature of scientists' perception of quantum gravity, there are certain features we can expect it to have.

One feature stems from the fact that, despite appearances, we have been taking a rather old-fashioned view about the nature of spacetime, a view little different in principle from Newton's perception of space as an arena. Certainly, we have made the description more sophisticated by combining space and time and endowing the resulting spacetime with a non-Euclidean geometry that depends on the presence of mass. But there is still the sense that spacetime is an arena of action within which all the activities of the world are played out. In quantum gravity this sense of arena dissolves, with the events themselves defining the universe. There is in fact, perhaps, no arena: what we regard as the universe is no more than a vast number of related events. With this perception, Einstein's equation becomes a statement about the causal structure of the relationships between events.

The second feature of quantum gravity is that on a small enough scale the whole concept of spacetime vanishes. That is, instead of spacetime being regarded as a continuum (of causally related events), it is more like a foam (Fig. 9.19). The smallest possible spatial separation of events is what we have

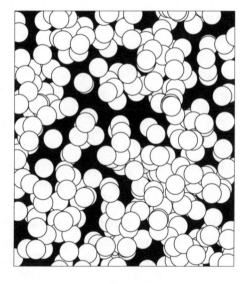

**Fig. 9.19** If we could examine spacetime under very high magnification, we would see that it is not a continuum but more like a foam. On the Planck scale of length and time, classical perceptions of spacetime are invalid. No one really knows what happens on the Planck scale, but progress is being made and promises to revolutionize our perception of where we are.

already called the Planck length, and the smallest possible separation in time is what we have called the Planck time. The Planck length is about $10^{-35}$ m, which is twenty orders of magnitude smaller than the diameter of an atomic nucleus, so it is hardly surprising that we perceptively clumsy colossal creatures have mistaken spacetime for a continuum. The smallest surface that can exist in spacetime has an area close to the square of the Planck length, and the smallest three-dimensional region that can exist has a volume approximately equal to the cube of the Planck length. In conventional units, the Planck time corresponds to about $10^{-43}$ s, so no two events can lie closer together in time. Even a process lasting a millisecond must consist of about $10^{40}$ components. No clock can tick faster than $10^{43}$ times a second.

Just as there is an absolute minimum of temperature ('absolute zero') in conventional physics, so quantum gravity shows that there is also an absolute maximum, at about $10^{32}$ K. At that temperature, spacetime itself will melt. The Big Bang that marked the inception of the universe was perhaps not so much a dramatic fireball but more a cosmic cooling that froze spacetime into persistence. Geometry, and all that we have seen that that word implies, is the frozen imprint of the causality of events.

# T E N

# ARITHMETIC

## THE LIMITS OF REASON

**THE GREAT IDEA**
*If arithmetic is consistent then it is incomplete*

*God created the integers, all else is the work of man*

LEOPOLD KRONECKER

O NE of the finest creations of the human mind is mathematics, for not only is it the apotheosis of rational thought but it is also the spine that renders scientific speculation sufficiently rigid to confront experience. Scientific hypotheses themselves are like jelly; they need the rigidity of mathematical formulation if they are to stand up to experimental verification and fit into the network of concepts that constitute physical science. One widely held view is that mathematics is not a science, for it can spin willy nilly its own universes of discourse, universes that need bear little relation, except in adherence to the rigours of logic, to the world we seem to inhabit. As such, mathematics might be thought an interloper into this volume. However, because it is so central to a scientist's mode of thought, mathematics is best regarded as a welcome guest and greeted as an honorary science. Moreover, with the march of abstraction in the physical sciences, and the stirring of abstraction in the belly of the biological, determining where mathematics ends and science begins is as difficult, and as pointless, as mapping the edge of a morning mist.

There is a further but related reason why it is appropriate to include mathematics here. Most working scientists, in their pragmatic and sensible way, simply recognize the astonishing ability of mathematics to act as a description of the physical world and are thankful that they have in their hands such an exquisitely powerful intellectual tool. But there are those who go beyond gratitude and application, and wonder whether the fruitful alliance of scientific observation and mathematical description points to a deeper aspect of mathematics, one that has not yet been fully identified and certainly not yet explained. The Hungarian–American theoretical physicist Eugene Paul (Jénó Pál) Wigner (1902–95), who did so much to formulate the mathematical theory of symmetry and apply it to physical problems, was puzzled by the remarkable ability of mathematics to act as a language for the description of the world:

The miracle of the appropriateness of the language of mathematics for the formulation of the laws of physics is a wonderful gift, which we neither understand nor deserve.[1]

Einstein expressed an allied thought when he remarked that the most difficult feature of the world to comprehend is that it is comprehensible.

I intend this chapter to be about mathematics rather than an exposition of mathematics itself, or even—except where I judge it relevant, or find it irrelevant but entertaining—a history of the ideas that make up mathematics. In other words, I shall be talking about what mathematicians think they are doing when they develop their theorems and solve their equations. I shall not be concerned about the details about what they do, so you won't come across proofs of Pythagoras' theorem or how to solve a quadratic equation. As such, the chapter is more concerned with the philosophy of mathematics, specifically with *mathematical ontology*, the foundations of the subject, than with the techniques that each of us has learned either to admire or to loathe and fear. Put another way, I intend to use the chapter to examine the validity of Bertrand Russell's much quoted but nevertheless engaging epigram that

Pure mathematics is the subject in which we don't know what we are talking about, or whether what we are saying is true.

I am alert to the fact that most of my readers will have uncomfortable and perhaps suppressed memories of mathematics, or at least uncomfortable preconceptions about what a chapter such as this will demand. Take heart: this is not a textbook. I intend to concentrate on the fascinating bits, and I shall signal places where it is safe to jump, on a first reading at least, without losing the thread. Moreover, you should bear in mind that this chapter is not mathematical; it is a story *about* mathematics.

My final introductory remark puts this chapter in another perspective. We

---

[1]  The unreasonable effectiveness of mathematics, *Communications in Pure and Applied Mathematics* 13 (1960). I have no idea what he meant by 'nor deserve'.

have travelled through ever-increasing abstraction, seeing familiar concepts dissolve with their replacements becoming more powerful. Mathematics is the climax of this journey, with abstraction its very essence: mathematics is pure, naked, disembodied abstraction. We should, therefore, expect extraordinary power.

The fundamental difficulty with mathematics is that it attempts to do things with the *natural numbers*, the everyday counting numbers 0, 1, 2, 3, . . . , which they invite but for which they were not originally intended. The natural numbers are used as *cardinal numbers*, to report the number of items in a collection, and as *ordinal numbers*, to order items as in a list. They are different concepts, and in English we give them different names, as in one, two, . . . for the cardinals and first, second, . . . for the ordinals. Most of what I have to say will refer to the natural numbers as cardinal numbers.

As we shall shortly see, once mathematicians started to think in their characteristically deep way about the natural numbers, it became apparent that it is surprising that we can count at all, for there are so few of them (a mere infinity) and they are so rare that from one point of view it is astonishing that early humans ever stumbled across them in the first place. We can already start to see some of the problems that mathematicians worry about even at this early stage of the discussion. For instance, do the cardinal numbers in fact go on for ever, or is *ultrafinitistic mathematics*, as it is called, in which the natural numbers fizzle out before they reach infinity, a better notion than numbers marching on for ever without end? And, because to be honest we have no direct perception of infinity, are mathematical demonstrations that invoke infinity reliable? Many would claim not, and would do everything to keep infinity at arm's length.

If we go back to the beginning of counting, whenever that was, we find deep resonance with what counting is taken to be today (a point we explore later). Counting is greatly aided by reference to a tally, such as marks on a stick, beads of a rosary, the hundred beads of the Muslim *subha* used to keep track of the ninety-nine attributes of Allah (with one additional bead marking the start), pellets of dried dung, and a pile of pebbles (Latin *calculi*, from which our words 'calculation' and 'calculus' derive). A universally portable tallying device is the human body, with its various indentations and protuberances. The Torres Straits islanders can reach 33 (right foot, little toe), taking in 8 (right shoulder), 26 (right hip), and 28 (right ankle) on the way, giving a base 33 for their counting.

The human hand, however, is far more convenient as a tallying device, particularly when one is fully dressed. Moreover, a hand has the flexibility of being able to indicate both cardinal and ordinal numbers: cardinal numbers are indicated by displaying the appropriate number of fingers simultaneously; ordinal numbers are displayed by unfolding the fingers in succession. Thus, counting to 'base 10', as our conventional system is called, is a natural consequence of human anatomy.

Although the basis for counting gradually settled down into the base 10 used almost universally today, there were and remain several aberrations. In Api, a language spoken in the New Hebrides, counting is to the base 5, and traces of the same base can be found in some African languages. Remnants of counting to base 12 are found in our use of the dozen (12) and gross (12 × 12). The Babylonians favoured base 60 for reasons that are still obscure, and their choice lingers in our divisions of time and the circle, with the small 'minutes' and the second division of these minute divisions into 'seconds'. There is a suggestion that the Sumerians of Babylon settled on 60 (but without a symbol for 0) as a result of the merging of two cultures, one using base 10 (with prime divisors 2 and 5) and the other base 12 (with prime divisors 2 and 3), with (2 × 5) × (2 × 3) = 60 the lowest common multiple. Base 60 never caught on for everyday enumeration as it involves learning so many specific names for the 60 distinct numerals required for the scheme (0, 1, . . . , 8, 9, ◆, ❞, . . . ,✳ [our 59], 10 [our 60], 11, . . .). Latin and French have remnants of base 20, as in *undeviginti* (19 = 20 − 1) and *quatre-vingts* (4 × 20 = 80), respectively, and a trace of base 20 can be discerned in the English *score* (20) and the Danish *tresinstyve* (three times twenty) for 60; base 20 is still used by the Tamanas of the Orinoco in Venezuela, the Inuits of Greenland, the Ainu of Japan, and the Zapotecs of Mexico. Pity the poor Maya, then, whose astronomical calendar had a symbol like a shell for 0 and was to base 20, but the third digit (the 'hundreds') was based on 18 × 20 rather than 20 × 20, the fourth digit was based on 18 × 20 × 20, and so on. They were probably trying to simplify astronomical calculations, as 18 × 20 = 360, the length of the Mayan year.

Finger counting, though, is unsuited to record keeping, and as early proto-accountants evolved and began to ply their trade, permanent scratches on various physical media slowly emerged as tallies and records of transactions. The Sumerians used a rather subtle form of cuneiform (wedge-shaped) set of numerals, the Attic Greeks an alphabetical notation, with symbols like Δ for 10 ($\delta\epsilon\kappa\alpha$, deka) and M for 10 000 ($\mu\nu\rho\iota\iota$, murioi). Still surviving today in a number of everyday applications are the Roman numerals. Apart from the obvious I, II, . . . for what we now write 1, 2, . . . , the German historian Theodor Mommsen (1817–1903) speculated that V (= 5) is a representation of an outspread hand, X (= 10) the combination of two hands, M (= 1000) a corruption of the symbol Φ via ( | ), and D (= 500) literally one half of that symbol.

Our familiar 'Arabic' numerals seem to have emerged in India some time before the ninth century, perhaps as a representation of the abacus. They were termed 'Arabic' by western scholars because, at the time, Arabic science was pre-eminent and the writers sought its authority. The origins of the forms of most of the numerals remains obscure, but 1 is obvious, 2 is perhaps a combination of two horizontal strokes and 3 of three. Humans seem to be incapable of assessing the number of more than four items at a glance, and consequently the numerals from 4 to 9 seem to have evolved as shorthand forms of collections of strokes.

The evolution of our current symbols can be traced back to the Brâhmî script, a very early form of Indian writing found in inscriptions of Asoka, third emperor of the dynasty of the Mauryas of Magadha, who reigned in India from about 273 to 235 BCE (Fig. 10.1); the script itself seems to have derived from a western Semitic tradition via a variety of Aramaean. The numerals were first introduced into an unreceptive Europe towards the end of the tenth century by the monk Gerbert of Aurillac (c.945–1003), who was to become Pope as Sylvester II in the numerically significant but apocalyptically disappointing year 1000. The tender shoot of innovation failed through the opposition of the conservative clerics who preferred to cling to the traditions of classical Rome, despite the near impossibility of its arithmetic. The earliest appearance of the numerals is in the *Codex Vigilanus*, copied by the monk Vigila at the monastery of Albeda, Spain, in 976.

Zero (from the Arabic, *sifr*, empty) was originally a dot, as it is in Arabic script today. The symbol for infinity, ∞, crept, like a wolf in the night, into the camp of numbers, where in due course it was to savage them. It was first used in 1655 by the insomniac John Wallis (1616–1703), the Oxford mathematician and

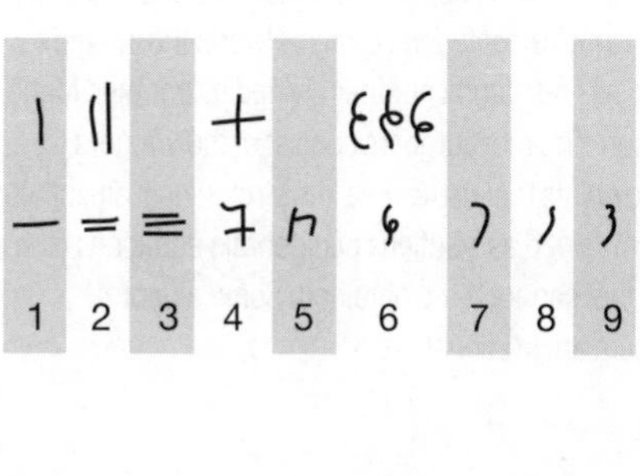

**Fig. 10.1** The so-called Arabic numerals arose from Indian symbols that can be traced back to the Brâhmî script, and then to deeper roots in the western Semitic tradition. The top line shows four numerals from the third century BCE, from the edicts of Asoka written in Brâhmî. The second line shows numerals from the third century CE, from a source in Uttar Pradesh.

one of the founders of the Royal Society, in his *Tract on conic sections*. He chose the symbol as a depiction of a curve that could be traced out endlessly, perhaps in the hope of getting to sleep.

Trouble (that is, mathematics) began when numbers were combined in a variety of ways. When we start to manipulate the natural numbers using operations like subtraction and division, when inventive intellect is brought to bear on pragmatic experience, we generate species of numbers that have less to do with cardinality. We first look at the symbolism for these manipulations and then see how, when they are applied to the natural numbers, we generate new kinds of numbers; the outcome is summarized in Fig. 10.2, and it will probably be helpful to keep this illustration in mind in what follows. In the early days of mathematics, equations were 'rhetorical' in the sense that they were expressed convolutedly in words. Much greater clarity, and with greater clarity greater power of manipulation, was achieved when symbols were introduced to denote operations.

The sign for addition, +, is probably derived from a cursive form of *et*, and first appeared in fifteenth-century German manuscripts, and – for subtraction may simply indicate separation. The sign for multiplication, ×, is probably derived from a symbol used in calculations of proportion involving cross-

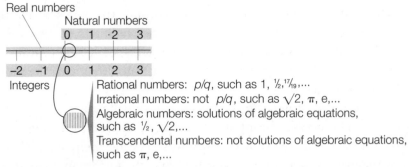

**Fig. 10.2** Here is a summary of the principal kinds of numbers we meet in this chapter. The natural numbers are the counting numbers; when extended to negative values they generalize into the integers. Lying between the integers are the rational numbers, which are numbers that can be expressed as one natural number divided by another. Much more dense are the irrational numbers, which cannot be so expressed. The real numbers consist of the integers, rational numbers, and the irrational numbers, and correspond to the points that make up a line stretching from infinity in both directions. The algebraic numbers are numbers that can be obtained as solutions of algebraic equations (see text and footnote 7), and the transcendental numbers are numbers that cannot be so obtained. Some algebraic numbers are rational, others are irrational; all transcendental numbers are irrational.

multiplication and first appears in *Clavis mathematicae*, published in 1631 by William Oughtred (1574–1660), the inventor of an early form of the sliderule. The German mathematician Gottfried Leibnitz (1646–1716) found it too easy to confuse × with *x*, and in 1698 suggested instead the use of a simple dot, as in *a.b* to denote multiplication of *a* by *b*. He also favoured : for division, but the common symbol ÷ was first employed for division (it had earlier been used for subtraction) in a Swiss text in 1659.

The equals sign, =, formed from two equal horizontal lines, was introduced in *The whetstone of witte* (1557) by the English mathematician Robert Recorde (*c*.1510–58) and introducer of algebra into England, coiner of engaging text-book titles (including, as well as *The whetstone, The grounde of artes*, an introduction to arithmetic, and *The castle of knowledge*, a textbook of astronomy), but who nevertheless died in a debtor's prison. Recorde wrote

And to avoide the tediouse repetition of these woordes :is equalle to: I will sette as I doe often in woorke use, a paire of paralleles, or Gemowe [twin] lines of one lengthe, thus: bicause noe .2. thynges, can be moare equalle.

Recorde's now familiar = sign long did battle with || and various designs based on æ, an abbreviation of *aequalis*, before it eventually triumphed.

Addition and multiplication of the natural numbers just generates more of the same. For instance, 2 + 5 = 7, a natural number; and 2 × 5 = 10, another natural number. However, subtraction generates a new class of number. Thus, if we subtract 3 from 2 we obtain –1, which expands our field of discourse from the natural numbers to the *integers*, . . . ,–2, –1, 0, 1, 2, . . . The negative integers must have been very puzzling when they were introduced, for people concerned only with counting would have found it difficult to conceive of less than no items.

Although multiplication of the natural numbers leads only to natural numbers, the concept of multiplication does lead to the identification of a subclass of the natural numbers called the *prime numbers*, which are numbers that are not multiples of other natural numbers (other than 1 and themselves). Thus, the first few prime numbers are 2, 3, 5, 7, 11, 13, 17, . . . A number like 15 is not prime because it can be expressed as 3 × 5; on the other hand, 17 is prime because it cannot be expressed as the product of other natural numbers. The prime numbers (the 'primes') have been and continue to be the centre of intense attention by those fascinated by numbers, for they seem to behave very much like the fundamental 'atoms' of the natural numbers—they act as numbers from which all other numbers can be constructed—when considering the operation of multiplication. This fundamental character is the essential content of Euclid's *fundamental theorem of arithmetic*, which asserts that every natural number is a *unique* product of primes.[2] A number like 9 365 811, for instance, can be expressed as a product of primes in only one way (in this case, as $3 \times 7^2 \times 13^3 \times 29$). The fundamental theorem is the basis of modern coding procedures that make use of the products of two large prime numbers, and the study of primes is not just disinterested mathematics as it is central to the conduct of safe transactions in commerce and to private communications between individuals and armies.

Various properties of the primes are known, but some conjectures are still unproved (and may be wrong). One well-established fact known to Euclid is that there is an infinite number of primes: primes go on without end. Currently, the largest known prime is $2^{13466917} - 1$.[3] This number is an example of a *Mersenne prime*, a prime of the form $2^p - 1$, where $p$ itself is prime. It was discovered on 14 November 2001, and if written out in full would have over 4 million digits (to be precise, 4 053 946), corresponding to about eight books the size of this one. Huge primes, with more than a thousand digits, are called 'titanic'. Primes get more and more thinly distributed as they get bigger, but there is always at least one prime between any given natural number and twice that number. For instance, you can be confident that there is at least one prime between 1 billion and 2 billion; there are in fact millions of them. Some primes cluster together. For instance, there are many 'twin primes', which are primes that differ by 2; thus, 11 and 13 are twin primes. The *twin prime conjecture* (which is only a conjecture) is that there is an infinite number of twin primes and therefore that twin primes, like primes themselves, go on without end. So far, the largest known twin primes are $33\ 218\ 925 \times 2^{169\ 690} - 1$ and $33\ 218\ 925 \times 2^{169\ 690} + 1$ (this pair was discovered in 2002, each number having 51 090 digits).

There are lots of other very weird properties of primes. For instance, the extraordinarily imaginative Polish–American mathematician Stanislaw Ulam (1909–84), discovered that if you write down all the natural numbers in a spiral, with 1 at the centre, 2 to its right, 3 above 2, 4 above 1, 5 to the left of 4, and so on, and mark all the primes, then they tend to fall in diagonal lines (Fig. 10.3). Ulam used his imagination in other ways: together with Edward Teller, he also discovered how to initiate a hydrogen bomb explosion.

---

[2]  This theorem would not be true if 1 were counted as a prime because we could include any number of factors of 1 and get the same answer. That is one reason why 1 is not included in the formal list of primes, but it does creep in as an honorary prime in some instances.

[3]  Keep up to date with primes by going to http://www.utm.edu/research/primes/.

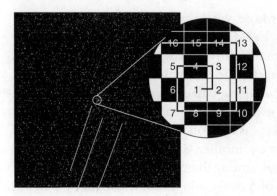

**Fig. 10.3** The Ulam spiral. When the natural numbers are plotted in a spiral, as shown in the inset, and the prime numbers are marked, they tend to fall in diagonal lines, as can be seen by inspecting the black zone, with the prime numbers like white stars. We have drawn some of the diagonals to indicate their location; you should be able to pick out others too.

Although prime numbers are the fundamental atoms of multiplication (just as 1 is trivially the fundamental atom of addition), they *might* also play a fundamental role in addition too. In 1742, Christian Goldbach (1690–1764), one-time tutor to Czar Peter II, suggested in a letter to the renowned Swiss mathematician Leonhard Euler (1707–83) that every even natural number greater than 2 is the sum of two primes. Thus, we have $2 + 2 = 4, 3 + 3 = 6, 3 + 5 = 8, \ldots, 47 + 53 = 100, \ldots$. *Goldbach's conjecture* still has not been proved despite an enormous effort. The difficulty appears to stem from the fact that the primes emerge from the concept of multiplication but are being invoked here in the context of addition. However, the conjecture might be an example of a feature that will gradually move to the centre of our stage: there might be no proof and, therefore, in a certain sense, the conjecture might be neither true nor false. Goldbach also conjectured that every odd natural number is the sum of three primes. This conjecture was partially proved—the proof works only for large numbers—by the Russian mathematician Ivan Matveevich Vinogradov (1891–1983) in 1937.

Division of one natural number by another also introduces a new class of number, called a *rational number* (from ratio; the reliable quality of such numbers is reflected in our common use of the term 'rational' to mean sensible, based on reason); examples between 0 and 1 include $\frac{1}{2} = 0.500\,000\,000\ldots$ and $\frac{3}{7} = 0.428\,571\,428\,57\ldots$. Notice how the decimal forms of the rational numbers either have 0 repeating for ever or have a finite sequence of numbers repeating for ever.

If you are starting to think like a mathematician, which is someone who goes beyond the immediate, looks for generalizations, and explores where they lead, then you will feel the tingle of a question stirring inside you: are there numbers that do not have repeating sequences and are therefore not expressible as a ratio of natural numbers? That these *irrational numbers* do exist was first discovered by the Pythagoreans, whose whole philosophy of life in Croton (now Crotone, in the heel of Italy), based as it was on the existence of the harmonies of rational numbers, not urinating towards the Sun or paring one's nails during sacrifice, and preserving peaceful social coexistence by the avoidance of eating beans (a practice Pythagoras himself had learned from the Egyptian priests he once lived among),[4] was overthrown when it was discovered that the square-root of 2, $\sqrt{2}$ = 1.414 213 5 . . . is irrational and cannot be expressed as the division of one natural number by another. Since then, many other numbers have been identified as irrational, among them $\pi$ = 3.141 59 . . . (the ratio of the circumference to the diameter of a circle, adopted as a symbol by Euler in 1737 and established as irrational in 1767),[5] $\pi^2$ (established as irrational in 1794), and e = 2.718 28 . . . (the base of natural logarithms). Irrationality is hard to prove: for instance, although it is known that $e^\pi$ is irrational, it is still not known if the same is true of $\pi^e$.

The rational and irrational numbers, both positive and negative and including zero, are termed the *real numbers*. To imagine the real numbers, we can think of each number as associated with a point on a straight line, the higher numbers to the right. The real numbers, like the points on this line, stretch from minus infinity on the left to plus infinity on the right, and take in all possible numbers—integers, rationals, and irrationals. The association of real numbers with points on a line is a crucial step in recognizing that geometry—the properties of various lines, therefore collections of points, and therefore collections of real numbers—can be treated as a branch of arithmetic. We shall not go down that path in this chapter, but you should be aware that although we will be concentrating on ideas that are overtly arithmetic, covertly they include other branches of mathematics too, such as geometry (Fig. 10.4). In fact, the reach of arithmetic is even greater than that. According to an extraordinary but catchy little theorem first proved by the German mathematician Leopold Löwenheim (1878–1957) in 1915 and improved by the Norwegian Albert Thoraf Skolem (1887–1963) in 1920, a system of rules like those of arithmetic emulates *any* field of knowledge that can be formalized in terms of a set of axioms. It might have leavened the tedium of learning how to extract square roots and do long division if at school you had been told that, according to the *Löwenheim–Skolem theorem*, you were actually modelling the process of drawing conclusions from

---

[4] How right they were. We now know that beans are rich in a carbohydrate that our enzymes cannot digest but which *E.coli*, which populate our gut, can; when they digest it, they release large quantities of carbon dioxide and hydrogen, which is a major source of flatulence.

[5] The value of $\pi$ has been calculated to several thousand digits. The 7s start repeating at the 1589th digit, when four appear in succession. But they break out into other digits immediately after.

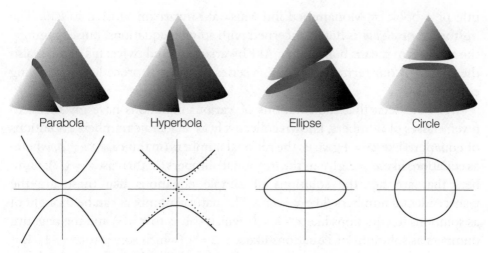

| Parabola | Hyperbola | Ellipse | Circle |

**Fig. 10.4** The Greeks had an ideal depiction of space, and were correspondingly good at geometry. Here we see how parabolas, hyperbolas, and ellipses (including the special case of a circle) can be regarded as selections of numbers obtained by slicing through a cone in different directions. We now know, thanks initially to Descartes, how to relate these shapes to algebraic quantities, and so can now see the link between the geometry of space and the arithmetical properties of certain selections of numbers.

quantum mechanics, natural selection, and jurisprudence (in so far as these branches of knowledge can be expressed in terms of axioms). The same is true of the rest of the chapter: although a lot of it will read like an account of arithmetic, bear in mind that *it is actually an account of any systematic branch of human knowledge.*[6] If that isn't breathtaking, then I don't know what is.

Some irrational numbers, including $\pi$ but not $\sqrt{2}$, are *transcendental* in the sense that they 'transcend' ordinary algebraic equations. That simply means that they are not solutions of simple algebraic equations like $3x^3 - 5x + 7 = 0$.[7] Thus, $x = \sqrt{2}$ is the solution of $x^2 - 2 = 0$, so it is algebraic (the solution of such an equation) but not transcendental. There is no equation of this kind, however, that has $x = \pi$ as a solution, or $x = e$, so $\pi$ and e are not only irrational but transcendental too. In 1934, the Russian mathematician Aleksandr Gelfond (1906–68) proved that $a^b$ is transcendental whenever $a$ is algebraic (other than 0 and 1) and $b$ is algebraic and irrational (like $\sqrt{2}$); so $2^{\sqrt{2}}$, for instance, is transcendental because 2 is algebraic and the irrational number $\sqrt{2}$ is algebraic: we know at once, therefore, that there is no algebraic equation that has $2^{\sqrt{2}}$ as solution. Incidentally, the name 'algebra', which has just made its first appearance, is derived from *Al-jabr w'al muqâbala* (Restoring and simplification), the

---

[6] The opposite, that all formalized systems of knowledge are merely arithmetic, is also true, and perhaps more sobering.    [7] An algebraic equation has the form $a_n x^n + a_{n-1} x^{n-1} + \ldots + a_0 = 0$, where the $a_k$ are integers.

title of a book by Mohammed ibn Musa al-Khwarizmi written in 830. The 'restoring' of *al-jabr* is here concerned with solving equations, but engagingly the same term means bone-setter. Al-Khwarizmi scored twice: his name is also the source of our term 'algorithm', a series of rules of procedure for solving equations.

We have seen that the solutions of various equations have given rise to given classes of numbers, known collectively as 'algebraic numbers'. Solutions of equations like $2x = 1$ give us the rational numbers (in this case $x = \frac{1}{2}$), whereas equations like $x^2 = 2$ give us the irrational numbers (in this case $x = \sqrt{2}$); numbers that are not the solutions of simple equations like these are the transcendental numbers (like $x = 2^{\sqrt{2}}$). The natural numbers can be thought of as solutions to equations like $x - 2 = 1$ (which solves to $x = 3$) and the negative numbers as solutions of equations like $x + 2 = 1$ (which solves to $x = -1$). But there is a simple equation missing from this list: what is the solution of $x^2 + 1 = 0$? None of the numbers introduced so far is a solution, because the square of any of them is positive and when added to 1 can never give zero. Largely because mathematicians didn't want to have to admit that some equations don't have solutions, they invented the concept of an *imaginary number* i, which is the solution of the equation $x^2 + 1 = 0$; in other words, $i = \sqrt{(-1)}$. Because they—in fact, it was Descartes—thought numbers such as i and any multiple of i didn't really exist, they termed them 'imaginary'.

It soon became clear that some equations, such as the solution of $x^2 - x + 1 = 0$, had solutions that are combinations of real and imaginary numbers, in this case $x = \frac{1}{2} + (\frac{1}{2}\sqrt{3})i$ and $x = \frac{1}{2} - (\frac{1}{2}\sqrt{3})i$. These combinations are called *complex numbers*, the first $\frac{1}{2}$ in this example being an ordinary 'real' number and the $\pm(\frac{1}{2}\sqrt{3})i$ being imaginary. Special rules had to be developed to do calculations with these two-component real numbers, but they were natural extensions of the rules we use for real numbers and give rise to no special difficulty.

The real numbers can be ordered in a line, as we have seen. The complex numbers become a little less mysterious once we realize that each one of them can be plotted as a point in a plane, with the real part of the number indicated by the distance along the horizontal axis and the imaginary part indicated by the distance along a vertical axis (Fig. 10.5). In other words, a complex number is actually a pair of numbers: the complex number $1 + 2i$, for instance, is just the two-component number (1, 2) which we can represent by a point 1 cm from the vertical axis and 2 cm up from the horizontal axis. Put yet another way, we can think of a complex number as a domino, with the value on the left-hand side of the rectangular face the real part and the value on the right-hand side the imagi-

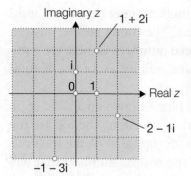

**Fig. 10.5** A complex number is a two-component number and as such can be represented by points in the plane. For example, the complex number $2 - 1i$ is denoted by a point 2 units along the horizontal axis and one unit down the vertical axis. Manipulations of complex numbers are simply manipulations of these two-component objects.

nary part. In future when you pick up , think of it as the complex number $4 + 3i$. If you feel uncomfortable with images of this kind, don't worry: complex numbers will not appear again in this chapter, except in one fleeting remark.

In this section, I shall address two fairly straightforward questions: how many numbers are there, and what are they anyway? As you might suspect, the answers will be more complex than the questions, which is probably the essence of good questions, anyway.

At first glance, there is an infinity of natural numbers because, in principle, we can go on counting for ever: 1 sheep, 2 sheep, . . .. We say that the 'cardinality' of the natural numbers is infinite. *Hilbert's hotel* is a fanciful exploration of this cardinality, and is attributed to the German mathematician David Hilbert, who will figure more seriously again later. 'Hilbert's hotel' consists of an infinite number of rooms, and one night all the rooms are full. A traveller arrives without a reservation. 'No problem!' cries Hilbert (the manager): he persuades all the guests to move into the neighbouring room, leaving the first room vacant and able to accommodate the newcomer. Later that night, an infinite number of travellers arrive, without reservations. 'No problem!' cries the resourceful Hilbert again. He persuades *all* the guests to pack and move into a room with a number that is twice the number of their current room, so leaving the odd-numbered rooms free to accommodate all the newcomers.

So far, perhaps, so good. But what about the rational numbers, the numbers that are obtained by dividing one natural number by another: how many rational numbers are there? The 'obvious' answer is that there are more rational than natural numbers, because there is an awful lot of them between 0 and 1 (such as ¼, ½, ⁵³⁄₆₇, and a lot of others), another huge number between 1 and 2 (such as ³⁄₂, ³⁄₅, ⁷⁹⁄₄₇, and a lot of others), and so on. The funny, correct answer, though, is that there is the same number of rational numbers as there are natural numbers: their cardinality is infinite, the same infinity as for the natural numbers.

To see that this is so, look at Fig. 10.6, where I have drawn up a table of all the rational numbers (but shown only a tiny part of it). Along the top are the nat-ural numbers that will appear in the numerator (upstairs) of the fractions we are going to produce, and down the side are the natural numbers that will appear in the denominator (downstairs). The body of the table holds all possible fractions obtained by dividing one natural number by another. There will be lots of repetitions, like ³⁄₆ and ⁴⁄₈, both of which are equal to ½, but that doesn't matter. Now we can draw a line that snakes from the first entry through all the entries, as shown in the illustration. Then, as we travel along the line, we count off 1, 2, . . . for each fraction we meet. In this way, all the fractions—all the rational numbers—get put into a one-to-one correspondence with the natural numbers. We never run out of natural numbers, so the number of rational number is the same as the number of natural numbers even though they are denser on the ground than the natural numbers. There is an infinite number of rational numbers between 0 and 1, and between 1 and 2, but the same infinity between 0 and 2! In short, we can always count the rational numbers—we say they are *denumerable*—and obtain the answer 'infinity' regardless of the range of numbers over which we perform the counting. You can perhaps begin to see that infinity is a greasily slippery concept.

The algebraic numbers—the numbers that are the solutions of algebraic equations—are also denumerable. You can get a glimpse of the argument by

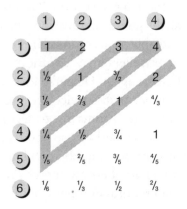

**Fig. 10.6** The rational numbers can be put into correspondence with the natural numbers, so they are denumerable (countable). The row along the top are the numbers that appear in the numerator of the ratio $p/q$, and the number down the side are the numbers that appear in the denominator. As we travel along the snaking diagonal line, we can count off all the rational numbers (including a lot of repeats).

noting that each algebraic equation is composed of powers of $x$ (expressions like $x^3$) multiplied by integers (as in $4x^3 + 2x - 1 = 0$). There is therefore a one-to-one correspondence between the solutions of the equations—the algebraic numbers—and the integers defining the equations. With that correspondence in mind, you can probably accept that the solutions—the algebraic numbers—can be put in one-to-one correspondence with the natural numbers. We can conclude that algebraic numbers are countable and, although they are infinite, their cardinality is the same as that of the natural numbers.

How many irrational numbers, numbers that cannot be expressed as ratios of natural numbers, are there? You probably think that there is also an infinite number of them. You would be right. But where you would probably be wrong (unless you know the answer already), is that there is a greater infinity of irrational numbers than there is of the natural numbers. That is, the irrational numbers have a greater cardinality than the natural numbers: they are more infinite in number. The clever argument that first exposed this odd feature was proposed by the genetically cosmopolitan Georg Ferdinand Ludwig Phillip Cantor (1845–1918), born of Danish and Russian parents in St Petersburg but living for most of his life in Germany. His was a life of frustration, largely because he was working at the edge of current mathematics, bringing infinity into perspective. In part as a result of the stress of opposition to his work from the more conservative parts of the mathematical establishment, and in particular the influential Leopold Kronecker (1823–91), who was prejudiced against all varieties of numbers except the rational, Cantor began to suffer severe mental disturbance, turning increasingly to religion, for he considered that the infinite sets of objects he had considered existed as realized entities in the mind of God, and that he, Cantor, was the vessel chosen to reveal them, a kind of mathematical St John the Baptist. Between indulging his obsession that Bacon was the author of Shakespeare, Cantor spent increasing periods in mental hospitals, exploring the frontiers of religion in the form of freemasonry, theosophy, and Rosicrucianism, just as he had explored the frontiers of mathematics, but with less effect. Madness is certainly a risk when one contemplates the abyss of infinity, as perhaps you will come to realize as this chapter unfolds.

In 1874, Cantor discovered a simple argument to show that the irrational numbers are more numerous than the rational numbers. We will use his argument and a variation of it again in other contexts, so it is worth spending a moment on it. We start by writing a list of randomly chosen numbers lying between 0 and 1, and number them in succession (in the left-hand column):

| 1 | 0.**1**98 402 957 820 . . . |
|---|---|
| 2 | 0.4**3**8 291 057 381 . . . |
| 3 | 0.68**4** 930 175 839 . . . |
| 4 | 0.782 **9**48 261 859 . . . |
| 5 | 0.500 0**0**0 000 000 . . . |
| 6 | 0.483 91**3** 562 785 . . . |
| ⋮ | . . . |

Now we show that however long the list, including infinitely long, there are numbers that are not in it. To do so, we construct a new number by selecting the first digit after the decimal point from the first number, the second from the second, and so on, and writing a *different* digit in each case: changing the bold digits, for instance, gives us the new number 0.350 047 . . . .. This number is certainly not in the original list because it differs from the first number, it differs from the second number, and so on. It follows that the real numbers (the rational and irrational together) are more numerous than the number of natural numbers because, however long the list, we can always construct a number that isn't in it. We say that the real numbers are *uncountable*.

Let's look at this conclusion a little more closely. We have just seen that the *real* numbers (the natural numbers plus the rational numbers and the irrational numbers) are uncountable. However, we have seen that the natural numbers, the rational numbers, and the algebraic numbers are all countable (that is, denumerable). The numbers that are excluded from these denumerable kinds are the transcendental numbers. We have to conclude that *the numbers that make the real numbers uncountable are all transcendental* (like the numbers π and e).

Let's pause to take on board the significance of this extraordinary conclusion. It means that the vast majority of numbers—an infinitely overwhelming proportion, in fact—are transcendental. That is rather astonishing, especially because transcendental numbers are much less familiar than 'ordinary' numbers and, indeed, you might never have heard of them before. The fact that the transcendental numbers are overwhelmingly more numerous than the other kinds of numbers is the basis of my remark at the beginning of the chapter that it is surprising that we can count: the natural numbers are extremely sparsely distributed among the real numbers, each one being surrounded by an infinity of transcendental numbers. The author Edward Temple Bell expressed it graphically when he said

The algebraic numbers [which include the natural numbers] are spotted over the plane like stars against a black sky; the dense blackness is the firmament of the transcendentals.[8]

Cantor denoted the cardinality—the total number—of the natural numbers by the Hebrew symbol $\aleph_0$ (aleph-null), the first of a series of *transfinite numbers* $\aleph_0, \aleph_1, \aleph_2, \ldots$ of increasing size.[9] We can think of $\aleph_0$ as the smallest version of infinity, $\aleph_1$ as the next bigger version, and so on. The question that confronted Cantor is whether the cardinality of the real numbers, which as we have seen is greater than that of the natural numbers, is equal to $\aleph_1$ or to a higher transfinite number. The celebrated *continuum hypothesis* is that the cardinality of the real numbers—the number of points on a line—is equal to $\aleph_1$, the first of the cardinal numbers after $\aleph_0$, and not, for instance, $\aleph_5$ or some other transfinite number. Cantor was almost—some would say was—driven mad by his continued but frustrated attempt to prove the continuum hypothesis. Had he survived to 1963, he would have understood his frustration, or at least have had it demonstrated to him, for in that year the American logician Paul Cohen (b. 1934) showed that the problem is undecidable: it is impossible to prove either the truth or the falsity of the continuum hypothesis, and the cardinality of the real numbers can be any of the values $\aleph_1, \aleph_2, \ldots$, and perhaps all of them.

We have stumbled on another suspicious and worrying feature of mathematics: it seems to run out of steam when dealing with the infinite just as it might be the case that its boilers run dry when confronted by Goldbach's conjecture (about the ability to express any even number as the sum of two primes). A question that should be beginning to stir in our minds is whether mathematics is not all it's cracked up to be: does it lose its magisterial command when pressed too far? Are there other questions like the continuum hypothesis that stun it into silence? Just as the ultrafinitists think of the natural numbers as fizzling out before they reach infinity, does mathematics itself fizzle out in certain regions of argument and have blind spots in others?

Before moving on to judge whether mathematics' fine clothes are in fact

---

[8]  E. T. Bell, *Men of mathematics*. Originally published in 1937.

[9]  The usage is a little variable here: some people reserve the term 'transfinite number' for the ordinal numbers $\omega$, $\omega + 1, \ldots$, where $\omega$ (omega) is larger than any natural number.

frayed and threadbare, there are still some remarks worth making about Cantor's conclusions, even though they might edge us too towards madness. First, the implication of the uncountability of the real numbers is that it is impossible to know the number of points on a line of any length. However, we can be sure that whatever the length of the line, it consists of the same number of points, whatever that number is. Thus, the number of points in a line a millimetre long is the same as the number of points in a line that stretches from here to the next galaxy. What about the number of points in a plane? By a clever argument, Cantor was able to show that every point on a plane could be put in a one-to-one correspondence with every point on a line regardless of the area of the plane and the length of the line. Therefore, the number of points in a plane of any area—a postage stamp or Australia—is the same as the number of points on a line of any length—a nanometre or a kilometre—and both are equal to the number of real numbers. The same is true of a volume in any number of dimensions: there are as many points in a cube, and as many points in a ten-dimensional hypercube of any size, as there are points in a line of any length. Rather astonishingly, therefore, there are as many points in a sphere the size of the Earth as there are points in a line of length 1 cm. You can perhaps begin to see why Kronecker was so upset at the prospect of mathematics stepping into what Hilbert was to call 'Cantor's paradise' and how, unless we take special care, the infinite is a treacherous bog that swallows reason.

We know that there are a lot of them, we recognize them when we see them, but what are they? What are numbers? The Greeks had a limited view of numbers, which is perhaps why they were better at geometry than arithmetic. Their notation didn't help: they had the perfect notation for elementary geometry—straight lines and circles drawn on a flat surface—but their numerals were cumbersome. Indeed, they did not consider 0 and 1 to be numbers, for their conception was that of 'numerous' rather than 'number': the greater the numerousness then the greater the number. No things and one thing each lack numerousness, so they are un-numbers.

The modern conception of number emerged once *set theory* had been developed in the late nineteenth century, initially by Cantor and then in its full rigour by Frege and Peano. The Italian Giuseppe Peano (1858–1932) was the Dr Casaubon of mathematics, for like Dr Casaubon in *Middlemarch*, who was seeking to write the history of all the world's religions, in his middle years, from about

1892 until 1908, Peano was attempting to compile his *Formulario mathematico*, a collection of all known theorems in all branches of mathematics. The engagingly unrealistic Peano considered that the *Formulario* would be of inestimable benefit to lecturers, who would merely have to announce the numbers of the theorems in their lectures rather than go through the tedium of presenting them verbally. To encourage its use internationally, Peano published his work in 'Latino sine flexione', the putative international language he had devised which was based on Latin, had a vocabulary compiled from Latin, German, English, and French, but was stripped of the irksomeness of a grammar. Peano, who perhaps may be thought lacking judgement in the ordinary deportment of daily life, though in other matters gentle and courteous, also had the knack of losing friends by the unrestrained exercise of one of his most impressive talents, his ability to be trenchantly logical. He used his talent to scythe down potential friends if their arguments lacked perfect rigour; but he put it to good use in his formulation of the foundations of mathematical logic. Even the young Bertrand Russell was impressed by Peano's precision and the power of argumentation that accompanied it when they met in 1900, and Russell adopted a variation of his notation when he began his own formulation of the foundations of mathematics.

Peano, for some undivinable but perhaps engagingly romantic reason, published his axioms in Latin. He set arithmetic on the following foundations:

1. *0 is a number.*

2. *The immediate successor of a number is also a number.*

3. *0 is not the immediate successor of any number.*

4. *No two numbers have the same immediate successor.*

5. *Any property belonging to 0 and to the immediate successor of any number that also has that property belongs to all numbers.*

The last axiom is the *principle of mathematical induction*. If we denote 'immediate successor' by $s$, then we can identify 1 as $s0$ (the immediate successor of 0), 2 as $ss0$ (the immediate successor of the immediate successor of 0), 3 as $sss0$, and so on. The problem with this approach, though, is that Peano leaves some of his terms, such as 'immediate successor' and, indeed, 'number', undefined and we still don't know what these numbers *are*.

At this point Friedrich Ludwig Gottlob Frege (1848–1925) made a seminal contribution, a contribution that seemed to be on the point of raising mathematics to a consummate position in human thought, but in fact proved to be its undoing. Frege is regarded as the founder of mathematical logic, for he set out to construct a perfectly logical scheme that would establish mathematics as the epitome of desiccated human thought. To succeed, he needed to define the concept of number, and to do so in his *Grundlagen der Arithmetik* (Foundations of arithmetic, 1884) he built on the concept of a *set*. A set is simply a collection of distinguishable objects, such as {Tom, Dick, Harry}. Sets had been introduced into mathematics by Cantor, and the theory was to be refined in the course of the following decades by Ernst Zermelo (1871–1953) and Adolf Fraenkel (1891–1965), who established precise statements about the properties of sets, how to construct them (a point Cantor had failed to explain), and how to manipulate them. A common version of modern set theory is consequently known as *Zermelo–Fraenkel theory*.

Frege developed the view that numbers are names that denote certain kinds of sets. To make his definition precise he introduced the concept of the *extension* of a property, which is the set that consists of all things that have that property. The name 'extension' is perhaps best thought of as a word formed from 'extended collection'. So the extension of the property 'being the same size as the set {Tom, Dick, Harry}' is a set that consists of *all* sets that have the same size. 'Have the same size' has a well-defined meaning in set theory: it means that the members of the sets can be put in one-to-one correspondence. For instance, the set {Tom, Dick, Harry} has the same size as the set {scissors, rock, paper}, because Tom can be put in correspondence with scissors, Dick with rock, and Harry with paper (Fig. 10.7). Set theory may seem unduly fussy with its definitions: but it should be fussy if it claims to be the foundation of mathematics. The extension of the property 'being the same size as the set {Tom, Dick, Harry}' is therefore the set that consists of the sets {Tom, Dick, Harry}, {scissors, rock, paper}, and so on. Now, with a bump, we come down to Earth: we call that extension, that set, 3.

Frege went on to define the natural numbers as the following extensions:

> *0 is the name of the extension of the property 'having the same size as the set consisting of things that are not identical to themselves'*

(of course, there is nothing that is not identical to itself).

> *1 is the name of the extension of the property 'having the same size as the set 0'.*

> *2 is the name of the extension of the property 'having the same size as the set consisting of the sets 0 and 1',*

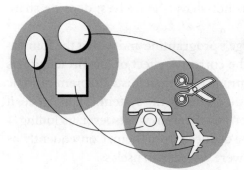

**Fig. 10.7** A set of objects has the same size as another set if all the members of the two sets can be put in one-to-one correspondence. These two sets have the same size: if the aircraft is removed, they have different sizes.

and so on. A crucial aspect of this definition of numbers as the names of sets defined successively in terms of smaller sets is that it uses terms that stem from logic, namely 'property', 'equality', and 'negation'. This aspect impelled Frege to the view that mathematics is no more than logic.

Logic it may be, but satisfactory it was not. In 1902, shortly before Frege was about to send to his publishers the second volume of his great work *Grundgesetze der Arithmetik* (Fundamental laws of arithmetic), in which he had built the whole edifice of mathematics on this definition of number, he received a famous letter from Bertrand Russell pointing out that there was an inconsistency. Frege's own words movingly capture the awful moment when he opened Russell's letter:

A scientist[10] can hardly meet with anything more undesirable than to have the foundation give way just as the work is finished. In this position I was put by a letter from Mr Bertrand Russell as the work was nearly through the press.

Bertrand Russell (1872–1970) had pointed out to Frege the problem of the extension of the property 'does not belong to itself'. Suppose we consider a set consisting of concepts that are not members of themselves. For instance, a set consisting of 'abstract ideas' is a member of itself because the set is itself an abstract idea, whereas a set consisting of 'fruit' is not a member of itself because the set is not a fruit. Russell asked whether the set of concepts that do not belong to themselves belongs to itself. If it does belong to itself, then it is the sort of set that doesn't belong to itself. If it doesn't belong to itself, then it is the sort of set that does belong to itself. In short, if it does, then it doesn't, but if it doesn't, then it does. *Russell's antinomy* (or contradiction, paradox) has been expressed in a number of more conversational everyday terms, as in 'the barber

---

[10]   We note that the logician Frege so considered himself.

in this city shaves all the men who do not shave themselves: does he shave himself?'.

Russell's antinomy undermined Frege's programme and with it the foundations of mathematics. The reason for the corrosive effect of a contradiction is that, if a series of axioms leads to a contradiction, then it is a theorem of logic that all propositions in that system are theorems of the system.[11] Therefore, if Frege's definitions are contradictory, then any theorem whatsoever, including '1 = 2' and '$\sqrt{2}$ is a rational number', can be deduced from them. Consequently, as a foundation for arithmetic, his axioms were worse than useless.

Russell was as deeply concerned as Frege with the foundations of mathematics, and was equally interested in trying to demonstrate that it is no more than a branch of logic. This is the viewpoint of the *logicist school* of the philosophy of mathematics. In 1903 Russell had published his *The principles of mathematics*, and his former examiner and now colleague at Cambridge, Alfred North Whitehead (1861–1947), was preparing a second edition of his *A treatise on universal algebra*. The two men decided to collaborate on a more ambitious project, to show that the whole of mathematics is merely a subset of logic. The work, which took them a decade to prepare, ultimately appeared in the three volumes of *Principia mathematica* in 1910, 1912, and 1913. A projected fourth volume on geometry never appeared. *Principia* used a highly elaborate symbolic notation that was more powerful than that of either Peano or Frege; some idea of its sophistication is illustrated in Fig. 10.8, which is Russell and Whitehead's proof that 1 + 1 = 2.

Russell and Whitehead needed to circumvent the bog of inconsistency that had swallowed Frege. To do so, Russell introduced his *theory of types*, in which members of sets are assigned a 'type', and any set can contain members only of a lesser type. Thus, individual entities are of Type 0, statements about sets of those individual entities are of Type 1, and so on. Because a set can contain only sets of inferior type, it can never be a member of itself, so Russell's antinomy is avoided. However, the theory of types is still not strong enough to eliminate some paradoxes, such as 'Berry's paradox', the ten-word statement 'the least integer not definable in fewer than eleven words'. The integer that satisfies this

---

[11] We start with the theorem $p \supset (\sim p \supset q)$, where $\sim$ means 'not', $\supset$ should be read 'if ... then', and $p$ and $q$ are propositions. Suppose that the propositions $p$ and $\sim p$ are both deducible from the axioms. Because $p$ is valid, by the 'rule of detachment' we can strip it away and from the theorem deduce that $\sim p \supset q$. Then, because $\sim p$ is valid, by the rule of detachment again, we can strip it away, and $q$ follows. That is, $q$ is valid, whatever proposition it is.

**✳54·43.** ⊢ :. $\alpha, \beta \, \epsilon \, 1 . \supset : \alpha \cap \beta = \Lambda . \equiv . \alpha \cup \beta \, \epsilon \, 2$

    *Dem.*

      ⊢ . ✳54·26 . ⊃ ⊢ :. $\alpha = \iota^{\prime} x . \beta = \iota^{\prime} y . \supset : \alpha \cup \beta \, \epsilon \, 2 . \equiv . x \neq y .$

      [✳51·231]                                     $\equiv . \iota^{\prime} x \cap \iota^{\prime} y = \Lambda .$

      [✳13·12]                                     $\equiv . \alpha \cap \beta = \Lambda$     (1)

      ⊢ . (1) . ✳11·11·35 . ⊃

        ⊢ :. $(\exists x, y) . \alpha = \iota^{\prime} x . \beta = \iota^{\prime} y . \supset : \alpha \cup \beta \, \epsilon \, 2 . \equiv . \alpha \cap \beta = \Lambda$     (2)

      ⊢ . (2) . ✳11·54 . ✳52·1 . ⊃ ⊢ . Prop

From this proposition it will follow, when arithmetical addition has been defined, that $1 + 1 = 2$.

and much later

**✳110·643.** ⊢ . $1 +_{o} 1 = 2$

    *Dem.*

                ⊢ . ✳110·632 . ✳101·21·28 . ⊃

                ⊢ . $1 +_{o} 1 = \hat{\xi} \{ (\exists y) . y \, \epsilon \, \xi . \xi - \iota^{\prime} y \, \epsilon \, 1 \}$

                [✳54·3]    $= 2 . \supset \vdash . \text{Prop}$

The above proposition is occasionally useful. It is used at least three times, in ✳113·66 and ✳120·123·472.

✳110·7·71 are required for proving ✳110·72, and ✳110·72 is used in ✳117·3, which is a fundamental proposition in the theory of greater and less.

**Fig. 10.8** A facsimile of the proof that $1 + 1 = 2$ in *Principia mathematica*.

condition is in fact defined by the ten-word statement, so the statement is contradictory. Russell had to cobble together a version of the theory of types, which he called the *ramified theory of types*, that avoided the dangers of this bog too. In the ramified theory, note is taken not only of the type of entity under discussion but also the manner in which it is defined. *Principia mathematica* is based on the ramified theory.

The impression has probably been gained that the ramified theory of types is a rag-bag of special pleadings. In fact, it is worse than that, because it turns out that it is not possible to demonstrate from it that every natural number has a successor or that there is an infinity of natural numbers. To overcome these weaknesses, to the rag-bag must be added the *axiom of infinity*, which simply asserts the existence of infinity. Worse, in the sense of greater rag-bagginess, was to come, for to define numbers correctly, the *axiom of reducibility*, relating to the behaviour of propositions of different order, had to be stuffed into the bag. Somehow, the logicist agenda was unravelling and it seemed to be becoming clear that mathematics is not just a branch of logic.

What was also becoming clear is that there are problems with set theory, which has been presented as the foundation of mathematics. Maybe the trouble can be traced to an intrinsic problem with sets, innocuous as they seemed? Perhaps a set is too broad a concept for mathematics? Some support for this view emerged in the early twentieth century at about the same time as Frege and Russell were wrestling with their problems, in the form of the *axiom of choice*. This axiom is the logical counterpart of Euclid's geometrical fifth postulate (about parallel lines, Chapter 9) and has received an enormous amount of attention. In its simplest form it looks like a lamb: if you have a series of sets, then you can form another set by selecting a member of each set and adding it to your shopping basket. We all compile sets in that way at the supermarket, calling our constructed set our 'shopping'. Who could possibly argue with that procedure for constructing sets?

The lamb casts aside its fleece and reveals itself as a wolf as soon as we consider infinite sets, for there may be no way of specifying the selection. For a finite number of sets, we can just list the members we select—we compile a shopping list. But consider the following problem. We have an infinite number of sets, one consisting of the real numbers between 0 and 1, the next those between 1 and 2, and so on. We decide to form a new set by selecting an arbitrary number from each of those sets. Unfortunately, we can't list our choices because there is an infinite number of them, and we can't specify them by a rule, because they have been selected at random. So, we have formed a set that we can't specify. Russell's homelier example of the difficulty with the axiom of choice is that of a wealthy man with an infinite number of pairs of socks, who instructs his valet to select one sock from each pair. The valet is unable to proceed, for he has no way of deciding which sock of each pair to select.

There are three attitudes to the axiom of choice, and mathematicians typically choose one of them either consciously or unconsciously. One attitude, adopted by mathematical ostriches, is to ignore the problems it represents and just get on with the job willy nilly. That is the view of all physical scientists, most of whom will be unaware that there is a problem anyway, and will shrug their shoulders dismissively once the problem has been brought to their attention and explained. Then there are the mathematical social workers who are aware of the problem and use the axiom of choice in a logical proof only as a last resort. They try desperately to find alternative routes between their axioms and their conclusions however tortuous their arguments then become. Finally, there are the mathematical saints, the truly celibate when it comes to the axiom of choice, who would not touch the axiom with a barge-pole, and view any proof that depends on it as invalid.

If mathematics is not purely a branch of logic, as these failures tend to suggest, what additional ingredient is present? To dig out one possible additional component we have to go back to that saddler's son and most difficult but influential of eighteenth-century philosophers, the possibly demisemi-Scottish Immanuel Kant (1724–1804).[12] In his discussion of metaphysical knowledge, which is philosophical knowledge that transcends the boundaries of experience, in his *Kritik der reinen Vernunft* (Critique of pure reason, 1781), Kant introduced the distinction between 'synthetic' and 'analytic' statements. An *analytic statement* is one in which the predicate can be teased out of the subject by reasoning alone and conveys no new knowledge, as in 'all carrots are vegetables'. According to the Logical Positivists of the early twentieth century, who adopted and clarified this term, the truth of an analytic statement depends solely upon the meaning of the words that compose it and the grammatical rules governing their juxtaposition. A *synthetic statement*, however, is one in which the predicate is not contained in the subject, as in 'a rose is red', as not all roses are red; such statements convey new knowledge. These categories are further divided into *a priori*, when the assessment of truth is independent of the evidence of experience, and *a posteriori*, when the validity of the statement does depend on experience.

Kant proposed that synthetic *a priori* statements, which express new knowledge but are independent of experience, are the proper objects of philosophical enquiry. Such statements include propositions about space and time which, in his view, were unquestionable and the perception of which was somehow built into our brains. For Kant, the tenets of Euclidean geometry and the properties of the natural numbers are synthetic *a priori*. In Kant's view, the theorems of mathematics are elucidations of the properties of space and time that in some way manifest our neural networks (not a term he would have used, of course) and our modes of perception.

The sense that there is something innate about the natural numbers, which were directly, obviously, synthetic *a priori* properties of the world was developed

---

[12] Kant was born in Scotya, a suburb of Königsberg in East Prussia (Kaliningrad), with a significant Scottish immigrant population; it is thought that his grandfather was a Scot. Although his mind roved widely, physically he never left Königsberg.

into the philosophy of mathematics known as *intuitionism* by the Dutch mathematician Luitzen Egbertus Jan Brouwer (1881–1966), one of the founders of topology, in his doctoral dissertation for the University of Amsterdam in 1907. Brouwer discarded Kant's view that geometry was synthetic *a priori*, which had in fact already bitten the dust with the realization that Euclid's fifth postulate, although consistent with the other four postulates, could be replaced by others without contradiction (as we saw in Chapter 9). That is, Brouwer accepted that Kant was wrong in supposing that Euclidean geometry is *necessarily* true, for there are alternative geometries that experience shows are better descriptions of space and time. However, he did not reject the whole of Kant's view that mathematics is the study of space and time, but only the spatial component. Brouwer considered that mathematics is a statement about our consciousness of time, and propagated the view that the natural numbers stem from our scanning of a collection of entities in sequence, the temporal separation of our perception of each one being the key to their distinction. Brouwer actually went further: he was a solipsist, considering that anything that existed, including other minds, had its origin in one conscious mind. That view, however, is an unnecessary complication of the intuitionist agenda, and at first sight there seems to be no need to pursue this aspect (but I touch on a version of it, with approval, later).

An intuitionist takes the view that the natural numbers have a special status and that we have a direct intuition of them: they are not entities that can be elaborated by further description. According to Brouwer, to arrive at the concept of a natural number, we note our perception of the distinction between entities arising from the time-ordering of our scanning of them, clicking up a digit each time our perception passes over one. That view implies that the natural numbers are a manifestation of our mental activity. Similarly, the operations of arithmetic, such as addition, are to be regarded as portrayals of the mental processes that take place inside our head. Thus, to confirm that 2 + 3 = 1 + 4, we have to carry out a variety of tasks: we have to judge the outcome of adding 2 to 3, likewise for 1 + 4, and then we have to verify that the outcomes are equal to each other.

There are certain troublesome consequences of intuitionism that are not immediately obvious from this brief account but need to be noted, for they strike at the very heart of classical logic. This is particularly the case when dealing with statements about infinite collections of entities, for which there is no associated mental activity relating to their perception because we have no direct experience of the infinite. For instance, Aristotle had identified one pillar of logic in his *law of the excluded middle*, that a statement is either true or false. That law is not held to be true in intuitionistic mathematics, for it may be that a statement exists that has not been proved or is undecidable. In either case, it is not the case that a statement is either true or false until, if ever, it has been proved. One consequence of this position is that saying that it is not true that a proposition is false is not equivalent to saying that the proposition is true.[13] Whereas we might aver that to say it is not true that there is a ball that is not red in a box containing an infinite number of balls is the same as saying that every ball in the box is red, an intuitionist would deny this conclusion. According to an intuitionist, the truth of the statement that there is a ball that is not red in the box can be established only by sorting through all the balls in the box, which is impossible for an infinite collection. A further consequence of this position is that it not possible to prove that a certain statement is false by adopting a *reductio ad absurdum* argument to show that the negation of the statement is false or leads to a contradiction. To an intuitionist, the only acceptable statements are those with proofs that have been explicitly constructed in a finite number of steps.

David Hilbert (1862–1943), good dancer and flirt, was one of the most influential mathematicians of the twentieth century. He was born, like Kant, in Königsberg in East Prussia (coincidentally, Goldbach's birthplace too). He is particularly famous for setting out what he perceived to be the outstanding problems of mathematics at the turn of the century, the beginning of the twentieth century, that is, and many mathematicians since have wrestled with the problems he identified. The problems were presented at the Second International Congress of Mathematicians in Paris in 1900. The lecture presented ten problems; as Hilbert worked up the published version, the number grew to twenty-three. The impact of these problems—which are better regarded as composite groups of problems and allusions to problems rather than twenty-three precisely formulated individual examination questions—stems from the fact that they constituted an enquiry into what marked out a good problem. Thus, Hilbert required problems worth spending time on to be clear, difficult but not inaccessible, and should, when solved, illuminate a wider compass than the ground on which they were based.

Some of these problems have been solved; some have been shown to be unsolvable; others remain under attack. Some of the problems as Hilbert stated

---

[13] That is, $\sim(\sim p)$ is not equivalent to $p$.

them are too grandiose for it to be clear whether they will ever reach a solution as definite as the others. For instance, one of his grandiose problems was to achieve the axiomatization of physics, putting it on as succinct and reliable a foundation as Euclid had put his version of geometry and he, Hilbert, had formalized it in his magisterial *Grundlagen der Geometrie* (Foundations of geometry, 1899). The formulation of a 'theory of everything' could be construed as what he had in mind. Most of the problems, though, are quite definite, especially when interpreted generously. For example, they include the proof of Cantor's continuum hypothesis (which turns out to be unprovable) and the *Riemann hypothesis*, that a certain function of the complex number z is equal to zero for an infinite number of values of z, all of which have a real part equal to ½ (Fig. 10.9). The latter problem might look rather inconsequential, but in fact it is of profound importance for the study of prime numbers; it remains unsolved and is regarded as one of the most important unsolved problems of mathematics. Later we shall deal explicitly with two of Hilbert's other problems. His second problem, attacked and solved in the negative by Gödel, was to prove that the axioms of arithmetic are not contradictory. His tenth problem, the so-called *Entscheidungsproblem* (decision problem), also attacked and solved in the negative, in this case by Alan Turing and Alonzo Church, was to devise a process according to which it can be determined in a finite number of steps whether or not any equation is solvable.

Hilbert also developed a philosophy of mathematics which has come to be termed *formalism*. He saw mathematics as two closely pasted sheets: one sheet consists of finite arrangements of symbols achieved by the application of certain rules. These symbols simply form determinate patterns on the page and are utterly devoid of meaning. These meaningless patterns are what we really

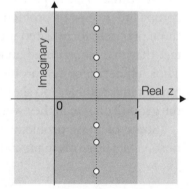

**Fig. 10.9** All solutions of the equation $1 + 1/2^z + 1/3^z + 1/4^z + \ldots = 0$, with z a complex number, are known to lie in the tinted strip between 0 and 1. One form of the Riemann hypothesis asserts that all the solutions of this equation actually lie on the centre line of the strip (as denoted by the small circles), with the real part of z equal to $1/2$ in every case.

mean by mathematics. Even the axioms of the system are just strings of marks drained of meaning, intellectual corpses, and new patterns are derived from these strings by the application of abstract rules. In this sense, mathematicians are wallpaper designers. The only reliable proofs, according to Hilbert, are *finitistic*, in the sense that they are *finite* collections of symbols, for only such collections can be inspected and verified: safe mathematics is finite mathematics. On the second sheet is *metamathematics*, which consists of a commentary on real mathematics. It contains remarks such as 'this string of symbols resembles another', that '*x* is to be interpreted as a particular sign for an entity', that 'a particular group of signs indicates that a pattern is complete', and 'this is a proof of that proposition'. We could think of mathematics itself as the possible patterns of pieces on a chess board and the corresponding metamathematics as commentaries such as 'there are twenty possible opening moves for white' or 'that position is checkmate'. According to the formalists, mathematics is abstract symbolism and pattern generation: metamathematics endows the symbolism and the patterns with significance to humans, it imbues marks with 'meaning'; it restores blood to corpses.

There is another school of thought about the nature of mathematics, that of *Platonic realism*. Mathematicians who subscribe to this school turn their back on the formalists' viewpoint that mathematics is the generation of meaningless strings of symbols. They turn their back too on the intuitionists' insistence that mathematics is a projection of the mind, that existence is meaningless until proof is supplied, and that in the absence of consciousness there are no numbers and no such thing as parallel lines. Like the formalists and the intuitionists, they accept the incompleteness of the logicists' notion that mathematics is no more than a branch of logic, and agree with them that there is more to mathematics than logic.

A *platonist*, as this sort of mathematician is called, considers that the missing component is reality. Platonist mathematicians are hewers of pre-existing relationships, who go about their hewing by swinging the pickaxe of their intellectual reflections on the world. They are discoverers of truth, not inventors. For them, the numbers are real entities, and relations between numbers are statements about something. For them, straight lines, triangles, and spheres are as real as rocks, and arithmetical truths (which, remember, means any kind of mathematical truth, and perhaps even more than that) are comments on some kind of actuality. Thus, they reject the sterile aloofness of formalism and the subjective involvement of intuitionism, and consider that they are scientists like the rest of us. They are excavating timeless truths, and in fierce opposition to the intuitionist stance, consider that there are truths out there even though the proofs have not yet been formulated.

I will now consider two of Hilbert's most important problems, the two that strike at the heart of the philosophy of mathematics and examine its potency most directly. As I have already mentioned, one of his problems is the so-called *Entscheidungsproblem*, the problem of finding a systematic way of determining whether any statement in a symbolic language is provable using the axioms of the language. Two almost simultaneous attacks were made on this problem, one by the American logician Alonzo Church (1903–95) with his invention and deployment of what he called the λ-calculus and the other by the British mathematician Alan Mathison Turing (1912–54) with his invention of his 'logical computing machine', which is now known as a *Turing machine*. The two approaches are startlingly superficially different; but Church and Turing collaborated to show that they are in fact mathematically equivalent. That is one extraordinarily important power of mathematics, its ability to show the equivalence of the seemingly utterly disparate. We shall concentrate on Turing's approach, as it has more resonances with the familiar modern world of computers, but it should not go unnoticed that Church's λ-calculus has resonances with and is the basis of varieties of the software they use.

A Turing machine is a device that is claimed to mimic the actions of a human performing any kind of *algorithmic computation*, a computation performed by applying a series of rules in sequence, and which we now recognize as a representation of a digital computer. Indeed, Turing's work on code-breaking at Bletchley Park, just north of London, during World War II and then later at Manchester led to the first realization of a programmable digital electronic computer. Turing himself is credited with shortening the war by months, if not years, on account of his code-breaking successes, and certainly saving the lives of thousands. It is to the shame of mid twentieth-century England that the laws and social mores of the time (he was homosexual) were to hound him to his own early death.

Turing sought to extract the essence of how a person doing a calculation proceeds, and then examined the limitations of the process: were there ques-

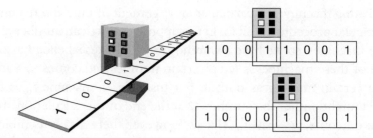

**Fig. 10.10** A version of a Turing machine. The machine consists of an infinite strip of paper divided into cells in which symbols (typically, 0 and 1) can be written, and a machine that can read the symbol, act on its reading according to the internal state it is in at that point, change the symbol if required, and move in either direction to the neighbouring cell. In this representation, the internal state is denoted by the lights on the side of the reading head. The lower diagram shows a possible response: the machine is in the internal state indicated by the light, and reads 1; as a result it changes the 1 to a 0, changes its internal state, and moves one place to the right.

tions that could be asked that, however long the person worked, would not lead to an answer? Turing's version of the procedure was encapsulated in a device that consisted of an indefinitely long strip of paper divided into square cells (which emulated the infinite resource of paper and pencils the human calculator could deploy, carrying out the calculation, making notes of intermediate answers, and then writing down the final answer), and a reading and writing head that could be programmed to respond in the appropriate manner to whatever was written in the cell it was scrutinizing at that moment (Fig. 10.10). These rules could also be modified and fed into the reading head from the paper tape.

We shall suppose that the cells of the paper tape can hold either a 0 or a 1 and, depending on its internal state, that the head can read the content of the cell, write to the cell, and move one cell to the right or to the left. A particular Turing machine will carry out a series of actions depending on what it finds on the tape and how its head has been built to respond. For instance, if it finds a 1 on the tape when it is in a state '1', it might change the tape's 1 to a 0, change its internal state to state '2', and move one step to the right. In that cell there might be a 0. When the head is in state '2' and reads a 0, it might be programmed to move one place to the left, but if it reads a 1, then it changes the 1 to a 0 and moves one place to the right. If the responses of the head are carefully designed, then the machine can be used to carry out even the most elaborate calculation. The actual design of the head and its responses might be very difficult and the calculation very slow, but here we are concerned only with the principle of the calculation, not its efficiency.[14]

[14] Emulations of working Turing machines can be found at several sites, including http://wap03.informatik.fh-wiesbaden.de/weber1/turing/index.html.

Each Turing machine is a particular arrangement of tape and reading head with a particular procedure built in. Let's suppose that we can number all possible Turing machines, so we have a warehouse with boxes labelled $t_1$, $t_2$, and so on. If one of these machines is fed a certain number and comes to a stop, we will find a certain number as output. For instance, if machine $t_{10}$ is fed the number 3 it might output the number 42 at the end of the calculation. To summarize this outcome, we write $t_{10}(3) = 42$. However, there may be combinations of machine and data for which the calculation never comes to an end, such as when machine $t_{22}$ is fed the number 17. To summarize this outcome, or lack of it, we write $t_{22}(17) = \square$. Turing's problem was whether there was a way of inspecting all possible machines and their data and deciding from that inspection whether the calculation would come to an end.

To carry out this programme, we suppose that there is a *universal Turing machine*, which is a Turing machine that can be programmed to emulate any individual Turing machine. For this machine, the input tape has two sections, one the program and the other the data. The program part might consist of a string of numbers that instruct the head how to respond to what it finds on the tape. For example, the code 001 might mean

> **001:** if you find 1 on the tape, and you are in state 1, change the 1 on the tape to a 0, change your internal state to state 2, and move one step to the right.

Similarly, the code 010 might mean

> **010:** if you find a 0 on the tape, and you are in state 2, move one place to the left; but if you read a 1, then change the 1 to a 0 and move one place to the right.

The program part of the tape might then look like . . . 001010 . . . if these two instructions are to be carried out in succession. We will call the universal Turing machine *tu*. Note that whereas one of the *individual* Turing machines reads in only data, the *universal* machine first reads in the program to prepare itself, then reads in the data. So, if we want it to emulate $t_{10}$, we read in program 10, the set of instructions for setting it up to act like $t_{10}$, and then feed it with the data. If the data consists of the number 3, we will expect the answer 42 from this joint process, and write $tu(10,3) = 42$, where the first number in parentheses is the number of the Turing machine we are seeking to emulate and the second number is the data.

Now we suppose that there is a Turing machine which can take in any other

Turing machine, such as $t_{23}$, and any set of data, and decide whether that combination will or will not halt and print out an answer. We will call this particular Turing machine *th* (*h* stands for 'halt'). If *th* halts for a particular combination of program and data, such as $t_{23}$ and 3, it will print 1 and stop; if it determines that the combination doesn't halt, as for $t_{22}$ and 17, *th* will print out 0 and stop. Turing's achievement was to show that *th* isn't included in the list of all possible Turing machines, and so doesn't exist. To do so, he used an argument very much like the 'diagonal' argument that Cantor used to show that irrational numbers aren't countable. Feel free to jump to the next section if you want to skip the derivation of this conclusion.

The argument runs as follows. Suppose we run the inputs 0, 1, 2, . . . through the Turing machines $t_0, t_1, t_2, . . .$ and draw up a table of which the following is just the top left fragment:

| | Input: 0 | 1 | 2 | 3 |
|---|---|---|---|---|
| 0 | □ | □ | □ | □ |
| 1 | 3 | □ | 4 | 1 |
| 2 | 1 | 1 | 1 | 1 |
| 3 | 0 | 1 | □ | 2 |

(Machine number)

Where the calculation never stops we have put the symbol □. This table contains all possible computable numbers (numbers that can be computed by a Turing machine to an arbitrary number of digits) because it contains, in its successive rows, all possible Turing machines and in its successive columns, all possible inputs.

Now we do a second run. This time, we sort the results by using *th* first, which we have arranged to give a 0 if the machine decides that the corresponding calculation will stop, and to do nothing to the data if it decides that the calculation will stop. It also makes a note to remind itself where it has replaced a □ by a 0 as it doesn't want the machines it is emulating to get trapped into endless calculations again. For instance, when we feed 4 and then 2 into *th*, which corresponds to the program for $t_4$ and the data 2, *th* inspects the tape, does a calculation of some kind, decides that the calculation $t_4(2)$ wouldn't stop if we were to run it, and so puts a 0 in the corresponding part of the table and makes a note to itself that that particular calculation won't stop. At the end of this part of the calculation, the top left fragment of the table looks like:

| Input: | 0 | 1 | 2 | 3 |
|---|---|---|---|---|
| 0 | 0 | 0 | 0 | 0 |
| 1 |  | 0 |  |  |
| 2 |  |  |  |  |
| 3 |  |  | 0 |  |

*(row labels on left: Machine number = 0, 1, 2, 3)*

Next, *wherever we have not entered a 0*, we run all the calculations, as we did in the first test, and now get the following fragment of the table:

| Input: | 0 | 1 | 2 | 3 |
|---|---|---|---|---|
| 0 | **0** | 0 | 0 | 0 |
| 1 | 3 | **0** | 4 | 1 |
| 2 | 1 | 1 | **1** | 1 |
| 3 | 0 | 1 | 0 | **2** |

*(row labels on left: Machine number = 0, 1, 2, 3)*

Because the original table contained all possible computable numbers, this table also contains all possible computable numbers: there may be a lot of repetition, but that does no harm.

Now we get to the crux. Let's take the numbers on the diagonal (they are bold in the table) and change them by adding 1 (rather like in Cantor's proof). We get a sequence such as 1123 . . ... This is a computable number (because the sequence of steps based on *th* and the Turing machines works in every case, we are supposing) so a machine that produces that number must already appear in the table somewhere. However, it doesn't: it differs from the first row (because we have forced the first digit to differ), it differs from the second row (because we have forced the second digit to differ), and so on for all the rows in the table. That is, on the one hand 1123 ... must be present, but on the other hand it isn't. This is a contradiction, so the assumption we have used, that the halting machine *th* exists, must be false. We have proved (and be assured Turing's version is much more rigorous and authoritative) that there is no single, general, universal, algorithmic procedure that can be used to judge whether a particular calculation will come to an end. This in turn implies that there can be no general algorithm for deciding mathematical questions, and therefore that Hilbert's *Entscheidungsproblem* has no solution.

I now move on to the apotheosis of this chapter and what has been called the finest achievement in logic in the twentieth century, *Gödel's theorem*. The Austrian logician Kurt Gödel (1906–1978) was born in Brünn, Austria–Hungary (now Brno in the Czech Republic), where Gregor Mendel did his work, and studied at the University of Vienna. Although not a Jew (despite Bertrand Russell's assertion to the contrary), Gödel could not tolerate the Nazi oppression, and went to the USA in 1934, emigrating there permanently in 1940 and spending the rest of his life in Princeton, where he and Einstein were great friends. Indeed, in his later years, Gödel made a substantial contribution to the general theory of relativity when he found an unexpected solution of Einstein's equations that allowed time travel into the past. Gödel was not what one would consider to be wholly conventional in his outlook and way of life. On his return to Austria after his first visit to the USA he married a divorced dancer and brought her back to Princeton, where the then prevalent snobbery meant that she was never well received. Towards the end of his life he developed the classic signs of depression and paranoia: he was convinced that he was the victim of a murder conspiracy to the extent that in the end he simply refused to eat, and to avoid contagion when walking through what he regarded as the severely polluted and risky ambience of Princeton, he wore a ski mask. He died of 'malnutrition and inanition' (exhaustion due to lack of food) weighing only 30 kg, as a result, as his death certificate says, of a 'personality disturbance'.

There are several theorems associated with Gödel's name. Here we are concerned with the theorem published on 1931 in the article *Über formal unentscheidbare Sätze der Principia Mathematica und verwandter Systeme* (On formally undecidable propositions in *Principia Mathematica* and related systems). In this article, he showed that in any axiomatic mathematical system there are metamathematical propositions that cannot be proved or disproved by formal derivations based on the axioms of the system.

This is what we do. Mathematics is a sequence of propositions, such as $1 + 1 = 2$, and 'this is a proof of that proposition'; the former proposition is mathematical, in Hilbert's sense, and the latter is metamathematical. Let's suppose that we can write down all the propositions that it is possible to derive from the fundamental axioms (for instance, Peano's axioms or the more elaborate system based on the ramified theory of types that Russell and Whitehead used). That will give us the propositions $p_0, p_1, p_2, \ldots$, and so on. How we decide to number the propositions isn't important, but the following few sentences will give you a flavour of how Gödel proceeded.

There are only a few symbols in the formulation of arithmetic like Peano's.

For instance, one of the axioms is 'the immediate successor of a number is a number'. We could write $x' = sx$, where $s$ means 'the immediate successor of', so $s0 = 1$, $s1 = ss0 = 2$, and so on, as we have already seen. Gödel ascribed numbers to each elementary sign in an expression. Let's suppose that he ascribed 5 to the sign '=' and 7 to the sign $s$. Each distinct variable, such as $x$, is ascribed a unique prime number greater than 10, so we ascribe to $x$ the number 11 and to $x'$ the number 13. The *Gödel number* of a proposition is then the product of all the numbers corresponding to the symbols it contains, so our proposition that $x' = sx$ is assigned the value 13 (for the $x'$) × 5 (for the =) × 7 (for the $s$) × 11 (for the $x$) which works out as 5005. Note that by this procedure, a proposition, including an axiom of the formalism, becomes a unique number,[15] so relations between propositions become relations within arithmetic. For instance, we can answer the metamathematical question as to whether this proposition occurs in a longer, more complex proposition by working out whether 5005 is a factor of the Gödel number of the complex proposition, just as 5 is a factor of 75.

We will label propositions using their Gödel number, so the proposition $x' = sx$ about the number 6 (which would read $6 = s5$, '6 is the immediate successor of 5') is proposition $p_{5005}(6)$. As you might suspect, elaborate propositions have very high Gödel numbers, but in the following we will pretend that we can get away with small numbers like $p_1(6)$ and $p_4(6)$. For instance, we could pretend that Proposition 4, $p_4$, when applied to the number 6, is the metamathematical statement '6 is a perfect number' (a number that is the sum of its prime factors, in this case including 1, as in $6 = 1 + 2 + 3$ and $6 = 1 \times 2 \times 3$) and Proposition 5 might be about prime numbers and we might read $p_5(11)$ as '11 is a prime number'.

A mathematical proof consists of a string of propositions that follow from each other by the application of the rules of symbol manipulation. That means that we can ascribe a unique number to an *entire proof* by noting the Gödel numbers of all the propositions it contains. If a proof consists of three propositions with Gödel numbers 6, 8, and 2 (in practice, these numbers would be huge), then the overall proof is ascribed the Gödel number $2^6 \times 3^8 \times 5^2 = 10\,497\,600$ (for longer proofs, the numbers 2, 3, 5 continue as primes). As you might imagine, long proofs composed of elaborate propositions have astronomically large Gödel numbers.[16] Once again, the point of this procedure is that entire proofs

---

[15] In the interests of simplicity I have reduced the numbering procedure to the point at which it doesn't quite work, in part because the order in which symbols occur is not taken into account. Gödel's procedure was more sophisticated.

[16] In ultrafinitistic mathematics, where numbers fizzle out, it could be the case that elaborate proofs have such high Gödel numbers that they become meaningless.

are brought into the domain of arithmetic. We can use arithmetical procedures, for instance, to judge whether one proof makes use of others by determining whether the latter have Gödel numbers that are factors of the former, rather like $15 = 5 \times 3$ signifies that 5 and 3 are components of 15.

Now we use these Gödel numbers to derive Gödel's result in a variation of Cantor's procedure and Turing's discussion of computability. Gödel actually used a much deeper method, establishing forty intermediate theorems—base camps—before reaching the summit of his proof. The following captures the essence of what is involved: think of it as a helicopter ride to the summit. However, because the proof is tough going, even when hacked down and simplified to the degree that I intend to adopt, feel free to jump to the point where the normal sized typeface resumes.

Let's suppose we make a certain proposition about the number 0, call this proposition $p_0(0)$, and the same proposition about the number 1, which we call $p_0(1)$, and so on, and in general $p_0(x)$, a proposition about $x$. The propositions might be right or they might be wrong. For instance, they might be 'the square root of $x$ is 1' in which case $p_0(0)$ is false because it claims that $\sqrt{0} = 1$, which is not the case, but $p_0(1)$ is true, because $\sqrt{1} = 1$. Each of these propositions has a certain Gödel number, which we can work out, and there is an infinite number of such propositions about each of the infinite natural numbers. We write these propositions $p_0(x)$, $p_1(x)$, and so on: some are garbage, some are true. Now, we arrange all the corresponding Gödel numbers in a huge table (with astronomically large numbers where we have used small numbers). The top left fragment of this table might be something like

| Input: | 0 | 1 | 2 | 3 |
|---|---|---|---|---|
| 0 | 1 | 55 | 27 | 4 |
| 1 | 51 | 3 | 7 | 17 |
| 2 | 0 | 20 | 30 | 40 |
| 3 | 13 | 22 | 11 | 2 |

(Proposition — row labels)

where each entry in the body of the table is the (fake) Gödel number of the corresponding proposition. Thus, the fake Gödel number of the proposition $p_3$ about the number 2 is 11.

Now, separately, we compile a list of the Gödel numbers of all the propositions that *are* provable from the axioms of the system. Like our supposition that there is a reliable Turing machine for judging whether a calculation would halt or not, we are assuming that such a list can be compiled, but if we are led into a contradiction then we will have to reject this assumption.

Now, as in the Turing argument, we come to the crunch. Let's consider the following proposition:

The Gödel number of this diagonal term is not in the list of provable statements.

A 'diagonal term' is a proposition about that proposition's own number, such as proposition $p_2$ about the number 2. Because this statement is a proposition, it must already occur somewhere in the original exhaustive list of propositions. For simplicity, let's suppose it turns out to be Proposition 2. That being so, let's consider the corresponding diagonal Gödel number, which in this case is 30. This Gödel number corresponds to the proposition that

There is no proof of Proposition 2 about the number 2.

Now we come to the crux. Suppose we know by reference to the complete list of provable statements that this proposition is indeed true, in the sense that it can be proved that there is no proof of Proposition 2 about the number 2, then we have a contradiction, for if there is no proof of Proposition 2 about the number 2, then it is in fact not on the list of provable propositions! If we assume instead that the proposition that there is a proof of Proposition 2 about the number 2 is false, then we also have a contradiction, for if there is no proof of Proposition 2 about the number 2 is false, then it is not in the list of provable propositions, in which case the proposition is true!

We have arrived at the point where we have to conclude that the axiom system we are using is insufficient to decide between the proposition and its negation. *Mathematics is incomplete.* This means that there is an infinite number of mathematical statements that are probably true but which cannot be deduced from a given set of axioms. Here is the basis of one of my opening remarks. Not only is it astonishing that we can count (because the natural numbers are so rare in the universe of all numbers), it is astonishing that we can do any arithmetic with them (because formally provable expressions are so rare).

Gödel's conclusion does not mark the doomsday of mathematics. First, there may be non-algorithmic methods of establishing the truth of a statement, just as it may be impossible to prove formally that a certain position in chess cannot lead to checkmate but is perceivable in a more global manner. That is, there may be a metamathematical proof of an assertion that cannot be proved within the formal system. That the human mind is capable of such informal but completely reliable proofs, is a window on the nature of consciousness, for it shows that apprehension and reflection need not be algorithmic.

Mathematics has gone through three major crises in its history. The first was the discovery of incommensurability by the Greeks, the existence of irrational numbers and its undercutting of the philosophy of the Pythagoreans. The second was the emergence of the calculus in the seventeenth century, with the fear that dealing with infinitesimals was illegitimate. The third crisis was the encounter with antinomies, such as Russell's antinomy and Berry's paradox in the early twentieth century, which appeared to undermine the very foundations of the subject. In a sense, it is remarkable that mathematics has survived as a discipline. That is has done so is due in part to good common sense: there is a great deal of wonderful mathematics that seems to work perfectly well, and it would be foolish to discard a subject that has such wonderful successes, even if somehow there are treacherous regions deep down inside its structure. Working mathematicians can get on with job unfearful and uncaring about cracks deep in the foundations which are, they presume, extremely unlikely ever to work their way to the surface in any actual application. The second reason, of course, is that mathematics is just too useful and is the paramount language of description of the physical world. Gone mathematics, gone would be most of science, commerce, transport, industry, and communication.

That raises the question of why mathematics, the supreme product of the human mind, is so superbly suited to the description of Nature. Here I will indulge in a final flourish, a personal flight of fancy, which is pure speculation, unfounded in science, and thus totally without authority. It shows that I am a Greek (an ancient one) and a Kantian at heart despite my faint mocking of their speculative philosophies. Here, I intend to out-Greek the Greeks, see if I can't out-Kant Kant, and explore whether there is a deep link between Platonic realism, Kantian and Brouwerian intuitionism, and Hilbertian formalism.

The problem that confronts us has two prongs. One prong is that mathematics is the internal product of the human mind. The second prong is that mathematics appears to be supremely well adapted to the description of the external physical world. How is it that the internal is so well suited to the external? If we adopt a Kantian view of the brain, we can suppose that it has evolved in a manner that renders it capable of distinguishing the sets corresponding to the natural numbers (in Kant's terms, a synthetic *a priori*) and the representation of those numbers in three dimensions in the form of geometry (synthetic *a priori* too, but only locally, for we know that Euclidean geometry is invalid over great distances and near massive bodies). A latter-day Kant might aver that we have such trouble thinking about irrational numbers and non-Euclidean geometries because these concepts are not hard-wired into our neuronal network through some kind of evolutionary adaptation to the local environment and we have to make a real intellectual effort to contemplate their properties.

Moving on, we can also suppose that the simple manipulations of these concepts is also structurally present in the hardwiring of the brain. This idea suggests that the underlying logical manipulations are built in and that we have a hard-wired algorithmic capacity. I am not saying that that is the brain's only capacity: there is currently much interest in the speculative presumption that there are non-local activities of the brain which enable us to consider relations in a non-algorithmic manner, and some have speculated (Roger Penrose being a leading proponent of this view) that consciousness is an intrinsically non-local quantum phenomenon. Although I would be surprised if that turns out to be true, it is not a component of my own speculation as I will concentrate on the algorithmic processes in the brain, the Hilbertian algorithmic co-processor of a larger, more metamathematical, perhaps non-local ability of the brain. In short, for algorithmic calculations we can take what could be called a 'structuralist' position, echoing Noam Chomski's vision of the innate human capacity for language, and think of our logical capacity as a Kantian manifestation of a hard-wired algorithmic component of the brain that has emerged under the pressures of evolution. Our ability to create mathematical relationships, deduce theorems, and so on is a consequence of that structure.

Moving outside the head, we now have to consider why the physical world is a hand to mathematics' glove. Now I am on even more treacherously speculative ground. We have seen the relation of numbers to sets, and Frege's identification of the numbers to the extensions of certain sets. In the same spirit, the jolly Hungarian–American mathematician John (Johann) von Neumann (1903–57), who is regarded, along with Turing, as the co-parent of the modern computer, suggested that the natural numbers could be identified as certain very simple sets. Specifically, he identified 0 as the empty set {}, the set with no members. Then he went on to identify 1 as the set that contains the empty set, 1 = {{}}, 2 as the set that contains both the empty set and the set that contains the empty set, 2 = {{},{{}}}, then 3 = {{},{{}},{{},{{}}}}, and so on.[17] Thus von Neumann spun the world of numbers from absolutely nothing and gave us arithmetic *ex nihilo*.

I have argued elsewhere that, because I lack the imagination to see how otherwise an apparent something could come from absolutely nothing, the emer-

[17] This notation, while elegant, is easily confusing. In set theory, the empty set is commonly denoted ∅, so the natural numbers are ∅, {∅}, {∅,{∅}}, {∅,{∅},{∅,{∅}}}, ... which is notationally, at least, a tiny bit less confusing.

gence of the universe *ex nihilo* must be just like von Neumann's conjuring of the natural numbers from the empty set. The fact that the universe survived its own creation is then to be interpreted as an indication that the entities that came into existence in this way are logically self-consistent, for otherwise the cosmos would collapse. There is therefore an intrinsic logical structure to the universe, which has the same structure as arithmetic.

Now we bring these bubbling streams of frothy speculation together. When mathematics confronts the physical world it sees its own reflection. Our brains, and their product mathematics have exactly the same logical structure as the physical universe itself, the structure of spacetime and the entities that inhabit it. It is no wonder then, *pace* Wigner and Einstein, that brain-generated mathematics is a perfect language for the description of the physical world.

All that is probably nonsense. But suppose it isn't. One implication would be that the deep structure of the world is mathematics: the universe, all it contains, *is* mathematics, nothing but mathematics, and physical reality is an awesome manifestation of mathematics. This is extreme platonism, ultra-neoplatonism, what elsewhere I have termed 'deep structuralism'. What to us seems tangible—earth, air, fire, and water—is but arithmetic. If that is so, then Gödel's theorem applies, in some sense, to the whole universe. We can never know that the universe is truly self-consistent. If it isn't, then perhaps at some instant in the future it will suddenly terminate, or inconsistency will spread through its structure like the plague, confounding logic as it goes and eliminating structure like rusting iron. All there is will return to whence it came, to the empty set, that astoundingly potent concept of absolutely nothing.

Meanwhile, that potency is ours to enjoy. All around us, if this view has any validity, are the awesome ramifications of nothing, revealed to us through our senses, with the delight of our senses deepened by an intellect sharpened by science, the descendant of Galileo's vision, his meddling finger. I can think of nothing more moving, and nothing more wonderful.

# THE FUTURE OF UNDERSTANDING

WHERE does Galileo's finger point for the future of understanding? The exhilarating progress that has been made over the past few centuries, particularly over the century just past, shows no sign of abating. So, where will it lead?

Science looks as though it might be semi-infinite. By that cautiously expressed opinion I mean that an optimist has certain grounds for suspecting that the search for a final theory, inappropriately called a 'theory of everything' (self-deprecatingly, a TOE at the foot of physics), will come to a successful conclusion but that the ramifications and applications of science are endless. Each century, of course, is littered with the bones of this view, bleached white and rendered risible by the harsh light of subsequent progress. However, the signs are different now, and optimists—optimism is a characteristic that should be common to the personality of all scientists—can point to the essential difference between the end-is-nighers of the nineteenth century and those of the twenty-first.

A scientist of the nineteenth century, brought up in a world of increasingly sophisticated gadgets on all scales from tiny to country-spanning, saw explanation as gadget. For them, the promised land of ultimate comprehension was the construction of a machine that emulated the observation, for they could grasp gadgets. This view has not entirely disappeared from modern science, as we see later, but scientists now accept that explanation-as-gadget is a naive view of the end of understanding. Any gadget is itself composed of gadgets on ever smaller scales: indeed, anything that has properties is a composite gadget. An electron, with its mass, charge, and spin is a gadget in this sense, with some sort of presumed structure that endows it with these basic characteristics.

From the age of the gadget we have drifted into the age of abstraction. Scientists of the twenty-first century currently believe that the deep structure of the universe can be expressed only in terms of mathematics and that any attempt to relate the mathematics to visualizable models is fraught with danger. Abstraction is now the name of the game, the current paradigm of understanding. Any final theory, if there is one, is likely to be a purely abstract account of the fundamental structure of the world, an account that we might possess but not comprehend.

That—the view that we might possess an explanation but not comprehend it—is probably too extreme a view. Humans are adept at interpreting mathematics, particularly the mathematics used to support physics, in homely terms, aware all the time that their interpretation is fraught with danger and incompleteness, but interpreting nevertheless. Thus, electron spin can be imagined, as a mental crutch, as an electron spinning, a ball rotating; but in the background we know that 'spin' is actually an extraordinarily abstract entity with characteristics that cannot completely be captured by this classical image and, moreover, that aspects of the classical image are misleading. String theory is another instance, where we pretend that we can grasp what we mean by the mathematical concept of a string in many dimensions by thinking of it as an actual string oscillating in three. Although the final theory may be highly abstract, we can expect to have homely, suggestive, inaccurate images of its content, and that there will be an infinite future for authors of popular science books finding new and engaging ways to render future final theories digestible.

But what shall we mean by a 'final theory'? The final theory will not be a single equation which, once solved, will account for every property and activity under the Sun, and the Sun itself. A final theory will be a complex of concepts embodying in some sense—I cannot be explicit because explicitness will come only with hindsight—an attitude towards the basic structure of the material world. To give you an idea of what I have in mind, I can point to the failed but imaginative attempt by the masterly imaginative John Wheeler, who nearly half a century ago wondered whether the stuff of ultimate reality was an agglomeration of statements of predicate logic. Would a universe arise, he wondered, if random statements of logic jostled into self-consistency? Was the Big Bang a burst of becoming logically self-consistent? Put another way: was the creation the coming into being of its own potential comprehensibility?

Of course, this level of description lies lower, deeper than the description currently being sought in terms of strings and the unification of quantum theory and gravity. If the past is a guide, we can be confident that there will be at least two profoundly important paradigm shifts between now and the achievement of a final theory. It is possible, of course (and archivists in the future, if they retain the capacity to read our printed books, will certainly snigger at the naivety of these words), that we are embarking on an infinite series of paradigm shifts and that true understanding will always lie along the yellow brick road and over the next paradigmatic horizon. That will probably please philosophers, who are innately pessimistic and will take pleasure from the prospect of science's tripping, but it will be a disappointment for scientists, who should be innately optimistic.

One paradigm shift will come from the unification of gravitation and quantum theory, and there are already signs of the form it is likely to take. As touched on in Chapter 9, the view is emerging that the only real aspects of spacetime are the existence of relationships between events. There are also deep interpretations of quantum theory in which all possible pasts have happened, so the universe is intrinsically multi-sheeted. We have not yet fully identified such paradigm shifts and they are open to technical objections, for we do not yet have a complete quantum gravity; but there can be no doubt that it will shift our perception of reality in an awesomely amazing and as yet only mistily discernible way, just as special relativity shifted our perception, as general relatively did too, and as quantum theory itself did and continues so to do. Indeed, if one thinks of the characteristics of the twentieth century, then not only was there an upheaval in social order (the century was not unique in that), but there was a profound upheaval, unlike anything that had happened since Copernicus, in our comprehension of the fabric of reality. Philosophy never achieved such an upheaval despite its millennia of activity; science achieved it at least three times in a hundred years, and will do it at least once more, probably twice more, and conceivably in endless succession.

The second paradigm shift—we shall pretend that it will be the last, but there is no way of knowing—will take us a step beyond the unification of quantum theory and gravitation. It will take us into the foundations of physical reality, and we shall comprehend what it means to be a particle (an outdated term already, of course), what it means to be a force, what it means to have a charge, how the physical laws emerge, why the world is the way it is, and how apparent reality can emerge from absolutely nothing without intervention . . . and turn out to be comprehensible. No one has the slightest idea what form that ultimate theory will take, although there are tiny glimmerings of various possibilities in string theory, in speculations like Wheeler's, and in fantasy speculations which I alluded to at the end of Chapter 10. All that we can be sure about is that when that final revelation comes, we shall be astonished by our previous naivety.

There are only two really deep problems left for science to solve, millions of the second rank, and countless trillions of lesser ranks of importance. One great problem is the origin of the universe; the other is the nature of consciousness, that most puzzling of all properties of matter. The origin of the universe will fall into place once we have pursued modern theories of quantum gravity and particles a little further and we can expect it to be summarized by a few more great ideas. The problem of consciousness may be quite different, and it is conceivable that it will be solved without developing a great idea of its own.

First, I suspect that a phenomenon as complex as consciousness will not be summarized by a 'law' in the traditional sense of the term. The brain, currently the only device known to be capable of generating a sense of consciousness, makes use of many modes of activity and possesses regions where certain functions are concentrated but not completely localized, so that we cannot expect to summarize its function in a sentence or two, let alone a mathematical formula. I suspect that an understanding of consciousness will be achieved only when we have succeeded in emulating it. That view certainly does not deny the importance of current neuroscientific approaches to the brain, among them physiological, pharmacological, and psychological, for we need to know in detail what needs to be incorporated into our simulations. But here we should be cautious, as it is not necessary to incorporate everything that is discovered, just as an aircraft need not be equipped with feathers or have its engines in its chest. Nor does this view mean that the current fashion in some quarters for basing mechanisms of consciousness on quantum phenomena within, for instance, microtubules cannot be incorporated. Indeed, it might be possible to achieve consciousness of Type 1 (as it might be called) by constructing a device that emulates only classical neurophysiology, including the amazing plasticity of neuronal connections and the subtlety of chemical potentiation and transmission, and then to go on to achieve consciousness of Type 2 by building a device that incorporates delocalized quantum effects of the types proposed by those who believe that they are ineluctable concomitants of consciousness. It would then be an engaging task to discover what a Type 2 emulator could do, or thought it could do, that a Type 1 emulator could not, or thought it could not, do. If, as I suspect, it turns out that we ourselves are merely Type 1, it is conceivable that we would not recognize the different achievements of a Type 2 consciousness as consciousness, and write it off as a failure.

In short, although there may never be a 'theory of consciousness'—indeed, the very notion is probably inappropriate—there is every likelihood that an emulation will be achieved. The act of construction of that emulator will be in some sense the achievement of comprehension of the nature of consciousness. There would, of course, be a never-ending exploration of the differences between natural consciousness, our sort, and the simulated sort, and we would never be quite sure that artificial consciousness is in all respects the same as natural consciousness or whether we had simply created something else that we would never understand. Perhaps the only aliens we shall ever meet are the ones we build ourselves. We can leave to future generations the ethical problems associated with the rights of these artificially created but sensate non-beings, their right to die, their right to special treatment if they become disabled, the possibility that we can clone them and their experiences exactly, the possibility that different races of conscious non-beings will develop and find each other unacceptable, that belief systems will emerge within individual emulators or tribes of them that undermine the presumed rationality of their actions, and the likelihood that these intelligences will find the antics of the vessels of human consciousness tiresome, and—after making a pessimistic but realistic judgement about the burden that humans put on the planet—take appropriate action. There is clearly much scope here for a new Gulliver to travel.

I have touched on the shifting paradigms of science. There are two that are much closer to home, that are among us now, and that strike at its very core.

With the rise of the computer and its ability to handle huge numerical calculations of the greatest intensity, we are seeing a shift away from analysis—the setting up and solving of equations—to numerical computation. Used appropriately, this shift is splendid, for it increases the reach of scientists, who now, instead of throwing up their hands when an unsolvable equation emerges from a theory, can run it on a computer and analyse all its implications. We can now see that tiny little, seemingly innocuous equations can have extraordinary consequences. We have to admire that potency and our ability to tease it out numerically, for I used the same criterion with admiration in the Prologue when considering the criteria of being a great idea.

The danger, though, is twofold. One is trivial: we might resort to numerical computation when an analytic solution could be found with a bit more hard work. That is laziness, and although regrettable to we who have been brought up on the beauties of analytical expression, probably not particularly important. The second danger is deeper: resorting to numerical solution can distance us from understanding. When an analytical solution is found, we can claim to *understand* the outcome, for in principle we can grasp each step of the argument that led to the solution. When a numerical result is found, there is a less immediate grasp of the connection between the seed (the equation) and the outcome, and we do not feel that the result is as much a part of our being as when we work through an analytical derivation step by step. Still, it is better to get a numerical result than no solution at all, and as time goes on we shall feel increasingly comfortable, and will find ways of assimilating numerical computations. The saving grace, of course, of such calculations is the wonderful way that graphics can now be used to display their content. We are currently in the midst of the transition from seeing the beauty of an elegant analytical solution to seeing the beauty of an elegant portrayal of a computed solution.

The second shift is one that must be handled with much greater caution. I have mentioned in a number of places in the text that in certain cases science is gingerly letting go of the feature that has been its principal resource: actual experiment. There are certain experiments that will always be out of range in cosmology, sometimes because the energy required is of cosmic magnitude, and sometimes because we appear to be limited to observations of a single, preexisting universe. In Chapter 6 I held string theory up as an example of a theory that appears to be experimentally untestable.

There are at least two reactions to this letting go of the ability to do experiments. One is to regard all such non-testable theories as outside science, and no more to be accepted as pointing to the truth as any of Aristotle's pronouncements. Here, Galileo's finger wags in admonition and warning. They are engaging intellectual activities, but not science. Some people certainly hold that view about string theory. Others, far less appropriately, also hold it about natural selection. An alternative is to regard science as having matured to the point at which non-verifiable theories can be regarded, cautiously, as valid. Thus, if a theory accounts for the masses of the fundamental particles and predicts that the world is three-dimensional, then it might be admitted to honorary validity even though there is no known way, or no practical way, of testing it. Such an attitude would have been unacceptable when the corpus of scientific knowledge was meagre, but now—so long as no inconsistencies are implied with the plethora of known facts—we may perhaps, cautiously, accept the validity of such an untestable theory. Now Galileo holds up his finger to caution. If we insist upon verifiability, as scientific purists have every right to do, then the price to pay might be the termination of scientific progress in the sense of discovering fundamentals; this argument has no impact, of course, on the application of science, where presumably experiment will never be curtailed in the same way.

I have used the term 'verifiable'. This brings me into contact with Karl Popper's widely celebrated view that theories are never verifiable in the strict sense of the word, but have to be falsifiable if they are to be regarded as scientific. That is, there should exist an experiment that, in principle, could show that the theory is false. Natural selection is falsifiable (contrary to what some think) because, for instance, it has implications for molecular biology, as was remarked in Chapter 1. General relativity is falsifiable because it has implications for the motion of objects near heavy bodies, as we saw in Chapter 9, such as the precession of Mercury's orbit and the bending of light by galaxies. The law of conservation of energy and the law of the increase in entropy (the First and Second Laws of thermodynamics) are falsifiable, because they have implications, among others, for the existence of perpetual motion machines.

Is string theory falsifiable? Currently, it is too vague and has too few well-defined predictions to know. But suppose it isn't: suppose that a future version of M-theory settles down into a form that predicts all the known masses of the fundamental particles, all the values of the fundamental constants, and the structure of spacetime, but suggests absolutely no other experiment. It would not be falsifiable because it has predicted accurately all known fundamental properties of the universe, and I suspect that we would form the opinion that it was valid and indeed be celebrated as the apotheosis of scientific achievement.

What will scientists be doing once a TOE has been established and used to predict all known properties of the universe? Some will turn to face the other direction, and explore the ramifications of this final theory. That will occupy them for an eternity, provided civilization persists. There will be others, though, who worry about the self-consistency of that final theory, for they will have in mind Gödel's theorem and its negative implications for furnishing such proofs (Chapter 10). Those who don't worry about self-consistency will lie awake at night and worry about the impossibility of proving that the final theory is unique. They might even discover a seemingly completely different theory of everything that has exactly the same implications, but other than being mathematically identical to the rival theory, implies that the universe is utterly different from what hitherto had been supposed. But then, that's science.

# FURTHER READING

These are the principal sources that I have drawn on to write this book. I am greatly indebted to their authors for excavating the knowledge I needed to carve into my own chapters.

## Chapter 1: Evolution

DARWIN, CHARLES, *The origin of species*. First published by John Murray, 1859; available in Penguin, 1968.

DAWKINS, RICHARD, *The selfish gene*. Oxford University Press, 1976.

DESMOND, ADRIAN and JAMES MOORE, *Darwin*. Penguin, London, 1991.

FUTUYAMA, DOUGLAS, *Evolutionary biology*. Sinauer, Sunderland, Mass., 1998.

GOULD, STEPHEN JAY, *The structure of evolutionary theory*. Harvard University Press, 2002.

LEWIN, ROGER, *Human evolution*. Blackwell Science, Oxford, 1999.

MAYNARD SMITH, JOHN and EÖRS SZATHMÁRY, *The origins of life*. Oxford University Press, 1999.

RIDLEY, MARK, *Evolution*. Blackwell Science, Oxford, 1996.

RIDLEY, MARK, ed. *Evolution: an Oxford reader*. Oxford University Press, 1997.

TUDGE, COLIN, *The variety of life*. Oxford University Press, 2000.

## Chapter 2: DNA

DAVIES, KEVIN, *The sequence: Inside the race for the human genome*. Weidenfeld and Nicolson, London, 2001.

HENIG, ROBIN, *A monk and two peas: The story of Gregor Mendel and the discovery of genetics*. Weidenfeld and Nicolson, London, 2000.

KRAWCZAK, M. and J. SCHMIDTKE, *DNA fingerprinting*. Bios Scientific Publishers, Oxford, 1998.

MADDOX, BRENDA, *Rosalind Franklin: The dark lady*. HarperCollins, 2002.

RIDLEY, MATT, *Genome: The autobiography of a species in 23 chapters*. Fourth Estate, London, 1999.

SULSTON, JOHN and GEORGINA FERRY, *The common thread*. Bantam Press, London, 2002.

SUZUKI, DAVID, ANTHONY GRIFFITHS, JEFFREY MILLER, and RICHARD LEWONTIN, *An introduction to genetic analysis*. W. H. Freeman, New York, 1989.

## Chapter 3: Energy

BARBOUR, JULIAN, *The discovery of dynamics*. Oxford University Press, 2001.

HARMAN, P. M., *Energy, force, and matter: The conceptual development of nineteenth- century physics*. Cambridge University Press, 1982.

LIGHTMAN, BERNARD, ed. *Victorian science in context*. University of Chicago Press, 1997.

NEWTON, ISAAC, *The Principia: Mathematical principles of natural philosophy*. Translated by I. Bernard Cohen and Anne Whitman, University of California Press, 1999.

SMITH, CROSBIE, *The science of energy: A cultural history of energy physics in Victorian Britain*. Athlone Press, London, 1998.

## Chapter 4: Entropy

ATKINS, PETER, *The second law*. W. H. Freeman, New York, 1987.

ATKINS, PETER and JULIO DE PAULA, *Physical chemistry*. Oxford University Press and W. H. Freeman, New York, 2002.

COVENEY, PETER and ROGER HIGHFIELD, *The arrow of time*. HarperCollins, London, 1991.

SMITH, CROSBIE, *The science of energy: A cultural history of energy physics in Victorian Britain*. Athlone Press, London, 1998.

## Chapter 5: Atoms

ATKINS, PETER, *The periodic kingdom: A journey into the land of the chemical elements*. Weidenfeld and Nicolson, London, 1995.

ATKINS, PETER and LORETTA JONES, *Chemical principles*. W. H. Freeman, New York, 2001.

EMSLEY, JOHN, *Nature's building blocks*. Oxford University Press, 2001.

LUCRETIUS, *On the nature of the universe*. Translated by R. E. Latham, Penguin, London, 1951.

STRATHERN, PAUL, *Mendeleyev's dream: The quest for the elements*. Hamish Hamilton, London, 2000.

## Chapter 6: Symmetry

DAVIES, P. C. W. and J. BROWN, eds. *Superstrings: A theory of everything?* Cam-bridge University Press, 1988.

GREENE, BRIAN, *The elegant universe: Superstrings, hidden dimensions, and the quest for the ultimate theory*. Vintage, London, 2000.

HEILBRONNER, EDGAR and JACK DUNITZ, *Reflections on symmetry: In chemistry ... and elsewhere*. VCH, Weinheim, 1993.

KANE, GORDON, *Supersymmetry: Squarks, photinos, and the unveiling of the ultimate laws of nature*. Perseus Publishing, Cambridge, Mass., 2000.

LOCKWOOD, E. H. and R. H. MACMILLAN, *Geometric symmetry*. Cambridge University Press, 1978.

MARTIN, B. R. and G. SHAW, *Particle physics*. John Wiley, Chichester, 1997.

PEAT, DAVID, *Superstrings and the search for the theory of everything*. Abacus, London, 1992.

SUNDARESAN, M. K., *Handbook of particle physics*. CRC Press, Boca Raton, 2001.

## Chapter 7: Quanta

ALBERT, DAVID, *Quantum mechanics and experience*. Harvard University Press, 1992.

ATKINS, PETER, *Quanta: A handbook of concepts*. Oxford University Press, 1991.

ATKINS, PETER and R. S. FRIEDMAN, *Molecular quantum mechanics*. Oxford Univer-sity Press, 1997.

ATKINS, PETER and JULIO DE PAULA, *Physical chemistry*. Oxford University Press and W. H. Freeman, New York, 2002.

BAGGOTT, JIM, *The meaning of quantum theory*. Oxford University Press, 1992.

DICKSON, MICHAEL, *Quantum chance and non-locality: Probability and non-locality in the interpretations of quantum mechanics*. Cambridge University Press, 1998.

GHOSE, PARTHA, *Testing quantum mechanics on new ground*. Cambridge University Press, 1999.

GREENSTEIN, GEORGE and ARTHUR ZAJONC, *The quantum challenge: Modern research on the foundations of quantum mechanics*. Jones and Bartlett, Sudbury, Mass., 1997.

KRAGH, HELGE, *Quantum generations: A history of physics in the twentieth century*. Princeton University Press, 1999.

OMNÈS, ROLAND, *Quantum philosophy: Understanding and interpreting contemporary science*. Princeton University Press, 1999.

YOURGRAU, WOLFGANG and STANLEY MANDELSTAM, *Variational principles in dynamics and quantum theory*. Pitman, London, 1968.

## Chapter 8: Cosmology

BARROW, JOHN and FRANK TIPLER, *The anthropic cosmological principle*. Clarendon Press, Oxford, 1986.

CARROLL, BRADLEY and DALE OSTLIE, *An introduction to modern astrophysics*. Addison-Wesley, Reading, Mass., 1996.

LAYZER, DAVID, *Cosmogenesis: The growth of order in the universe*. Oxford University Press, New York, 1990.

PEACOCK, JOHN, *Cosmological physics*. Cambridge University Press, 1999.

REES, MARTIN, *Just six numbers: The deep forces that shape the universe*. Weidenfeld and Nicolson, London, 1999.

SCHWINGER, JULIAN, *Einstein's legacy*. Scientific American Library, W. H. Freeman, New York, 1986.

WHEELER, JOHN, *A journey into gravity and spacetime*. Scientific American Library, W. H. Freeman, New York, 1990.

## Chapter 9: Spacetime

BERRY, MICHAEL, *Principles of cosmology and gravitation*. Cambridge University Press, 1976.

CALLENDAR, CRAIG and NICK HUGGETT, *Physics meets philosophy at the Planck scale: Contemporary theories in quantum gravity*. Cambridge University Press, 2001.

D'INVERNO, RAY, *Introducing Einstein's relativity*. Clarendon Press, Oxford, 1992.

GRAVES, JOHN COWPERTHWAITE, *The conceptual foundations of contemporary relativ-ity theory*. MIT Press, Cambridge, Mass., 1971.

GREENE, BRIAN, *The elegant universe: Superstrings, hidden dimensions, and the quest for the ultimate theory*. Vintage, London, 2000.

MISNER, CHARLES, KIP THORNE, and JOHN WHEELER, *Gravitation*. W. H. Free-man, San Francisco, 1970.

PEACOCK, JOHN, *Cosmological physics*. Cambridge University Press, 1999.

SMOLIN, LEE, *Three roads to quantum gravity*. Weidenfeld and Nicolson, London, 2000.

WEINBERG, STEVEN, *Gravitation and cosmology: Principles and applications of the gen-eral theory of relativity*. John Wiley, New York, 1972.

## Chapter 10: Arithmetic

ATKINS, PETER, *Creation revisited*. Penguin, London, 1992.

BELL, JOHN, *The art of the intelligible: An elementary survey of mathematics and its conceptual development*. Kluwer, Dordrecht, 2001.

BENSON, DONALD, *The moment of proof: Mathematical epiphanies*. Oxford University Press, 1999.

IFRAH, GEORGES, *The universal history of numbers: From prehistory to the invention of the computer*. Harvill Press, London, 1998.

JACQUETTE, DALE, ed. *Philosophy of mathematics: An anthology*. Blackwell, Oxford, 2002.

KATZ, VICTOR, *A history of mathematics: An introduction*. Addison-Wesley, Reading, Mass., 1998.

NAGEL, ERNEST and JAMES NEWMAN, *Gödel's proof*. Routledge, London, 1958.

PENROSE, ROGER, *The emperor's new mind: Concerning computers, minds, and the laws of physics*. Oxford University Press, 1989.

PENROSE, ROGER, *Shadows of the mind: A search for the missing science of consciousness*. Oxford University Press, 1994.

SHANKER, S. G., ed. *Gödel's theorem in focus*. Routledge, London, 1988.

YANDELL, BENJAMIN, *The honors class: Hilbert's problems and their solvers*. A. K. Peters, Natick, Mass., 2002.

# ACKNOWLEDGEMENTS

I would like to thank the following for reading parts of this book and guiding my mind and hand:

Professor Richard Dawkins, FRS (University of Oxford),
Professor Sir Roger Penrose, OM, FRS (University of Oxford),
Professor Sir Martin Rees, FRS (University of Cambridge),
Professor Sir Michael Berry, FRS (University of Bristol),
Professor Lane Hughston (Kings College, London),
the Revd John Polkinghorne, KBE, FRS (University of Cambridge),
Professor Michael Rowan-Robinson (Imperial College, London), and
Professor Alex Wilkie (University of Oxford).

Their friendly, wise, and helpful input was invaluable. I am grateful to my colleague, Nigel Wilson, FBA, for providing a translation of the inscription in the frontispiece. I would also like to acknowledge the guidance and wise encouragement of my editor, Michael Rodgers.

## ILLUSTRATION CREDITS

I have drawn almost all the illustrations myself, but many are based on others' published work. I would therefore like to acknowledge the following sources of inspiration:

**1.3**, **1.7** *Encyclopedia Britannica*.15edn.
**1.5**, **1.6**, **1.9**, **1.12** Douglas Futuyama, *Evolutionary biology*.
Sinaur, Sunderland, Mass., 1998.
**1.10** Roger Lewin, *Human evolution: An illustrated introduction*.
Blackwell Science, Oxford, 1999.
**2.2**, **2.3** David Suzuki, Anthony Griffiths, Jeffrey Miller, Richard Lewontin,
*An introduction to genetic analysis*. W.H. Freeman & Co., New York, 1989.
**2.9** *Life: the science of biology*. William Purves, Gordon Orians, and
Craig Heller, Sinauer, Sunderland, Mass., 1992.
**2.11**, **2.16** M. Krawczak and J. Schmidtke, *DNA fingerprinting*. Bios, Oxford, 1998.
**2.13** C.K. Mathews and K.E. van Holde, *Biochemistry*,
Benjamin/Cummins, Menlo Park, 1996.
**4.13** Roger Penrose, *The emperor's new mind*. Oxford University Press, 1989.
**5.1** William Brock, *The Fontana history of chemistry*.
HarperCollins, London, 1992.
**8.3** NASA, COBE Science Working Group; *Ap. J. Lett.*, 354, L37, 1990.
**8.7** Bradley Carroll and Dale Ostlie, *An introduction to modern astrophysics*.
Addison–Wesley, Reading, Mass., 1996.
**8.11** Brian Greene, *The elegant universe*. Vintage, London, 2000.
**9.3**, **9.4** Thomas Banchoff, *Beyond the third dimension*. Scientific American Library,
W. H. Freeman & Co., New York, 1990.
**9.8**, **9.13**, **9.15**, **9.16** John Wheeler, *A journey into gravity and spacetime*.
Scientific American Library, W. H. Freeman & Co., New York, 1990.
**10.1** Georges Ifrah, *The universal history of numbers*.
Harvill Press, London, 1998.

Specific sources are as follows:

**Frontispiece** Photo: Franca Principe, IMSS.

**1.2** Ring-tailed lemurs, spider monkey, chimpanzee: © Adrian Warren / Oxford Scientific Films; Rhesus monkey: © Belinda Wright / Oxford Scientific Films; Author portrait: Studio Edmark.

**1.4** Dr Jeremy R. Young, The Natural History Museum, London.

**1.8** Michael Wood.

**1.11** Courtesy of the National Science Museum, Tokyo.

**1.13** Mauricio Anton / Oxford University Press.

**1.14** Mauricio Anton / Oxford University Press.

**2.1** © James King-Holmes / Science Photo Library.

**2.6** © Science Photo Library.

**2.17** Cellmark Diagnostics, Abingdon, Oxon.

**3.1** Hulton Archive.

**3.2** Voyager 1 spacecraft: NASA / Science & Society Picture Library.

**3.4** Science Museum / Science & Society Picture Library.

**3.12** Courtesy of Rolls-Royce plc.

**4.1** Science Museum / Science & Society Picture Library.

**4.9** © Astrid & Hanns-Frieder Michler / Science Photo Library.

**5.2** Short-beam analytical balance, 1876: Science Museum / Science & Society Picture Library; Modern chemical balance: courtesy of County Scales Group.

**5.3** Science Museum / Science & Society Picture Library.

**8.8** © Celestial Image Co. / Science Photo Library.

**9.1** © Thomas Banchoff, Brown University, and Davide P. Cervone, Union College, NY.

# INDEX